Lecture Notes in Computer Scie

Commenced Publication in 1973
Founding and Former Series Editors:
Gerhard Goos, Juris Hartmanis, and Jan van Leeuwen

Editorial Board

Alexander A. Shvartsman   Pascal Felber (Eds.)

Structural Information and Communication Complexity

15th International Colloquium, SIROCCO 2008
Villars-sur-Ollon, Switzerland, June 17-20, 2008
Proceedings

 Springer

Volume Editors

Alexander A. Shvartsman
University of Connecticut
Department of Computer Science and Engineering
Storrs, CT 06269, USA
E-mail: aas@cse.uconn.edu

Pascal Felber
Université de Neuchâtel
Institut d'informatique
Rue Emile-Argand 11
2009 Neuchâtel
E-mail: pascal.felber@unine.ch

Library of Congress Control Number: 2008928716

CR Subject Classification (1998): F.2, C.2, G.2, E.1

LNCS Sublibrary: SL 1 – Theoretical Computer Science and General Issues

ISSN   0302-9743
ISBN-10   3-540-69326-2 Springer Berlin Heidelberg New York
ISBN-13   978-3-540-69326-0 Springer Berlin Heidelberg New York

Springer is a part of Springer Science+Business Media

springer.com

© Springer-Verlag Berlin Heidelberg 2008
Printed in Germany

Typesetting: Camera-ready by author, data conversion by Scientific Publishing Services, Chennai, India
Printed on acid-free paper   SPIN: 12321964   06/3180   5 4 3 2 1 0

Preface

The Colloquium on Structure, Information, Communication, and Complexity (SIROCCO) is an annual research meeting focused on the relationship between information and efficiency in decentralized (distributed, parallel, and network) computing. This year, SIROCCO celebrated its 15th anniversary. Over the years, the colloquium has become a widely recognized forum bringing together researchers interested in the fundamental principles underlying the interplay between local structural knowledge and global communication and computation complexity. SIROCCO covers topics such as distributed algorithms, compact data structures, information dissemination, informative labeling schemes, combinatorial optimization, and others, with potential applications to large-scale distributed systems including global computing platforms, peer-to-peer systems and applications, social networks, wireless networks, and network protocols (such as routing, broadcasting, localization). SIROCCO 2008 was held in Villars-sur-Ollon, in the Swiss Alps, June 17–20, 2008.

There were 52 contributions submitted to SIROCCO 2008. All papers underwent a thorough refereeing process, where each submission was reviewed by at least 3, and on average 3.4, Program Committee members. After in-depth discussions, the Program Committee selected 22 high-quality contributions for presentation at the colloquium and publication in this volume. We thank the authors of all the submitted papers, the Program Committee members, and the external reviewers. Without their dedication, we could not have prepared a program of such quality.

There were two invited speakers at SIROCCO 2008: Nicola Santoro (Carleton University) and Boaz Patt-Shamir (Tel-Aviv University).

We express our gratitude to the SIROCCO Steering Committee, and in particular to Pierre Fraigniaud for his enthusiasm and his invaluable help throughout the preparation of this event.

We are also grateful to all the local people who were instrumental in making SIROCCO 2008 a success, notably Peter Kropf, the many students who volunteered on the organization team, and the IT service of the University of Neuchatel.

We acknowledge the use of the EasyChair system for handling the submission of papers, managing the refereeing process, and generating these proceedings.

June 2008

Alexander Shvartsman
Pascal Felber

Conference Organization

Steering Committee

Paola Flocchini	University of Ottawa, Canada
Pierre Fraigniaud (Chair)	CNRS and University Paris 7, France
Leszek Gasieniec	University of Liverpool, UK
Leszek Kirousis Lefteris	University of Patras, Greece
Rastislav Kralovic	Comenius University, Slovakia
Evangelos Kranakis	Carleton University, Canada
Danny Krizanc	Wesleyan University, USA
Bernard Mans	Macquarie University, Australia
Andrzej Pelc	Université du Québec en Outaouais, Canada
David Peleg	Weizmann Institute, Israel
Giuseppe Prencipe	Pisa University, Italy
Michel Raynal	IRISA, University of Rennes, France
Nicola Santoro	Carleton University, Canada
Pavlos Spirakis	CTI, Greece
Shmuel Zaks	Technion, Israel

Program Committee

Amotz Bar-Noy	City University of New York, USA
Costas Busch	Louisiana State University, USA
Bogdan Chlebus	University of Colorado Denver, USA
Andrea Clementi	University "Tor Vergata" of Rome, Italy
Roberto De Prisco	University of Salerno, Italy
Stefan Dobrev	Slovak Academy of Sciences, Slovakia
Michael Elkin	Ben-Gurion University, Israel
Pascal Felber	University of Neuchatel, Switzerland
Eli Gafni	UCLA, USA
Goran Konjevod	Arizona State University, USA
Dariusz Kowalski	University of Liverpool, UK
Adrian Kosowski	Gdansk University of Technology, Poland
Shay Kutten	Technion, Israel
Dahlia Malkhi	Microsoft Research, USA
Marios Mavronicolas	University of Cyprus, Cyprus
Achour Mostefaoui	University of Rennes, France
Alex Shvartsman (Chair)	University of Connecticut, USA
Sebastien Tixeuil	Pierre and Marie Curie University, France
Masafumi Yamashita	Kyushu University, Japan

Local Organization

Pascal Felber (Chair)   University of Neuchatel, Switzerland
Peter Kropf   University of Neuchatel, Switzerland

Sponsoring Institutions

University of Neuchatel

External Reviewers

Mohamed Ahmed
Chen Avin
Leonid Barenboim
Janna Burman
Andrea Clementi
Richard Cole
Jurek Czyzowicz
Shantanu Das
Seda Davtyan
Dariusz Dereniowski
Stéphane Devismes
Yoann Dieudonné
Yefim Dinitz
Pierre Fraigniaud
Cyril Gavoille
Chryssis Georgiou
Maria Gradinariu Potop-Butucaru
Ted Herman
David Ilcinkas
Erez Kantor
Haim Kaplan
Sotiris Kentros
Ralf Klasing
Alex Kravchik
Peter Kropf
Piotr Krysta
Lukasz Kuszner

Zvi Lotker
Adam Malinowski
Fredrik Manne
Sumit Narayan
Alfredo Navarra
Yen Kaow Ng
Nicolas Nicolaou
Hirotaka Ono
Christos Papadimitriou
Katy Paroux
Francesco Pasquale
Paolo Penna
Andrzej Pelc
Andrea Pietracaprina
Adele Rescigno
Adi Rosen
Gianluca Rossi
Etienne Rivière
Tim Roughgarden
Nicola Santoro
Riccardo Silvestri
Nir Tzachar
Ugo Vaccaro
Carmine Ventre
Ivan Visconti
Qin Xin
Pawel Zylinski

Table of Contents

Mobile Entities Computing:
Models and Problems

Nicola Santoro

School of Computer Science, Carleton University, Canada
santoro@scs.carleton.ca

Abstract. By *mobile entity computing* (MEC) we refer to the study of the computational and complexity issues arising in systems of autonomous computational entities located in a spatial universe $\mathcal{U}$ in which they can move. The entities have computational capabilities (i.e., storage and processing), can move in $\mathcal{U}$ (their movement is constrained by the nature of $\mathcal{U}$), exhibit the same behavior (i.e., execute the same protocol), and are autonomous in their actions (e.g., they are not directed by an external controller). Depending on the context, the entities are sometimes called *agents*, other times *robots*.

Depending on the nature of $\mathcal{U}$, two different settings are identified. The first setting, sometimes called a *graph world* or *discrete universe* or *netscape*, is when the $\mathcal{U}$ is a simple graph and the entities can move from node to neighbouring node. An instance of such setting is that of mobile software agents in a network. The other setting, called sometimes *continuous universe*, is when $\mathcal{U}$ is a geometric space which the entities, endowed with wireless sensorial/communication capabilities, can perceive and can move in. Instances of such settings are autonomous mobile robots, and autonomous vehicular networks.

These settings have been long the subject of separate intensive investigations in fields as diverse as AI, robotics, and software engineering, and only recently by the distributed computing community. Indeed, the use of mobile software agents is becoming increasingly popular when computing in networked environments, ranging from Internet to the Data Grid, both as a theoretical paradigm and as a system-supported programming platform but the theoretical research had focused mainly on the descriptive and semantic concerns. This situation has drastically changed in recent years, as an increasing number of algorithmic investigations have started to examine the setting. Similarly, in the last few years the problems related to the coordination and control of autonomous mobile robots have been investigated not only in the traditional fields of AI, robotics and control but also by researchers in distributed computing).

In both settings, the research concern is on determining what tasks can be performed by the entities, under what conditions, and at what cost. In particular, a central question is to determine what minimal hypotheses allow a given problem to be solved.

The purpose of this talk is to introduce the computational models and the fundamental problems in MEC; although the focus is on the discrete universe, several of the introduced concepts extend, mutata mutandis, to the continuous case.

A. Shvartsman and P. Felber (Eds.): SIROCCO 2008, LNCS 5058, p. 1, 2008.
© Springer-Verlag Berlin Heidelberg 2008

Reputation, Trust and Recommendation Systems in Peer-to-Peer Systems

Boaz Patt-Shamir*

School of Electrical Engineering
Tel Aviv University
Tel Aviv 69978, Israel
boaz@eng.tau.ac.il

The Internet has brought about the notion of peer-to-peer computing, whose reliance on a central authority (let alone a central server) is minimal. It seems fair to say that one of the Great Promises of the Internet is that such unmoderated direct interaction between users will reduce much of the traditional overhead due to the "man in middle" taking his share: either financially (in the context of e-commerce) or conceptually (in the context of opinion shaping, say). The flip side of this prospect, of course, is the danger that the system will deteriorate into a lawless jungle: some users in a peer-to-peer system, possibly coordinated, might exploit honest users to their advantage, since it appears that there is no effective way to enforce rules in this game.

Consider eBay for example, where users can buy and sell stuff. The immense success of this application of peer-to-peer computing is quite surprising for people who are used to consider worst-case scenarios, since eBay is "ripe with the possibility of large scale fraud and deceit" [6]. The answer, according to most experts, lies in eBay's closely watched *reputation system* [9]: after every transaction, the system invites each party to rate the behavior of the other party (the grades are "positive," "negative" and "neutral"). The system maintains a public billboard that records all feedbacks received for each user. It is common knowledge that consulting the billboard is a key step before making a transaction. A user with more than a few negative feedbacks has very little chance of getting any business in eBay. While empirical evidence show that the system is successful, many questions are unanswered. For starters, one annoying problem today is that new users with little or no feedback will find it quite difficult to attract any interest (either as buyers or sellers). Another clear weakness of the system is that it allows, at least theoretically, for a clique of players to accumulate a lot of positive reputation by praising each other, and then defraud naïve users. Above all, it is not clear how to interpret the raw data on the billboard, or, in other words, *what algorithm should an honest user follow?*

Another example of relying on reputation is *recommendation systems*. These are systems that provide the users with recommendations about new products (books and movies are a prime example, because of their relatively fast publishing rate). In this case the difficulty is that users have differing tastes. Obviously, not all users can expect to get good recommendations (if I am the only one who

* This research was supported in by the Israel Science Foundation (grant 664/05).

A. Shvartsman and P. Felber (Eds.): SIROCCO 2008, LNCS 5058, pp. 2–4, 2008.

likes a certain book, who can recommend it to me?). Rather, recommendation systems try to help users who share their taste with others. Recommendation systems are usually based on professional critics (like book reviews in newspapers) or on public opinion (like best-seller lists). More sophisticated systems cluster users [8] or products [10], and try to predict the response of a user to a new object based on the responses of similar users, or based on previous responses of that user to similar objects. However, these systems can be easily fooled by malicious users (see, e.g., [7]). Another disadvantage of current methods is that they are self-perpetuating: by definition, they highlight—and thus promote—popular products; in other words, this approach sets a high "entry barrier" a new product must overcome to reach the users.

These examples demonstrate a central problem in peer-to-peer systems: how to use the "good" information that exists in the system without being exploited by the potentially bad information out there? Using reputation, i.e., making judgments based on reports of interactions by others, seems to be the way to go. We seek to explore the question of how exactly to make these judgments, what types of guarantees are possible, and what are the reasonable models to be considered and implemented in the context of reputation systems. In other words, our goal is to study an algorithmic approach to reputation systems. Our approach is based on a few surprisingly powerful theoretical results we have obtained recently [1,2,3,4,5]. These results lead us to the somewhat counterintuitive conclusion that peer-to-peer systems can enjoy effective collaboration without elaborate notions of trust, even when any portion of the peers exhibit arbitrarily malicious (Byzantine) behavior. Moreover, our goal is that algorithms will incur only a relatively modest overhead to the honest players, when compared to algorithms for the case where all players are honest.

References

1. Alon, N., Awerbuch, B., Azar, Y., Patt-Shamir, B.: Tell me who I am: an interactive recommendation system. In: Proc. 18th Ann. ACM Symp. on Parallelism in Algorithms and Architectures (SPAA), pp. 1–10 (2006)
2. Awerbuch, B., Nisgav, A., Patt-Shamir, B.: Asynchronous active recommendation systems. In: Tovar, E., Tsigas, P., Fouchal, H. (eds.) OPODIS 2007. LNCS, vol. 4878, pp. 48–61. Springer, Heidelberg (2007)
3. Awerbuch, B., Patt-Shamir, B., Peleg, D., Tuttle, M.: Collaboration of untrusting peers with changing interests. In: Proc. 5th ACM conference on Electronic Commerce (EC), pp. 112–119. ACM Press, New York (2004)
4. Awerbuch, B., Patt-Shamir, B., Peleg, D., Tuttle, M.: Adaptive collaboration in synchronous p2p systems. In: Proc. 25th International Conf. on Distributed Computing Systems (ICDCS), pp. 71–80 (2005)
5. Awerbuch, B., Patt-Shamir, B., Peleg, D., Tuttle, M.: Improved recommendation systems. In: Proc. 16th Ann. ACM-SIAM Symp. on Discrete Algorithms (SODA), pp. 1174–1183 (2005)
6. Kollock, P.: The production of trust in online markets. In: Thye, S.R., Macy, M.W., Walker, H., Lawler, E.J. (eds.) Advances in Group Processes, vol. 16, pp. 99–124. Elsevier Psychology, Amsterdam (1999)

7. O'Mahony, M.P., Hurley, N.J., Silvestre, G.C.M.: Utility-based neighbourhood formation for efficient and robust collaborative filtering. In: Proc. 5th ACM Conf. on Electronic Commerce (EC), pp. 260–261 (2004)
8. Resnick, P., Iacovou, N., Suchak, M., Bergstrom, P., Riedl, J.: Grouplens: an open architecture for collaborative filtering of netnews. In: Proc. 1994 ACM Conf. on Computer Supported Cooperative Work (CSCW), pp. 175–186. ACM Press, New York (1994)
9. Resnick, P., Zeckhauser, R., Friedman, E., Kuwabara, K.: Reputation sytems. Comm. of the ACM 43(12), 45–48 (2000)
10. Sarwar, B., Karypis, G., Konstan, J., Reidl, J.: Item-based collaborative filtering recommendation algorithms. In: Proc. 10th International Conf. on World Wide Web (WWW), pp. 285–295 (2001)

Gathering Problem of Two Asynchronous Mobile Robots with Semi-dynamic Compasses

Nobuhiro Inuzuka, Yuichi Tomida, Taisuke Izumi,
Yoshiaki Katayama, and Koichi Wada

Nagoya Institute of Technology, Gokiso-cho Showa, Nagoya 466-8555, Japan
inuzuka@nitech.ac.jp, u1tomida@phaser.elcom.nitech.ac.jp,
t-izumi@nitech.ac.jp, katayama@nitech.ac.jp, wada@nitech.ac.jp
Phone: +81-52-735-5050, Facsimile: +81-52-735-5408

Abstract. Systems of autonomous mobile robots have been attracted as distributed systems. Minimal setting for robots to solve some class of problems has been studied with theoretical concerns. This paper contributes discussion on relationship between inaccuracy of compasses which give axes of coordinate systems of robots and the possibility of gathering robots. The gathering problem is to make all robots meet at a single point which is not predefined. The problem has been shown to be solvable for two robots with dynamically variable compasses within the difference of $\pi/4$ between two robots. This paper improves the limit of difference to $\pi/3$. This is shown with the fully asynchronous robot model CORDA. Configurations and executions of robot systems for CORDA with dynamic compasses are formalised. In order to deal with behaviors of robots a concept of relative configurations is also introduced.

1 Introduction

Systems of autonomous mobile robots have been gathering attention as distributed systems that are expected effective for activity in deep sea or outer space, where robots have to act autonomously. The system consisting of a large number of simple robots is expected to be effective to work with fault-tolerance. Hence minimal setting for robots which cope with this kind of difficult environment is worth studying from the practical and also theoretical points of view. This paper focuses on a theoretical point. As an interesting problem this paper studies the gathering problem. It is to make all robots meet at a single point which is not predefined, normally without agreement of their coordinate systems. The gathering problem is a typical consensus problem and is important as a preliminary stage of cooperative tasks by robots.

Literatures deal with this problem under various settings, which moderate the setting by giving additional ability, such as a sense of multiplicity of robots, limited use of memory and a limited sense of direction or some inaccurate compasses. We focus the gathering problem of two robots with inaccurate and variable compasses. Compasses are indispensable for autonomous robots in a distant space but it is large merit if our algorithm requires less accurate compasses. Prencipe

A. Shvartsman and P. Felber (Eds.): SIROCCO 2008, LNCS 5058, pp. 5–19, 2008.

Table 1. Results on gathering among robots with inaccurate compasses

	models	compasses	tilt angle	#robots
Souissi et al.[7] Imazu et al.[8]	asynchronous	fixed	$\pi/4$	2
Katayama et al.[9]	asynchronous	fixed semi-dynamic	$\pi/3$ $\pi/4$	2
Yamashita et al.[12]	asynchronous	fixed	$\pi - \varepsilon$	2
Izumi at al.[13]	semi-synchronous	semi-dynamic	$\pi/2 - \varepsilon$	n
this paper	asynchronous	semi-dynamic	$\pi/3 - \varepsilon$	2

showed that even for two robots the gathering can not be solved when robots are oblivious and share no global compass and do not have additional ability[1] on his asynchronous model called CORDA[2,3]. Even on semi-synchronous model given by Suzuki and Yamashita[4,5] two oblivious robots with no means of agreeing their orientation can not solve the problem. On the other hand Flocchini et al.[6] showed that the problem is solvable for any number of robots when their compasses are identical. Since then our interest moves to ease the perfect agreement of compasses to more realistic inaccurate compasses and to explore the minimal requirement to compasses to solve the problem.

Here an inaccurate compass means that it does not necessarily point the correct north direction and robots may not agree on direction. Souissi et al.[7] and Imazu et al.[8] independently proposed models of inaccurate compasses, where the inaccuracy is modelled by degrees of disagreement of compasses. The two papers showed algorithms that solve the gathering problem with at most $\pi/4$ of disagreement of two robots. Afterwards, inaccurate compasses are totally modelled by Katayama et al.[9]. In addition to disagreement of compasses the paper discusses time variance of compasses (called a dynamic compass) and whether an absolute north exists or not. R. Cohen and D. Peleg[10] proposed another models of inaccuracy of compasses and sensors. In their model robots acquire erroneous coordinates which are different from the correct but robots share the global direction. Souissi et al.[11] discusses gathering of robots in another setting, where compasses are inaccurate but eventually become consistent.

Katayama et al.[9] also improved the disagreement level to allow gathering to $\pi/3$ and also showed that it is possible for two robots with dynamic compass with at most $\pi/4$ disagreement (each compass may differ $\pi/8$ from absolute north) in full asynchronous model CORDA. In the case of fixed compasses this disagreement level is improved to $\pi - \varepsilon$ for any small positive ε[12]. Izumi et al.[13] showed a gathering algorithm for dynamic compasses of at most $\pi/2 - \varepsilon$ disagreement on the semi-synchronous model for any positive ε. This is not restricted for two robots. It is also shown that this is optimal on this model. The results on the gathering problem among robots are summarized in Table 1.

This paper is on the line of [7,8,9]. The main target is to improve limit of disagreement for dynamic compasses to solve gathering up to $\pi/3 - \varepsilon$ ($\pi/6$ from the absolute north) for any positive ε on CORDA model. Difference of model causes the difficulty in design of algorithms. While in the semi-synchronous model a cycle consisting of observation, calculation and movement is atomic, in CORDA events are occurred in separate. Consequently a robots may take several cycles while another robot starts for a movement after observation. We give formalization of configurations and executions of the robot system on CORDA with dynamic compasses. Then we also introduce the concept of relative configuration to understand behavior of robots.

The proposing algorithm is a generalized version of Imazu's algorithm[8,9]. In order to show the result we introduce a couple of concepts. Using the concepts we first show the algorithm accomplishes the problem for fixed compasses with $\pi/3 - \varepsilon$ disagreement although this is not new.[1] We show that the discussion is extended to dynamic compasses by using the concept of relative configurations and a range of change of relative configurations by change of compasses.

The following section introduces models of robots, their system, and compasses equipped. Section 3 defines the gathering problem after clarifying configurations and executions of a system. Section 4 proposes an algorithm and shows its correctness. At first its correctness is given for fixed compasses with a difference less than $\pi/3$ by using a concept of relative configuration. This will lead us to the correctness for dynamic compasses with a difference less than $\pi/3$.

2   Models of the System

In this section we review models of robots, their system, and compasses.

2.1   CORDA — A Model of Robots

We consider a system which consists of a set of robots $\mathcal{R} = \{r_1, r_2, \cdots, r_n\}$. Robots move on a two-dimensional plane, on which we assume a global Cartesian coordinate. The global coordinate of r_i at a time t is denoted by $r_i(t)$.

We assume CORDA, the model of robots, which is described as follows:

- A robot has no volume and its position is specified by a coordinate.
- A robot has its own rectangular coordinate system with whose origin is its current position.
- Robots are uniform and anonymous. That is, they use an identical algorithm and do not have any kinds of identifiers distinguishing a robot from others.
- A robots asynchronously repeats cycles consisting of the four states, LOOK, COMPUTE, MOVE and WAIT, in this order.
- A robot equips a sensor, with which at an instant during LOOK a robot acquires a set (or multiset) of coordinates of robots on the local coordinate system. The set is called an *observation*. Since the local origin is at the own position it distinguishes its own coordinate from others in spite of anonymity.

---

[1] Indeed the proposing algorithm solves the gathering problem for robots with $\pi/2 - \varepsilon$ disagreement for any $\varepsilon > 0$.

– During COMPUTE a robot performs a local computation by a given algorithm only using the observation as an input, and then results a coordinate, which is intended to be a destination with respect to the local coordinate system. A robot is oblivious. It remembers no information of the previous cycles.
– A robot moves on the line from the position at the beginning of MOVE to the destination, the result of COMPUTE. The schedule of movement is non-deterministic. At any instant during MOVE a robot is on the line and always goes forward but its position on the line at an instant is not prescribed. It is not assumed to reach the destination during a single MOVE.
– Throughout WAIT a robot does nothing.
– It has no direct means of communication. Observing others is only the mean.

A time period taken in a state or a cycle is finite but nondeterministically variable and unpredictable. COMPUTE may take unexpectedly long time and it causes the observation to become old and yields the large difference between the destination and the ideal destination that is to be determined by robots' positions at the time when the robot is really going to move.

From the aspect of fairness we make two assumptions. The first is that a robot travels at least the distance of $\min(d, \delta)$ during a single MOVE, where d is the distance to the destination given by COMPUTE and $\delta > 0$ is a constant distance which is unknown but there exists for each robot. The second is that the length of each cycle is at least a constant time $\varepsilon(> 0)$ and it is finite.

Each time a robot has a local coordinate system, which is relatively determined to the global one by three parameters, the tilt angle of axes ψ, the scale ratio sc, and the origin o. Given a global coordinate p its coordinate in terms of the local one is got by the function, $Z_{\psi,sc,o}(p) = \frac{1}{sc} \begin{pmatrix} \cos\psi & \sin\psi \\ -\sin\psi & \cos\psi \end{pmatrix} (p - o)$.
The function is called an *observation function*. Since the scale ratio is fixed for a robot we often do not care about a scale and denote the function by $Z_{\psi,o}$.

2.2 Compasses

We assume a robot to equip a compass, a device indexing a direction. The local y-axis of a robot is directed by its compass. The direction of compass of a robot r_i is given as an angle with respect to a global coordinate system and denoted by ψ_i. We identify the compass with its direction ψ_i.

To argue the effect of inaccuracy of compass we review models of compasses. Models are totally presented in Katayama et al.[9] in different aspects, including the variance range of compasses among robots, time variance of compasses, and existence or inexistence of an absolute direction, say the *absolute north*. The first factor is argued by giving an upper limit of tilt angles and combing with the other factors. We briefly reviews the second and the third points.

We imagine three types of time variance. A compass which may always vary through time is called a *full dynamic compass (FDC)*. A compass which does not change through a cycle is a *semi dynamic compass (SDC)*. An SDC may change when a cycle switches to another. A *fixed compass (FXC)* is one which never changes through execution. FXC $\subseteq$ SDC and SDC $\subseteq$ FDC, where FXC, SDC, and FDC denote the classes of compasses of the types.

Assuming the absolute north or the global compass and an upper limit of tilt angles from it the range of variance is restricted. A compass keeping the difference from the absolute north within an upper limit $\alpha/2$ is said α-*absolute error*. If we only assume the upper bound α for the difference of tilt angles between any pair of robots, we call the compass α-*relative error*. An α-absolute error compass is also α-relative error.

These three inaccuracy are to be combined. For example, an α-relative error SDC is of relative error, of semi-dynamic variance and with an upper limit α. In the rest of paper we deal with relative-error FXC and absolute-error SDC.

For the rest of this paper, we prepare a couple of notations. A line on two different points x and y is denoted by xy. A line through x with the direction which is tuned counterclockwise by θ from the positive direction of x-axis is denoted by $\overline{x}^{\theta}$. When we write $\overline{xy}$ it is a segment from x to y. $\overline{xy}$ also means the length of the segment. $\triangle xyz$ is a triangle whose apexes are x, y, and z, and it means the set of points inside and the border of the triangle. For two points $p = (x_1, y_1)$ and $q = (x_2, y_2)$, $\angle \overline{pq}$ means the degree ρ s.t. $(x_2 - x_1, y_2 - y_1) = (r \cos \rho, r \sin \rho)$ for a positive r.

3   Configurations, Executions and the Gathering Problem

We formalize executions taking account of the situations that (1) we consider only FXC and SDC which do not change direction during a cycle, and that (2) only instants of observation and the schedule of movement affect the system.

Let us call a robot *active* at t when it observes at t. We call a time instant t *active* when there is at least an active robot at t. A *step* is a time duration between an active time and its immediately following active time. We also define a *robot step* as the time duration between an active time t_a and the first active time until which all robot performs at least one cycle after t_a.

A *configuration* of the system at t is $\langle R(t), \Psi(t) \rangle$, where $R(t)$ is an n-tuple $(r_1(t), \ldots, r_n(t))$ of global coordinates of n robots at t, and $\Psi(t)$ is an n-tuple $(\psi_1(t), \cdots, \psi_n(t))$ of compass angles of robots.

COMPUTE determines a destination. After an observation even before the end of COMPUTE a destination is already determined no matter the robot knows. That is, we define the *destination* $dest_i(t)$ of r_i as the destination given by COMPUTE of r_i using an observation at t or the last observation of r_i before t. For t before the first observation we let $dest_i(t)$ be the initial position of the robot. Let denote the n-tuple $(dest_1(t), \ldots, dest_n(t))$ by $Dest(t)$.

We do not care *scales* of local coordinates.

The *set of active robots* at t, denoted by $\Gamma(t)$, is also a factor determining an execution. Now we formalize an execution. For a given algorithm $\mathcal{A}$ an *execution* of the system is a sequence of $\Gamma(t)$, $R(t)$, $\Psi(t)$ and $Dest(t)$ for all active time $t_0, t_1, \ldots, t_k, \ldots,$

$$\langle \Gamma(t_0), R(t_0), \Psi(t_0), Dest(t_0) \rangle, \cdots, \langle \Gamma(t_k), R(t_k), \Psi(t_k), Dest(t_k) \rangle, \cdots,$$

satisfying the following conditions for $j = 0$,

$$
\left(
\begin{array}{l}
\Gamma(t_0) \neq \emptyset \\
r_i(t_0) = \text{the initial position of } r_i \\
dest_i(t_0) = \begin{cases} r(t_0) & \text{if } r_i \notin \Gamma(t_0) \\ Z^{-1}(\mathcal{A}(\{Z(r_1(t_0)), \ldots, Z(r_n(t_0))\})) & \text{if } r_i \in \Gamma(t_0) \end{cases}
\end{array}
\right.
$$

and for $j \geq 1$,

$$
\left(
\begin{array}{l}
\Gamma(t_j) \neq \emptyset \\
r_i(t_j) = (1 - \gamma)r_i(t_{j-1}) + \gamma \, dest_i(t_{j-1}) \quad \text{for some } \gamma \in [0, 1] \\
\quad \text{i.e. } r_i(t_j) \in \overline{r_i(t_{j-1}) \, dest_i(t_{j-1})} \\
dest_i(t_j) = \begin{cases} dest_i(t_{j-1}) & \text{if } r_i \notin \Gamma(t_j) \\ Z^{-1}(\mathcal{A}(\{Z(r_1(t_j)), \ldots, Z(r_n(t_j))\})) & \text{if } r_i \in \Gamma(t_j), \end{cases}
\end{array}
\right.
$$

where $Z = Z_{\psi_i(t_j), r_i(t_j)}$ the observation function of r_i at t_j and Z^{-1} is the inverse mapping of it, and $\mathcal{A}$ is the function determined by the given algorithm.

When a robot r_i becomes active at t, i.e. $r_i \in \Gamma(t)$, r_i changes its destination to new one, determined by $\mathcal{A}$ with an input of observation. A robot moves on the line to the destination. Of course a robot does not really acquire a destination at the time of observation before finishing COMPUTE. The robot will start MOVE at a time after the observation, possibly after a couple of steps. During steps in which the robot does not start the γ is 0. When a robot is inactive at t_j it is still on the line $\overline{r_i(t_{j-1}) \, dest_i(t_{j-1})}$ and it holds $\overline{r_i(t_{j-1}) \, dest_i(t_{j-1})} \supseteq \overline{r_i(t_j) \, dest_i(t_j)}$ because $r_i(t_j) \in \overline{r_i(t_{j-1}) \, dest_i(t_{j-1})}$ and $dest_i(t_j) = dest_i(t_{j-1})$.

Depending on the model of compasses $\Psi(t_j)$ is subjected the following conditions, for any robots r_i and $r_{i'}$ $(i \neq i')$ and any time t_j:

$$
\begin{cases}
|\psi_i(t_j) - \psi_{i'}(t_j)| \leq \alpha & \text{in the case of } \alpha\text{-relative error compasses} \\
\psi_i(t_j) \in [-\alpha/2, \alpha/2] & \text{in the case of } \alpha\text{-absolute error compasses} \\
\psi_i(t_0) = \psi_i(t_1) = \cdots & \text{in the case of FXC} \\
\psi_i(t_j) = \psi_i(t_{j-1}) & \text{if } r_i \notin \Gamma(t_j) \text{ in the case of SDC}
\end{cases}
$$

Executions are also restricted from the fairness. That is, when a robot r_i is active at t_a and also at $t_b > t_a$ it must satisfy $\overline{r_i(t_a) \, r_i(t_b)} \geq \min(d, \delta)$ where $d = \overline{r_i(t_a) \, dest_i(t_a)}$ and $\delta > 0$ is the constant distance depending on r_i. It must also satisfy $\varepsilon \leq t_b - t_a < \infty$ where $\varepsilon >$ is a constant shared by robots. It is also subjected that any robot is included in $\Gamma(t_j)$ for infinitely many j. A fair execution must satisfy all of them. We only consider fair executions.

When we have an algorithm $\mathcal{A}$ and the initial positions $R(t_0)$ of robots an execution is determined by three factors, that is, choices of active robots $\Gamma(t_j)$, the changes of compasses $\Psi(t_j)$ and the degree of movement represented by γ for each t_j. These make a *schedule*.

Throughout this paper we consider the system of two robots, i.e. $\mathcal{R} = \{r_1, r_2\}$. An algorithm $\mathcal{A}$ is said to *solve the gathering problem* when for any fair execution of $\mathcal{A}$ there is a time instant $\hat{t}$ s.t. $r_1(t) = r_2(t)$ for any $t \geq \hat{t}$.

4   A Gathering Algorithm $\mathcal{A}_{\alpha,\theta}$

For our algorithm we prepare words. A robot is said to observe the other at the direction ϕ when its observation includes a point $(x, y) = (r\cos\phi, r\sin\phi)$ for a positive r. For a given upper limit α of difference of compasses we color angles of seeing a robot as follows (Fig. 1 (a)), where π is classified to blue for $\alpha = 0$.

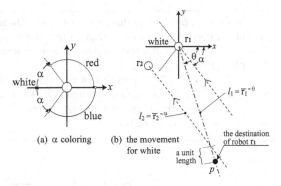

Fig. 1. α-coloring and the movement for white

- an angle in $[0, \pi - \alpha)$ is called a *red angle*.
- an angle in $[\pi - \alpha, \pi + \alpha)$ is called a *white angle*.
- an angle in $[\pi + \alpha, 2\pi)$ is called a *blue angle*.

When a robot observes the opponent at an angle in a color it is said to *see* the color. This coloring for angles and for observation is called α-*coloring*.

Table 2 shows the algorithm $\mathcal{A}_{\alpha,\theta}$. It takes two parameters α the upper limit of difference between compasses, and θ, given as greater than α. We will see that this algorithm works for robots with FXC and also SDC under conditions. Basically red and blue placed opposite. When a robot sees red the opponent is expected to see blue. If this is true a robot approaches to another waiting and they have a successful rendezvous. Because of disagreement of compasses the white region is prepared. Seeing white it moves to a place to be expected to see red. The place is calculated by the two lines, the line l_1 through itself with a tilt angle $-\theta$ and the line l_2 through the opponent with angle $-\alpha$. Because $\alpha < \theta$ the lines always cross below the robot. It sets the destination to the point p on l_1 a unit length further from the crossing point of l_1 and l_2 (see Fig. 1 (b)).

4.1   Relative Configurations and Relative Configuration Map

In order to see the behavior of robots we define a view from a robot. For $\langle R(t), \Psi(t)\rangle = \langle (r_1(t), r_2(t)), (\psi_1(t), \psi_2(t))\rangle$, a *relative configuration* (*RC* for short) from $r_i \in \{r_1, r_2\}$ is $\langle \phi(t), \delta(t)\rangle$, s.t. r_i observes $r_j \in \{r_1, r_2\} - \{r_i\}$ at the direction $\phi(t)$ or $Z(r_j(t) - r_i(t)) = (r\cos\phi(t), r\sin\phi(t))$ for a positive r, and $\delta(t) = \psi_j(t) - \psi_i(t)$, where Z is the observation function of r_i at t. We call a robot from which RC's are considered a *base robot* and the other an *opponent robot*. In the rest of paper r_1 denotes a base robot. The vector from a base robot to its opponent is called a *base-vector*. The color that a base robot sees is called a *base-color* of the RC. An *opponent-color* is the color being seen by the opponent.

Fig. 2 shows a map of RC (called the *RC-map*). It is a two-dimensional plane of parameters ϕ and δ and shows opponent-colors of places on the plane. Base-colors are shown on the ϕ axis. It is helpful to understand robots' behavior. For

Table 2. A pseudo-code of an algorithm $\mathcal{A}_{\alpha,\theta}$

1.	**if** observes only a coordinate at the origin **then** make a null movement
2.	**elseif** observes a *red* **then** make a null movement
3.	**elseif** observes a *blue* **then** make a movement toward the coordinate of the opponent
4.	**else** (/* observes a *white* */) **do**
5.	$l_1 := \overline{o}^{-\theta} = \overline{r_1}^{-\theta}$ (the line of tilt angle $-\theta$ through the origin or itself);
6.	$l_2 := \overline{r_2}^{-\alpha}$ (the line of tilt angle $-\alpha$ through the opponent);
7.	$p :=$ the point a unit length further from the cross point of l_1 and l_2 on l_1;
8.	make a movement toward p.

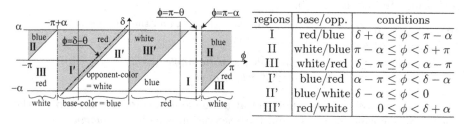

Fig. 2. The relative configuration map, regions and the conditions to divide regions

example a compass of an α-relative or -absolute FXC is fixed within the range $(-\alpha, \alpha)$ then the RC on the map is always on a line parallel to the ϕ axis.

4.2 Regions in RC-Map for $\mathcal{A}_{\alpha,\pi/3}$

In this paragraph let the upper limit α of a disagreement angle between two robots be less than $\pi/3$ and consider $\mathcal{A}_{\alpha,\theta}$ for $0 \le \alpha < \theta \le \pi/3$. Here we give a correctness of it for α-relative FXC.

The RC-map is used to classify initial configurations of robots. Then we see behavior of robots in each region. We have six regions as shown in Fig. 2. A region with a Roman numeral and another with the same numeral plus a prime (') are symmetric. Because of symmetry we need understand three situations.

The table in Fig. 2 shows colors and conditions for the regions. For example, Fig. 4 shows a situation of robots in Region II. The region is for base-color white and opponent color blue. As shown in Fig. 4-(a), in order to see blue from r_2 it has to hold $\phi < \delta + \pi$. Fig. 4-(b) illustrates the range of ϕ from the restriction to see white. The angle ϕ has to be equal or greater than $\pi - \alpha$. It has not to exceed the lower limit $\pi + \alpha$ but it is subsumed by $\phi < \delta + \pi$ because $\delta \le \alpha$.

We state the correspondence between regions and conditions without proofs:

Lemma 1. *When the conditions of each region are satisfied by an RC the base- and opponent-colors of the RC are those stated in Fig. 2. The conditions cover all cases and are exclusive.*

In advance we summarize the correctness proof of algorithm for robots with FXC and also SDC compasses. The correctness will be given by tracing the transition of regions of RC. We can easily see the gathering is achieved in Regions I and I' and for other cases robots transit to these regions. Our strategy is to show that RC always moves right to left monotonically and it moves at least a constant in a robot step. In the case of FXC RC only moves in parallel to ϕ axis. Then RC always reaches I or I' unless robots gather before they reach the regions. We also show that RC stops before certain lines.

This proof is a preparation for the SDC case. When robots have SDC compasses they change RC even if they do not make any movement. First we understand the area of possible RC without movement. We call the area the mobility range. (See Figs. 7 and 8.) In this case RC does not necessarily move right to left but the mobility range itself always moves right to left. Instead of RC we can argue the transition of the center of the mobility range. We show that the whole area of mobility range can fill in the safe region I or I'. It will be certified by showing that even with SDC compasses the movement of RC is characterized in the same way as FXC in a period of a robot step and RC always stops before the certain lines where the mobility range falls into Regions I or I'.

4.3 Correctness of $\mathcal{A}_{\alpha,\theta}$ for α-Relative FXC with $\alpha < \theta \le \pi/3$

In this section we show lemmata arguing movement of robots in the regions and transition of regions. We use the local coordinate system of a base-robot r_1 for all reference of coordinates and angles. Lemmata are given for α-relative FXC and $0 \le \alpha < \theta \le \pi/3$. Then we show the correctness of $\mathcal{A}_{\alpha,\pi/3}$ for $\alpha < \pi/3$.

Lemma 2. *For any execution of α-relative FXC using $\mathcal{A}_{\alpha,\theta}$ starting an RC in region I or I' includes a time t s.t. $r_1(t') = r_2(t')$ for any $t' \ge t$, where $0 \le \alpha < \theta \le \pi/3$. In the execution it holds $\phi(t_{k-1}) = \phi(t_k)$ unless $r_1(t_k) = r_2(t_k)$.*

Proof. By Lemma 1 the RC $\langle \phi(t_0), \delta(t_0) \rangle$ at t_0 in region I satisfies the condition given in Fig. 2 and its base- and opponent-colors are red and blue, respectively.

We claim that for any allowed execution starting at t_0 it holds for all t_k unless $r_1(t_k) = r_2(t_k)$,

1. $r_1(t_k) = dest_1(t_k) = r_1(t_0)$,
2. $dest_2(t_k) = r_2(t_0)$ before r_2 becomes active and $dest_2(t_k) = r_1(t_0)$ after that.
3. $r_2(t_k) \in \overline{r_2(t_0) r_1(t_0)}$, and
4. $\langle \phi(t_k), \delta(t_k) \rangle = \langle \phi(t_0), \delta(t_0) \rangle$ and the colors seen by robots do not change.

We found that the claim yields the lemma considering,

- r_2 approaches to r_1 during its robot step by at least a constant distance (or the distance to r_1 when it is less than the constant), and
- r_1 and r_2 take null movement after the time gathering together.

Then we show the claim by induction on k. It is trivial for t_0. Assuming it for t_k we check for t_{k+1}. The first two are direct from $\langle \phi(t_k), \delta(t_k) \rangle = \langle \phi(t_0), \delta(t_0) \rangle$, i.e. the two robots see red and blue at t_k. The third one is also direct from the definition of execution, because $r_2(t_{k+1}) \in \overline{r_2(t_k) \, dest_2(t_k)} = \overline{r_2(t_0) \, r_1(t_0)}$. $r_2(t_{k+1}) \in \overline{r_2(t_0) \, r_1(t_0)}$ means that r_2 does not leave from the initial line through $r_1(t_0)$ and $r_2(t_0)$. It yields the last one, that is, RC does not change.   □

Lemma 2 can be extended for the case which does not start at but enters I or I' during execution. We define an imaginable RC from the position of the destination of r_1 to the destination of r_2, which we call a *destination RC*. That is, for a time t a destination RC is $\langle \phi'(t), \delta(t) \rangle$ s.t. $Z(dest_2(t) - dest_1(t)) = (r \cos \phi'(t), r \sin \phi'(t))$ for $\exists r > 0$ and $\delta(t) = \psi_2(t) - \psi_1(t)$ and Z is the observation function of r_1 at t.

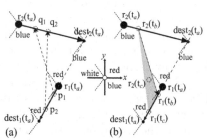

Fig. 3. (a) The situation of Lemma 3 and (b) Case (II) of the proof

Lemma 3. *Consider any execution of α-relative FXC using $\mathcal{A}_{\alpha,\theta}$ where $0 \leq \alpha < \theta \leq \pi/3$. If there is a time (not necessarily an active time) t_a satisfying (the situation is illustrated in Fig. 3),*

- *the RC $\langle \phi(t_a), \delta(t_a) \rangle$ and the destination RC $\langle \phi'(t_a), \delta(t_a) \rangle$ are in I;*
- *$dest_2(t_a)$, $r_1(t_a)$ and $dest_1(t_a)$ are on a line in this order; and*
- *for any $p_1, p_2 \in \overline{r_1(t_a) \, dest_1(t_a)}$ s.t. $\overline{p_1 \, dest_1(t_a)} \leq \overline{p_2 \, dest_1(t_a)}$ and $q_1, q_2 \in \overline{r_2(t_a) \, dest_2(t_a)}$ s.t. $\overline{q_1 \, dest_2(t_a)} \leq \overline{q_2 \, dest_2(t_a)}$, it holds $\angle \overrightarrow{p_1 \, q_1} \geq \angle \overrightarrow{p_2 \, q_2}$.*

then it holds the followings, where $\hat{\phi} = \min(\angle \overrightarrow{dest_1(t_a) \, dest_2(t_a)}, \angle \overrightarrow{dest_1(t_a) \, r_1(t_a)})$.

1. *There is a time t s.t. $r_1(t') = r_2(t')$ for any $t' \geq t$; and*
2. *$\phi(t_{k-1}) \geq \phi(t_k) \geq \hat{\phi}$ for any active time t_k after t_a,*

Proof. Let t_b denote the first active time after t_a. Because the RC $\langle \phi(t_a), \delta(t_a) \rangle$ and the destination RC $\langle \phi'(t_a), \delta(t_a) \rangle$ are both in Region I, the two robots are in I during between t_a and t_b.

(I) Case that r_1 is active at t_b When r_1 stops at t_b it sees red and makes null movements. The situation is the same until r_2 becomes active. At the first active time of r_2 it sees blue and it is the situation of Lemma 2.

(II) Case that r_2 is active at t_b When r_2 stops at t_b it sees blue and decides the destination at $r_1(t_b)$ which is somewhere between $r_1(t_a)$ and $dest_1(t_a)$. The first active time t_c of r_1 after t_b r_2 is at somewhere in $\triangle r_2(t_b) r_1(t_b) r_1(t_c)$. ($r_2(t_c)$ is not necessarily on the segment $\overline{r_2(t_b) r_1(t_b)}$ because r_2 may have several active times before t_c.) Then at t_c robots are in Region I and at the next active time of r_2 robots are in the situation of Lemma 2.   □

Lemma 4. *Consider any execution of $\mathcal{A}_{\alpha,\theta}$ starting an RC in regions II, III, II' or III', where $\alpha < \theta \leq \pi/3$. Then,*

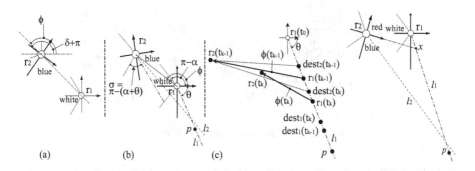

Fig. 4. Behavior of robots in Region II   **Fig. 5.** Region III

i It includes a time t s.t. $r_1(t') = r_2(t')$ for any $t' \geq t$; and
ii In the execution it holds $\phi(t_{k-1}) \geq \phi(t_k) \geq \hat{\phi}$ unless $r_1(t_k) = r_2(t_k)$,

where $\hat{\phi} = \pi - \theta$ when the execution starts from II or III and $\hat{\phi} = \delta - \theta$ when the execution starts from II' or III'.

Proof. We consider II. We can have a symmetric argument for II'.

From the assumption and Lemma 1 the base- and opponent-colors of $\langle\phi(t_0), \delta(t_0)\rangle\rangle$ are white and blue, respectively. If r_1 becomes active first then the algorithm sets $dest_1(t_0)$ to p as shown in Fig. 4. $dest_1$ remains p until r_2 becomes active. If r_2 is activated first $dest_2$ is set to the position of r_1 and r_2 approached to r_1 until r_2 meets r_1 or r_1 becomes active.

Let us consider the time t_a when both r_1 and r_2 are activated without completing gather. We claim to hold the followings for all $t_k \geq t_a$ unless $r_1(t_k) = r_2(t_k)$ or the RC goes out from the region II. This is shown by induction on k.

1. $dest_2(t_k)$, $r_1(t_k)$ and $dest_1(t_k)$ are in $\overline{r_1(t_{k-1})\,dest_1(t_{k-1})}$ on l_1 and lie in this order from $r_1(t_{k-1})$ to $dest_1(t_{k-1})$ (see Fig. 4-(c)).
2. $r_2(t_k) \in \triangle r_2(t_{k-1})dest_2(t_{k-1})dest_2(t_k)$, and
3. $\phi(t_{k-1}) \geq \phi(t_k) \geq \pi - \theta$,

Robots go forward at least the constant, i.e. r_1 goes along l_1 and r_2 approaches to r_1. Then there is a time t_b at which it becomes $r_1(t_b) = r_2(t_b)$ and r_1 is active at t_b or the RC is in Region I at t_b. The former happens in the case that r_2 reaches r_1 while r_1 stay somewhere it sees r_2 in Region II. In this case t_b is the time at which r_1 awakes and it realizes gathering. Since gathering is achieved within Region II it keeps $\phi(t) > \delta + \pi > \pi - \theta$.

For the latter the situation matches Lemma 3. It guarantees that robots gather and it keeps satisfying $\phi(t) \geq \min(\angle\overrightarrow{dest_1(t_b)\,dest_2(t_b)}, \angle\overrightarrow{dest_1(t_b)\,r_1(t_b)}) = \min(\angle\overrightarrow{dest_1(t_b)\,dest_2(t_b)}, \pi - \theta) = \pi - \theta$ before gathering. The symmetric arguments for the case starting from II' guarantees the angle from r_2 to r_1, $\phi(t) + \pi - \delta \geq \pi - \theta$ which implies $\phi(t) \geq \delta - \theta$.

We consider III. See Fig. 5. In this case only r_1 moves toward its destination designated as p in the figure while r_2 sees red. It continues until r_1 goes across

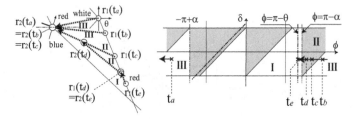

Fig. 6. An example movement of robots

the point x which is the cross point of l_1 and the horizontal axis of r_2. Let t_c to denote the active time of r_2 after the time when r_1 reaches x. Then the situation at t_c is the same as at t_a of the discussion for Region II.   □

Then the correctness of the algorithm is direct from Lemmata 2, 3 and 4.

Theorem 1. *The gathering problem for α-relative FXC for $\alpha < \pi/3$ is solved by $\mathcal{A}_{\alpha,\theta}$ for θ s.t. $\alpha < \theta \leq \pi/3$.*

For the following section we prepare another lemma, which we do not give a proof but it is direct from the arguments in the previous lemmata.

Lemma 5. *Consider any execution of α-relative FXC using $\mathcal{A}_{\alpha,\theta}$, where $\alpha < \theta \leq \pi/3$.*

- *Then it holds $\phi(t_{k-1}) \geq \phi(t_k)$ unless $r_1(t_k) = r_2(t_k)$ for any active time t_k;*
- *it keeps $\phi(t_k) \geq \hat{\phi}$ for any active time t_k where $\hat{\phi} = \pi - \theta$ when the execution starts from II, or III or $\hat{\phi} = \delta - \theta$ when it starts from II', or III'; and*
- *For active times t and $t' > t$ between which a robot step is passed if $\langle \phi(t), \delta(t) \rangle$ is not in I nor I', then $\phi(t) - \phi(t')$ is at least $\min(\xi, \eta)$, where ξ is a constant depending on only $\langle R(t_0), \Psi(t_0) \rangle$ and η is the minimal angle to enter I or I' from $\langle \phi(t), \delta(t) \rangle$.*

We show an example movement of robots in Fig. 6. Robots are in Region III at t_a then only r_1 can move downward. When r_1 stops and observes r_2 at t_b they are still in Region III. At t_c, r_2 awakes in Region II and it finds blue then goes to r_1. When r_2 stops on the way to its destination at t_d robots are in Region II. After that at t_e r_2 reaches to the position where r_1 had been at t_d. At that time r_1 is downward on the line and robots are in Region I. In the figure arrows show $\phi(t)$ and it can be seen that $\phi(t_a) > \phi(t_b) > \phi(t_c) > \phi(t_d) > \phi(t_e) = \pi - \theta$.

4.4   Correctness of $\mathcal{A}_{\alpha,\pi/3}$ for α-Absolute SDC with $\alpha < \pi/3$

This paragraph shows the correctness of $\mathcal{A}_{\alpha,\pi/3}$ for α-absolute SDC with $\alpha < \pi/3$ using the RC-map. Variance of compasses allows change of an RC even the robots do not move. However we can see that there is an RC of which the range of variance by compasses is included in the safe regions I or I' when $\alpha < \pi/3$. We also see that the algorithm $\mathcal{A}_{\alpha,\pi/3}$ will bring the RC to the point.

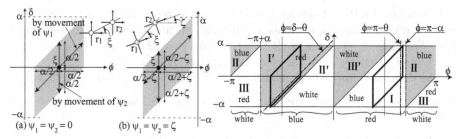

Fig. 7. Ranges of config. change of α-absolute error compasses

Fig. 8. Mobility ranges of α-absolute error compasses for $\alpha < \pi/3$

First let us imagine robots r_1 and r_2 with α-absolute SDC compasses with $\psi_1 = \psi_2 = 0$, i.e. both compasses have no bias. Let r_2 be in a direction ξ from r_1. The RC of this situation is plotted at $(\xi, 0)$ on the ϕ axis (Fig. 7-(a)).

Now we consider change of compasses. At first we think of change of ψ_2. It directly affects the change of RC on the vertical δ coordinate. When ψ_2 turns an angle η the RC moves from $(\xi, 0)$ to (ξ, η). Since ψ_2 has mobility in $[-\alpha/2, \alpha/2]$ their RC takes in range $\{(\xi, \eta) \mid -\alpha/2 \leq \eta \leq \alpha/2\}$. As for change of ψ_1, it affects both coordinates. When ψ_1 rotates η the relative angle δ decreases by η and ϕ also decreases by η. Then their RC takes in $\{(\xi - \eta, -\eta) \mid -\alpha/2 \leq \eta \leq \alpha/2\}$.

Let us think of robots on the ϕ axis, i.e. $\delta = 0$. If $\delta = 0$ and $\psi_1 = \psi_2 = 0$ the range of RC without move of their positions is $\{(\xi - \eta_1, -\eta_1 + \eta_2) \mid \eta_1, \eta_2 \in [-\alpha/2, \alpha/2]\}$ (Fig. 7-(a)). $\delta = 0$ does not mean $\psi_1 = \psi_2 = 0$ but $\psi_1 = \psi_2$. When $\psi_1 = \psi_2 = \zeta$ the mobility range of the RC is $\{(\xi - \eta_1, -\eta_1 + \eta_2) \mid \eta_1, \eta_2 \in [-\alpha/2 - \zeta, \alpha/2 - \zeta]\}$ (Fig. 7-(b)).

Then absolute-SDC robots are treated as their RC and its mobility range. Generally robots in $\langle R(t), \Psi(t) \rangle = \langle (r_1(t), r_2(t)), (\psi_1(t), \psi_2(t)) \rangle$ have their RC, $\langle \phi(t), \delta(t) \rangle$ with a mobility range $\{(\phi(t) - \eta_1, -\eta_1 + \eta_2) \mid \eta_1, \eta_2 \in [-\alpha/2 - \psi_1(t), \alpha/2 - \psi_1(t)]\}$, where $Z_{\psi_1(t), r_1(t)}(r_2(t) - r_1(t)) = (\rho \cos \phi(t), \rho \sin \phi(t))$ for $\rho > 0$ and $\delta(t) = \psi_2(t) - \psi_1(t)$.

Fig. 8 shows the RC-map for $\alpha < \pi/3$ and the regions. When we try to fit mobility ranges to the map we have Lemma 6. We also confirm behavior in I and I' in Lemma 7. When for the configuration $\langle R(t), \Psi(t) \rangle = \langle (r_1(t), r_2(t)), (\psi_1(t), \psi_2(t)) \rangle$ of robots, $\psi_1(t) = \alpha$, the RC is in the most right edge of the mobility area. The mobility range is taken the left hand side from the RC. We can observe the following lemma from Fig. 8.

Lemma 6. *(1) When $\psi_1(t) = \alpha$ and RC satisfies $\pi - \theta \leq \phi(t) \leq \pi - \alpha$ the mobility range of the RC is inside Region I.*
(2) When $\psi_2(t) = \alpha$ and RC satisfies $\delta - \theta \leq \phi(t) \leq \delta - \alpha$ the mobility range of the RC is inside Region I'.

Lemma 7. *While RC is in I or I' changing compasses has no effect for the movement of robots.*

Lemmata 6 and 7 allow robots to bring their RC and mobility range to I or I' and to gather. Within the mobility range let us call a point satisfying $\psi_1 = \psi_2 = 0$

the *center* of the RC. When we have $\langle \phi(t), \delta(t) \rangle$, the center is $\langle \phi(t) + \psi_1(t), 0 \rangle$, which is the RC when the compasses of both robots are coincident with the global one. We define $\phi_g(t)$ as the angle to see r_2 from r_1 with the global compass, i.e. $\phi_g(t) = \phi(t) + \psi_1(t)$. Then the center of $\langle \phi(t), \delta(t) \rangle$ is $\langle \phi_g(t), 0 \rangle$.

The discussion for FXC is valid for each instant, which yields the following.

Lemma 8. *Consider any execution of α-absolute SDC using $\mathcal{A}_{\alpha,\theta}$, where $\alpha < \theta \leq \pi/3$. Then,*

 i *It holds $\phi_g(t_{k-1}) \geq \phi_g(t_k)$ unless $r_1(t_k) = r_2(t_k)$ for any active time t_k;*
 ii *For active times t and $t' > t$ between which a robot step is passed if $\langle \phi(t), \delta(t) \rangle$ is not in I nor I', then $\phi_g(t) - \phi_g(t') \geq \min(\xi, \eta)$, where ξ is a constant depending only on $\langle R(t_0), \Psi(t_0) \rangle$ and η is the minimal angle to enter I or I' from $\langle \phi(t), \delta(t) \rangle$; and*
 iii *It holds $\phi_g(t) \geq \pi - \theta - \alpha/2$ when the execution starts from I, II or III and $\phi_g(t) \geq \delta - \theta - \alpha/2$ when it starts from I', II' or III'.*

Proof. Since during a single robot step change of compass does not affect behavior of the robot, it holds (i) and (ii) if ψ_1 does not change. Although ψ_1 may change, if we take $\phi_g(t_k)$ it holds (i) and (ii) because changing compass does not make the center of range change.

The $\phi(t)$ of RC is restricted by $\phi_g(t)$ and it satisfies $\phi_g(t) - \alpha/2 \leq \phi(t) \leq \phi_g(t) + \alpha/2$. Since $\phi(t) \leq \pi - \theta$ when the execution starts in Region II or III it has to be $\phi_g(t) \leq \pi - \theta - \alpha/2$. When the execution starts in Region II' or III' the symmetric arguments gives $\phi_g(t) \geq \delta - \theta - \alpha/2$. $\qquad\square$

Lemmata 6–8 yield the following theorem.

Theorem 2. *The gathering problem for α-absolute SDC for $\alpha < \pi/3$ is solved by $\mathcal{A}_{\alpha,\theta}$ for $\theta = \pi/3$.*

Proof. Lemma 7 tells that in I and I' robots gather as same as FXC. For other region Lemma 8 tells that the algorithm makes $\phi_g(t)$ left by a constant angles every robot step and it must stop before the line $\phi = \pi - \theta - \alpha/2$ or $\phi = \delta - \theta - \alpha/2$. Then the whole mobility range fits in Region I or I' by lemma 6. When it happens Lemma 7 guarantees to have gathering. $\qquad\square$

Impossibility of gathering for $\pi/2$-absolute SDC robots even on the semi-synchronous model is proven in [13]. The optimal angles where gathering is solvable for CORDA with absolute-SDC is still unclear between $\pi/3$ and $\pi/2$.

References

1. Prencipe, G.: On the feasibility of gathering by autonomous mobile robots. In: Pelc, A., Raynal, M. (eds.) SIROCCO 2005. LNCS, vol. 3499, pp. 246–261. Springer, Heidelberg (2005)
2. Prencipe, G.: CORDA: distributed coordination of a set of autonomous mobile robots. In: ERSADS 2001, pp. 185–190 (2001)

3. Prencipe, G.: Distributed coordination of a set of autonomous mobile robots. PhD thesis, Università di Pisa (2002)
4. Suzuki, I., Yamashita, M.: Distributed anonymous mobile robot – formation and agreement problems. In: SIROCCO 1996. Carleon Scientific, pp. 313–330 (1996)
5. Suzuki, I., Yamashita, M.: Distributed anonymous mobile robots: formation of geometric patterns. SIAM Journal of Computing 28(4), 1347–1363 (1999)
6. Flocchini, P., Prencipe, G., Santoro, N., Widmayer, P.: Gathering of asynchronous oblivious robots with limited visibility. Theoretical Computer Science 337(1-3), 147–168 (2005)
7. Souissi, S., Défago, X., Yamashita, M.: Gathering asynchronous mobile robots with inaccurate compasses. In: Shvartsman, M.M.A.A. (ed.) OPODIS 2006. LNCS, vol. 4305, pp. 333–349. Springer, Heidelberg (2006)
8. Imazu, H., Itoh, N., Katayama, Y., Inuzuka, N., Wada, K.: A gathering problem for autonomous mobile robots with disagreement in compasses. In: 1st Workshop on Theoretical Computer Science in Izumo (in Japanese), pp. 43–46 (2005)
9. Katayama, Y., Tomida, Y., Imazu, H., Inuzuka, N., Wada, K.: Dynamic Compass Models and Gathering Algorithms for Autonomous Mobile Robots. In: Prencipe, G., Zaks, S. (eds.) SIROCCO 2007. LNCS, vol. 4474, pp. 274–288. Springer, Heidelberg (2007)
10. Cohen, R., Peleg, D.: Convergence of autonomous mobile robots with inaccurate sensors and movements. In: Durand, B., Thomas, W. (eds.) STACS 2006. LNCS, vol. 3884, pp. 549–560. Springer, Heidelberg (2006)
11. Souissi, S., Défago, X., Yamashita, M.: Using eventually consistent compasses to gather oblivious mobile robots with limited visibility. In: Datta, A.K., Gradinariu, M. (eds.) SSS 2006. LNCS, vol. 4280, pp. 484–500. Springer, Heidelberg (2006)
12. Yamashita, M., Souissi, S., Defago, X.: Tight bound on the gathering of two oblivious mobile robots with inconsistent compasses. Research Report, JAIST, IS-RR-2007-006 (2007)
13. Izumi, T., Katayama, Y., Inuzuka, N., Wada, K.: Gathering Autonomous Mobile Robots with Dynamic Compasses: An Optimal Result. In: DISC 2007. LNCS, vol. 4738, pp. 298–312. Springer, Heidelberg (2007)

Locating and Repairing Faults in a Network with Mobile Agents*

Colin Cooper[1,**], Ralf Klasing[2], and Tomasz Radzik[1]

[1] Department of Computer Science, King's College, London WC2R 2LS, UK
{Colin.Cooper,Tomasz.Radzik}@kcl.ac.uk
[2] LaBRI – Université Bordeaux 1 – CNRS, 351 cours de la Libération,
33405 Talence cedex, France
Ralf.Klasing@labri.fr

Abstract. We consider a fixed, undirected, known network and a number of "mobile agents" which can traverse the network in synchronized steps. Some nodes in the network may be faulty and the agents are to find the faults and repair them. The agents could be software agents, if the underlying network represents a computer network, or robots, if the underlying network represents some potentially hazardous physical terrain. Assuming that the first agent encountering a faulty node can immediately repair it, it is easy to see that the number of steps necessary and sufficient to complete this task is $\Theta(n/k + D)$, where n is the number of nodes in the network, D is the diameter of the network, and k is the number of agents. We consider the case where one agent can repair only one faulty node. After repairing the fault, the agent dies. We show that a simple deterministic algorithm for this problem terminates within $O(n/k + D \log f / \log \log f)$ steps, where $f = \min\{n/k, n/D\}$, assuming that the number of faulty nodes is at most $k/2$. We also demonstrate the worst-case asymptotic optimality of this algorithm by showing a network such that for any deterministic algorithm, there is a placement of $k/2$ faults forcing the algorithm to work for $\Omega(n/k + D \log f / \log \log f)$ steps.

Keywords: Distributed computing, Graph exploration, Mobile agents.

1 Introduction

The *black hole search* problems, which have been recently extensively studied, assume that a mobile agent traversing a network is terminated immediately upon entering a faulty node. The task for a group of agents is to explore the network to identify all such faulty nodes, called black holes. In this paper we

* Partially supported by the project ALPAGE of the ANR "Masse de données: Modélisation, Simulation, Applications", the project CEPAGE of INRIA, the EC COST-TIST Action 293 "Graphs and Algorithms in Communication Networks" (GRAAL) and Action 295 "Dynamic Communication Networks" (DYNAMO), and the Royal Society IJP grant "Algorithms to find dense clusters in large networks".
** Partially supported by the UK EPSRC grant EP/D059372/1.

A. Shvartsman and P. Felber (Eds.): SIROCCO 2008, LNCS 5058, pp. 20–32, 2008.
© Springer-Verlag Berlin Heidelberg 2008

consider a weaker type of faults: agents are able to repair them, but one agent can repair only one faulty node. The task for a group of agents is now to explore the network to repair all faulty nodes. In this section we first review the black hole search model and introduce in more detail our repairable hole model. Then we summarise our results and compare them with previous relevant results.

Black-hole faults. The black hole search problems have been motivated by the following scenario. Mobile software agents can move through a network of computers, but some hosts (black holes) terminate any agent visiting it. The problem of protecting mobile agents from such malicious hosts has been studied in [11,12,15,16], with the focus on protecting sensitive information which mobile agents may carry.

From a more theoretical point of view, the researchers have investigated the problem of agents co-operatively exploring the network to identify the locations of black holes. Some of the agents might die, but the surving ones should learn where the black holes are (to avoid them in the future). This proplem was initially considered in the *asynchronous* model, and the majority of the results are for this model. An efficient protocol for two agents locating one black hole in an asynchronous ring network was presented in [6] and was extended to aribtrary two-connected networks in [7]. In subsequent research special network topologies were considered [4] and restrictions on communication between the agents were investigated, replacing the initially used whiteboard model with communication by means of a limited number of identical pebbles [5,8].

In the *synchronous* model, which we consider in this paper, the agents traverse the network in globally timed *steps*. In each step each agent can performed (unlimitted) local computation, which includes exchanging information with other agents who are at the same step in the same node, and can then move to a neighbouring node (or remain in the same node) [1,2,3,13,14]. Initially, all agents are at the same *start node s* and know the topology of the whole network, but do not know the number and the location of black holes. Also, no information about black holes is available in the *safe* nodes of the network (the nodes which are not black holes). Thus, in order to locate a black hole, at least one agent must visit it, but an agent entering a black hole disappears completely. The agents can communicate only when they meet, not being allowed to leave any messages in any form at the nodes. An agent learns that a node v is not a black hole either by visiting it (and surviving) or by meeting another agent who has already visited that node (and survived). An agent may deduce that a node v is a black hole if this agent is suppose to meet another agent, at some node other than v, but that other agent does not show up. The objective is to design a communication algorithm for the agents to identify all black holes reachable from the start node, minimizing the loss of agents and the number of steps required.

Most of the research on the synchronous black hole search problems has been concerned so far with the special case of two agents exploring a network which may have at most one black hole. The problem is to compute an "exploration schedule" for the agents which has the minimum possible number of steps. The first results regarding the computational hardness of this problem and approximation algorithms were presented in [2,3] and were subsequently improved in [13,14]. A more general case of locating black holes with k agents, where $k \geq 2$ is a parameter, was considered in [1]. The recent survey paper [9] discusses both asynchronous and synchronous models, various variants of the black hole search problem and solution techniques.

Repairable faults. In reality many types of faults can be fixed after some amount of trying, and with the expenditure of some effort. Thus there is a spectrum of problems with unfixable faults (black holes) at one end, and fixable faults (holes) at the other. If faults can be fixed, then we have to decide what is the appropriate way of modeling the cost of repairing a fault. For example, an agent fixing a fault may not be able to do anything else: the fault has been fixed but the agent has to be sacrificed. In other scenarios only the "content" part of the agent (the "repair kit") may have to be sacrificed to fix a fault, while the "carrier" part can remain mobile and can communicate with other agents, or can return to the starting node (the "depot") to pick up another repair kit.

Scenarios when the agent is sacrificed include robots traveling a road network, seeking to trigger land-mines, and software agents moving from node to node in a computer network to repair faults. In the latter example, the software agent is executable by the runtime environment at a faulty node, fixing that fault but permanently remaining at the node. A physical example where the contents of the agent are consumed, but the carrier survives, is that of trucks with loads of gravel travelling a road network in order to fill holes caused by, for example, flash flooding.

In this paper we consider the following synchronous model of repairable holes. All agents are initially in the start (or *source*) node s and move through the network synchronously, traversing one edge in one time step. If an agent encounters a hole at a vertex v, it will sacrifice itself to repair it. After the repair, which is completed instantenously, the node functions normally and other agents can pass through it as if the fault never existed. The first agent encountering a given hole must repair it. If two or more agents encounter the same hole at the same time, one of them repairs it, while the other agents can continue with their exploration. Given a network (the whole topology is known in advance), node s and k agents, the aim is to design an *exploration algorithm*, which repairs all holes in the network within a small number of steps.

Our results. Since an exploration algorithm must ensure that each node is visited by at least one agent, then $\Omega(n/k + D)$ is an obvious lower bound on the required number of steps, where n and D are the number of nodes and the diameter of the network, respectively. We show in Section 3 that a simple algorithm completes exploration in $O\left(n/k + D \log f / \log \log f\right)$ steps, where $f = \min\{n/k, n/D\}$, if there are at most $k/2$ holes. In Section 4 we prove that this algorithm is asymptotically optimal in the worst case. We do this by showing a tree network $T(n,d)$ with $\Theta(n)$ nodes and diameter $\Theta(d)$, for any $n \geq d \geq 1$, such that for any deterministic exploration algorithm, there is a placement of at most $k/2$ holes in $T(n,d)$, for any $n \geq k \geq 2$, which forces the algorithm to run for $\Omega\left(n/k + d \log f / \log \log f\right)$ steps. The assumption that the number of holes is at most $k/2$ and possible generalisations are discussed in Section 5.

The previous work which is most closely related to our results are the multi-agent black-hole search algorithms which we gave in [1]. Our repairable-hole model is stronger than the black-hole model, so multi-agent black-hole search algorithms can be adapted to the repairable hole model in a natural way. The two black-hole search algorithms from [1] adapted to the repairable-hole model run in $O((n/k)\log n/\log\log n + kD)$ and $O(n/k + D\sqrt{n})$ steps. No lower bounds were given in [1] other than the obvious $\Omega(n/k + D)$.

2 Preliminaries

The input is an undirected graph $G = (E, V)$ representing the network, a source node $s \in V$ and an integer $k \geq 2$, which is the number of available agents. We assume that n, the number of nodes in G, is sufficiently large, and the diameter of G is $D \geq 5$. The latter assumption can be dropped by replacing in the algorithm and its analysis D with $\bar{D} = \max\{D, 5\}$. We denote by b the (unknown) initial number of holes and assume that $b \leq k/2$.

A connected subgraph S of a tree T is called a subtree of T. If T is a rooted tree, then a subtree S is rooted at the node of S closest to the root of T. The size of a subtree S is defined as the number of nodes in S and is denoted by $|S|$. If T is rooted, then for a node v in T, *the* subtree of T rooted at v is the subtree of T spanning node v and all its decendants. During exploration of a network, the *active agents* are the agents which are still alive, that is, the agents who have not repaired any hole yet.

3 Exploration Algorithm

A natural approach to exploring a network with multiple agents is first to compute a cover of the network with connected regions, and then to have separate agents explore separate regions in parallel. Considering regions of size $\Theta(D)$ seems to be a good idea since an agent may need anyway $\Theta(D)$ steps to reach a region. We describe below a simple way of computing $q = O(n/D)$ regions, each of size $O(D)$, which cover the whole network.

For the problem of exploring a network which does not have any faults/holes, with the objective of having each node visited by at least one agent, such regions are the basis of a simple, asymptotically optimal $\Theta(n/k + D)$-steps algorithm. If $k < q$, then each agent explores $q/k = O(n/(kD))$ regions and exploration of one region takes $O(D)$ time (that is, each agent explores $O(n/k)$ nodes, and whole exploration is completed in $\Theta(n/k)$ steps). If $k \geq q$, then q agents explore the q regions in parallel in $\Theta(D)$ steps. The other agents may remain idle since this asymptotic bound cannot be improved.

Our algorithm, for our model where each agent can repair only one hole, consists of a sequence of rounds, and each round is similar to the exploration scheme sketched above. After each round, the active agents return to the start node s and learn which regions may still contain (unrepaired) holes. These are the regions from which no agent returned. In the next round, the active agents are sent in equal size groups to explore these remaining regions in parallel. We describe below the details of the algorithm and give in Theorem 1 a worst-case bound on its running time (the number of steps).

Computation of regions. The algorithm first computes a breadth-first tree T of G rooted at s. Then it computes subtrees $S_1, S_2, \ldots, S_q$ of tree T with the following properties:

1. each subtree S_i has size at most D;
2. subtrees $S_1, S_2, \ldots, S_q$ cover tree T, that is, each node of T belongs to at least one subtree S_i;
3. $q = O(n/D)$.

These subtrees define the regions of the network and can be computed by the following straightforward iterative process. Let $T_1 = T$. At the beginning of iteration i, $i \geq 1$, tree T is covered by subtrees $S_1, \ldots, S_{i-1}$ and T_i. The subtree T_i is rooted at s and represents the remaining part of tree T which has yet to be covered. If the size of T_i is at most D, then we set $S_i = T_i$, $q = i$, and the covering of T has been computed. Otherwise, when the size of T_i is greater than D, let v_i be a node in T_i such that the size of the subtree of T_i rooted at v_i is at least D but the size of each subtree of T_i rooted at a child of v_i is less than D. Let $w_1, w_2, \ldots, w_j$ be the children of node v_i ordered according to the sizes of the subtrees of T_i rooted at these nodes, starting from the largest. Let r be the first index such that the sum of the sizes of the subtrees rooted at nodes $w_1, w_2, \ldots, w_r$ is at least $D/2$. Set S_i to the subtree of T_i comprising node v_i and the subtrees rooted at nodes $w_1, w_2, \ldots, w_r$. If S_i includes all children of v_i in T_i (that is, if S_i is the subtree of T_i rooted at v_i), then tree T_{i+1} is obtained from T_i by cutting off subtree S_i (observe that in this case v_i cannot be the root of T_i, so $T_{i+1} \neq \emptyset$). Otherwise, tree T_{i+1} is obtained from T_i by cutting off the subtrees rooted at nodes $w_1, w_2, \ldots, w_r$. Node v_i and its subtrees rooted at nodes $w_{r+1}, \ldots, w_j$ remain in T_{i+1}.

It should be clear that the subtrees $S_1, S_2, \ldots, S_q$ constructed in this way satisfy Properties 1 and 2. If $i \leq q - 1$, then S_i has at least $D/2 + 1$ nodes. If a node in T belongs to two different subtrees S_r and S_j, $r \neq j$, then it must be the root of at least one of them. Thus sets $V(S_i) \setminus \{v_i\}$, for $i = 1, 2, \ldots, q$, are pairwise disjoint and each of them other than the last one has at least $D/2$ nodes. This implies $(D/2)(q - 1) < n$, so $q < (2n)/D + 1$ and Property 3 is satisfied as well.

Exploration. In our exploration algorithm, the agents move through the graph only along the edges of tree T. The exploration consists of two phases, and each of them consists of rounds. At the beginning of one round all active agents are at the source node s. They are partitioned into groups and each group explores during this round one of the *active trees* S_i. Initially all trees S_i are active. A group of l agents explores a tree S_i by first walking together along the path in T from s to the root of S_i, then fully traversing S_i in $O(|S_i|)$ steps (say, by following an Euler tour around tree S_i, which passes twice along each edge of

S_i), and finally walking back to s. All agents in one group keep moving together from node to node. If they encounter a hole which has not been repaired yet (either on the path from s to the root of S_i or within S_i), then one of them repairs it, while the other agents move on. If two or more groups meet at a hole, then one agent from one group (an arbitrary agent from an arbitrary group) repairs it.

If at least one agent from the group exploring tree S_i returns back to s, then all holes in S_i and on the path in T from s to S_i have been repaired, and tree S_i is no longer active. If no agent from this group returns to s, then tree S_i remains active, but we know that at least l additional holes must have been repaired.

During phase 1 of the algorithm, each tree S_i is explored with only one agent (single-agent groups). More specifically, in round 1 of phase 1, the k agents explore in parallel trees $S_1, S_2, \ldots S_k$, one agent per tree. In round 2, the $k' \leq k$ agents active at the beginning of this round explore in parallel the next k' trees $S_{k+1}, S_{k+2}, \ldots, S_{k+k'}$, and so on, until each tree has been explored once. If $k \geq q$, then there is only one round in phase 1. At the end of phase 1, there are at most b remaining active trees, because for each of these trees at least one hole must have been repaired (either in this tree or on the path from s to this tree).

We now describe phase 2 of the exploration. Let k_j, m_j and b_j denote the number of active agents, the number of active trees and the (unknown) number of remaining holes, respectively, at the beginning of round j, $j \geq 1$. During round j, the k_j active agents are partitioned into groups of size $\lfloor k_j/m_j \rfloor$ (some active agents may be left idle, if k_j/m_j is not an integer), and the groups explore in parallel the remaining m_j active trees, one group per tree. If a tree S_i remains active at the end of this round, then each of the $\lfloor k_j/m_j \rfloor$ agents assigned to this tree must have repaired one hole. Thus at least $m_{j+1}\lfloor k_j/m_j \rfloor$ additional holes are repaired during round j. The exploration ends when no active tree is left.

Theorem 1. *Assuming $k \geq 2b$, the exploration algorithm runs in*

$$O\left(\frac{n}{k} + D\frac{\log f}{\log \log f}\right) \tag{1}$$

steps, where $f = \min\{n/k, n/D\}$.

Proof. Each round consists of $O(D + \max\{|S_i|\}) = O(D)$ steps. The assumption that $k \geq 2b$ implies that there are always at least $k/2$ active agents. Thus during phase 1, at least $k/2$ new trees are explored in each round other than the last one. Therefore the number of rounds in phase 1 is at most $q/(k/2) + 1 = O(1 + n/(kD))$.

Now we bound the number of rounds in phase 2 using numbers k_j, m_j and b_j introduced above. Consider any round j other than the last one ($m_{j+1} \geq 1$). In this round at least $m_{j+1}\lfloor k_j/m_j \rfloor$ additional holes are repaired, so we have

$$b_{j+1} \leq b_j - m_{j+1}\left\lfloor \frac{k_j}{m_j} \right\rfloor \leq b_j - \frac{m_{j+1}}{m_j}\frac{k_j}{2} \leq b_j - \frac{m_{j+1}}{m_j}\frac{k}{4}. \tag{2}$$

The second inequality above holds because $k_j \geq m_j$ (we have $m_j \leq m_1 \leq b \leq k/2 \leq k_j$). The third inequality holds because $k_j \geq k/2$. Using (2) we get

$$0 \leq b_{j+1} \leq b_1 - \frac{k}{4}\left(\frac{m_2}{m_1} + \frac{m_3}{m_2} + \cdots + \frac{m_{j+1}}{m_j}\right)$$

$$\leq b_1 - \frac{jk}{4}\left(\frac{m_{j+1}}{m_1}\right)^{1/j}$$

$$\leq \frac{k}{2} - \frac{jk}{4}\left(\frac{m_{j+1}}{m_1}\right)^{1/j}. \tag{3}$$

The third inequality above holds because the geometric mean $(m_{j+1}/m_1)^{1/j}$ of numbers m_{i+1}/m_i, $i = 1, 2, \ldots, j$, is not greater than their arithmetic mean. Inequality (3) implies

$$\frac{m_1}{m_{j+1}} \geq \left(\frac{j}{2}\right)^j. \tag{4}$$

We have

$$m_1 \leq q \leq \frac{4n}{D}. \tag{5}$$

Let round $j + 2$ be the last one. We have $m_{j+1} \geq m_{j+2} \geq 1$. It also must hold that $k_{j+1}/m_{j+1} \leq 2D$, or otherwise round $j + 1$ would be the last one. Indeed, if $k_{j+1}/m_{j+1} > 2D$, then each tree S_i active at the beginning of round $j + 1$ would be explored during this round with $\lfloor k_{j+1}/m_{j+1} \rfloor \geq 2D \geq |S_i| + D$ agents, so all holes in S_i and on the path from s to S_i would be repaired in this round. Therefore,

$$m_{j+1} \geq \max\left\{1, \frac{k_{j+1}}{2D}\right\} \geq \max\left\{1, \frac{k}{4D}\right\}. \tag{6}$$

Inequalities (5) and (6) imply

$$\frac{m_1}{m_{j+1}} \leq \frac{4n/D}{\max\{1, k/(4D)\}} \leq 16\frac{n}{\max\{k, D\}} = 16\min\left\{\frac{n}{k}, \frac{n}{D}\right\} \leq 16f. \tag{7}$$

Inequalities (4) and (7) imply

$$\left(\frac{j}{2}\right)^j \leq 16f,$$

so

$$j = O\left(\frac{\log f}{\log \log f}\right).$$

Thus the total number of steps throughout the whole exploration algorithm is $O(D(n/(kD) + j))$, which is the bound (1). □

4 Lower Bound

For positive integers $n \geq d \geq 1$, we define $T(n, d)$ as the tree which is rooted at node s and has the following $\lfloor n/d \rfloor$ subtrees. Each subtree of the root is a d-node path with d leaf nodes attached to its end. Thus tree $T(n, d)$ has $1 + 2d\lfloor n/d \rfloor = \Theta(n)$ nodes and diameter $\Theta(d)$.

We show that for any (deterministic) exploration algorithm for this tree network, we (as the adversary) can place holes in such a way that the agents will be forced to go up and down the tree $\Omega(\log f / \log \log f)$ times. To decide where holes should be placed, we keep watching the movement of the agents and decide *not* to put any additional holes in the subtrees of $T(n, d)$ where currently relatively many agents are. These agents will have to go back up to the root of $T(n, d)$ and then down into other subtrees, where relatively few agents have gone before, to look for further holes.

Theorem 2. *For integers $n \geq d \geq 1$ and $n \geq k \geq 2$, and any algorithm exploring tree $T = T(n, d)$ with k agents starting from the root, there exists a placement of $b \leq k/2$ holes in T which forces the algorithm to run in*

$$\Omega\left(\frac{n}{k} + d\frac{\log f}{\log \log f}\right). \tag{8}$$

steps, where $f = \min\{n/k, n/d\}$.

Proof. Let $\mathcal{A}$ be any algorithm exploring tree $T = T(n, d)$ with k agents, and let

$$\alpha = \max\left\{i - \text{positive integer}: i^i \leq f\right\}. \tag{9}$$

We have

$$\alpha = \Theta\left(\frac{\log f}{\log \log f}\right),$$

and we assume in our calculations that f is sufficiently large, so that $\alpha \geq 4$. Since for any exploration algorithm and for any network the number of steps must be $\Omega(n/k)$, it suffices to show that algorithm $\mathcal{A}$ requires in the worst case $\Omega(d \log f / (\log \log f))$ steps.

We place the holes only at leaves of T. We simulate algorithm $\mathcal{A}$ to decide which subtrees get how many holes. We look at intervals of d consecutive steps and call them *rounds* of the exploration. We will place the holes in such a way that the algorithm needs at least α rounds to complete the exploration.

We look first at round 1, and observe that the way the agents move during this round is independent of the distribution of holes in the leaves. This is because no agent can reach further than distance d from the root, so cannot reach any leaf of T. Let $n_0 = \lfloor n/d \rfloor$ and let $\mathcal{S}^{(0)}$ be the set of the n_0 subtrees of the root s in T. We sort these subtrees into the ascending sequence $S_1^{(0)}, ..., S_{n_0}^{(0)}$, according to the number of agents in the subtrees at the end of the round. We take the $n_1 = \lfloor n_0/\alpha \rfloor$ lowest subtrees in this order to form the set $\mathcal{S}^{(1)}$ (where α is given in (9)). We decide that there are no holes in the subtrees in $\mathcal{S}^{(0)} \setminus \mathcal{S}^{(1)}$, that is, all holes are in the subtrees in $\mathcal{S}^{(1)}$. Informally, if many agents have gone into a subtree, then we decide not to put any holes in this subtree (this would be a subtree in $\mathcal{S}^{(0)} \setminus \mathcal{S}^{(1)}$). Instead we will be putting holes in the subtrees where relatively few agents have gone (the subtrees in $\mathcal{S}^{(1)}$).

Now we observe what the algorithm does during round 2. Note that at the beginning of this round, there must be a subtree in $\mathcal{S}^{(0)} \setminus \mathcal{S}^{(1)}$ with at most $\lfloor k/(n_0 - n_1) \rfloor$ agents, so each subtree in $\mathcal{S}^{(1)}$ has at most $\lfloor k/(n_0 - n_1) \rfloor$ agents. Whenever an agent visits during this round a new leaf of a subtree in $\mathcal{S}^{(1)}$ (not visited before by any agent), we place a hole there. Only the agents which are in a subtree at the beginning of a round can explore a leaf of this subtree during this round. Thus at the end of round 2, in each subtree in $\mathcal{S}^{(1)}$, all but at most

$$\left\lfloor \frac{k}{n_0 - n_1} \right\rfloor$$

leaves are still unexplored. We sort the subtrees in $\mathcal{S}^{(1)}$ into the ascending sequence $S_1^{(1)}, ..., S_{n_1}^{(1)}$, according to the number of agents in the subtrees at the end of the round, and we take the $n_2 = \lfloor n_1/\alpha \rfloor$ lowest subtrees in this order to form the set $\mathcal{S}^{(2)}$. We decide not to put more holes in the subtrees in $\mathcal{S}^{(1)} \setminus \mathcal{S}^{(2)}$, that is, all additional holes will be placed in the subtrees in $\mathcal{S}^{(2)}$.

We now generalise round 2 into a round $i \geq 2$. At the beginning of this round, we have (by induction) a set $\mathcal{S}^{(i-1)}$ of n_{i-1} subtrees of the root of T such that each of these subtrees has at most $\lfloor k/(n_{i-2} - n_{i-1}) \rfloor$ agents. We have already placed holes at some leaves of T in previous rounds. We will not put more holes at leaves of the subtrees in $\mathcal{S}^{(0)} \setminus \mathcal{S}^{(i-1)}$, but we may put additional holes in subtrees in $\mathcal{S}^{(i-1)}$. We observe how the agents explore tree T during this round, and whenever an agent visits a new leaf of a subtree in $\mathcal{S}^{(i-1)}$ (not visited before by any agent), we place a hole there. Only the agents which are in a subtree at the beginning of a round can explore a leaf of this subtree during this round. Thus at the end of round i, in each subtree in $\mathcal{S}^{(i-1)}$, all but at most

$$\left\lfloor \frac{k}{n_0 - n_1} \right\rfloor + \left\lfloor \frac{k}{n_1 - n_2} \right\rfloor + \cdots + \left\lfloor \frac{k}{n_{i-2} - n_{i-1}} \right\rfloor \tag{10}$$

leaves are still unexplored, and the number of holes we have placed in the network so far is at most

$$n_1 \left\lfloor \frac{k}{n_0 - n_1} \right\rfloor + n_2 \left\lfloor \frac{k}{n_1 - n_2} \right\rfloor + \cdots + n_{i-1} \left\lfloor \frac{k}{n_{i-2} - n_{i-1}} \right\rfloor \tag{11}$$

We sort the subtrees in $\mathcal{S}^{(i-1)}$ into the ascending sequence $S_1^{(i-1)}, ..., S_{n_{i-1}}^{(i-1)}$, according to the number of agents in the subtrees at the end of round i, and take the $n_i = \lfloor n_{i-1}/\alpha \rfloor$ lowest subtrees in this order to form the set $\mathcal{S}^{(i)}$. We decide not to put more holes in the subtrees in $\mathcal{S}^{(i-1)} \setminus \mathcal{S}^{(i)}$, that is, all additional holes will be placed in the subtrees in $\mathcal{S}^{(i)}$. At the end of the round there must be a subtree in $\mathcal{S}^{(i-1)} \setminus \mathcal{S}^{(i)}$ with at most $\lfloor k/(n_{i-1} - n_i) \rfloor$ agents, so each subtree in $\mathcal{S}^{(i)}$ has at most $\lfloor k/(n_{i-1} - n_i) \rfloor$ agents.

Round i is not the last one, if the following three conditions hold.

1. $n_i \geq 1$ (that is, there is at least one subtree in the set $\mathcal{S}^{(i)}$).
2. The upper bound (10) on the number of explored nodes in a subtree in $\mathcal{S}^{(i)}$ at the end of round i is less than d, the number of leaves in one subtree.
3. The upper bound (11) on the number of holes placed in the network by the end of round i is less than $k/2$.

We show that the above three conditions hold for $i = \alpha/4$, assuming for convenience that $\alpha/4$ is integral. We have $n_0 = \lfloor n/d \rfloor \geq f \geq \alpha^\alpha$, and for $i = 0, 1, \ldots, \alpha$,

$$n_i = \lfloor n_{i-1}/\alpha \rfloor, \tag{12}$$

so

$$n_i \geq \alpha^{\alpha-i}. \tag{13}$$

In particular, $n_{\alpha-1} \geq \alpha \geq 1$. Thus Condition 1 holds for $i = \alpha - 1$, so must also hold for $i = \alpha/4$.

Now we bound (10) for $i = \alpha - 1$:

$$\left\lfloor \frac{k}{n_0 - n_1} \right\rfloor + \left\lfloor \frac{k}{n_1 - n_2} \right\rfloor + \cdots + \left\lfloor \frac{k}{n_{\alpha-3} - n_{\alpha-2}} \right\rfloor$$
$$\leq \frac{2k}{n_0} + \frac{2k}{n_1} + \cdots + \frac{2k}{n_{\alpha-3}}$$
$$\leq \frac{2k}{n_{\alpha-3}} \left(\frac{1}{\alpha^{\alpha-3}} + \frac{1}{\alpha^{\alpha-4}} + \cdots + 1 \right)$$
$$\leq \frac{4k}{n_{\alpha-3}}. \tag{14}$$

The first inequality above holds because $n_i \leq n_{i-1}/\alpha \leq n_{i-1}/2$, $(\alpha \geq 4)$. The second inequality holds because $n_{i+j} \leq n_i/\alpha^j$, so $n_i \geq n_{\alpha-3}\alpha^{\alpha-3-i}$. If $k \leq d$, then we continue (14) in the following way:

$$\frac{4k}{n_{\alpha-3}} \leq \frac{4d}{\alpha^3} < d,$$

where the first inequality follows from (13) and the second one from the assumption that $\alpha \geq 4$. If $k > d$, then $\alpha^\alpha \leq \lfloor f \rfloor = \lfloor n/k \rfloor$, so

$$n_0 = \left\lfloor \frac{n}{d} \right\rfloor \geq \left\lfloor \frac{k}{d} \right\rfloor \left\lfloor \frac{n}{k} \right\rfloor \geq \left\lfloor \frac{k}{d} \right\rfloor \alpha^\alpha,$$

and (13) becomes

$$n_i \geq \left\lfloor \frac{k}{d} \right\rfloor \alpha^{\alpha-i}. \tag{15}$$

Thus in this case we can continue (14) in the following way:

$$\frac{4k}{n_{\alpha-3}} \leq \frac{4k}{\lfloor k/d \rfloor \alpha^3} \leq \frac{8d}{\alpha^3} < d,$$

where the last inequality follows from $\alpha \geq 4$. Since in both cases the bound (10) is less than d for $i = \alpha - 1$, then Condition 2 holds for $i = \alpha - 1$, so it must also hold for $i = \alpha/4$.

Now we bound (11) for $i = \alpha/4$:

$$
\begin{aligned}
n_1 \left\lfloor \frac{k}{n_0 - n_1} \right\rfloor &+ n_2 \left\lfloor \frac{k}{n_1 - n_2} \right\rfloor + \cdots + n_{i-1} \left\lfloor \frac{k}{n_{i-2} - n_{i-1}} \right\rfloor \\
&\leq \frac{k}{n_0/n_1 - 1} + \frac{k}{n_1/n_2 - 1} + \cdots + \frac{k}{n_{i-2}/n_{i-1} - 1} \\
&\leq \frac{\alpha}{4} \cdot \frac{k}{\alpha - 1} < \frac{k}{2}
\end{aligned}
$$

The second inequality above follows from (12) and the fact that we consider $i = \alpha/4$. Thus also Condition 3 holds for $i = \alpha/4$.

We conclue that algorithm $\mathcal{A}$ requires in the worst case $d\alpha/4 = \Omega(d \log f/(\log \log f))$ steps. □

5   Conclusions

We have introduced a variation of the black hole search model by enabling the agents to repair the faulty nodes. We have shown matching worst-case upper and lower bounds on the running time of a k-agent exploration algorithm in this new model. These bounds imply that the trivial lower bound of $\Omega(n/k + D)$ is not always tight.

Our arguments assume that the number of holes b is at most $k/2$. This assumption can be weakened to $b \leq ck$ for an arbitrary constant $c < 1$ without affecting the asymptotic bounds given in Theorems 1 and 2. To repair all holes and to have at least one agent left at the end of the exploration, we only need to assume that $b \leq k - 1$. An adaptation of our upper and lower bound arguments to this general case would change the bounds by replacing n/k with $n/(k - b)$.

Our algorithm is *worst-case* asymptotically optimal. If we view it as an *approximation* algorithm, then its ratio bound is $O(\log f/\log \log f)$ and it can be shown that it is not constant. An interesting question is whether one can find a better approximation algorithm. The problem of designing a k-agent exploration with the minimum number of steps for the special case when there are no faults (or when each agent can repair any number of faults) is equivalent to the following k-TSP problem. Find k cycles which cover all nodes of a given undirected graph, minimising the length of the longest cycle. A $(5/2 - 1/k)$-approximation algorithm for this problem is shown in [10]. It is not clear, however, whether this or other approximation algorithms for the k-TSP problem could be used effectively in our model, when there may be many faults and agents die after repairing one fault.

Most of the work on the synchronous black hole search problems, and this paper as well, assume that the topology of the whole network is known in advance and only the locations of faults are unknown. An interesting direction for further research is to consider the case when the topology of the network is not known or only partially known.

References

1. Cooper, C., Klasing, R., Radzik, T.: Searching for black-hole faults in a network using multiple agents. In: Shvartsman, M.M.A.A. (ed.) OPODIS 2006. LNCS, vol. 4305, pp. 320–332. Springer, Heidelberg (2006)
2. Czyzowicz, J., Kowalski, D.R., Markou, E., Pelc, A.: Searching for a black hole in tree networks. In: Higashino, T. (ed.) OPODIS 2004. LNCS, vol. 3544, pp. 67–80. Springer, Heidelberg (2005)
3. Czyzowicz, J., Kowalski, D., Markou, E., Pelc, A.: Complexity of searching for a black hole. Fundamenta Informaticae 71(2-3), 229–242 (2006)
4. Dobrev, S., Flocchini, P., Kralovic, R., Ruzicka, P., Prencipe, G., Santoro, N.: Black hole search in common interconnection networks. Networks 47(2), 61–71 (2006); (Preliminary version: Black hole search by mobile agents in hypercubes and related networks. In: Proceedings of the 6th International Conference on Principles of Distributed Systems, OPODIS 2002, pp. 169–180 (2002))
5. Dobrev, S., Flocchini, P., Kralovic, R., Santoro, N.: Exploring an unknown graph to locate a black hole using tokens. In: Navarro, G., Bertossi, L., Kohayakwa, Y. (eds.) Fourth IFIP International Conference on Theoretical Computer Science, TCS 2006. IFIP International Federation for Information Processing, vol. 209, pp. 131–150. Springer, Heidelberg (2006)
6. Dobrev, S., Flocchini, P., Prencipe, G., Santoro, N.: Mobile search for a black hole in an anonymous ring. Algorithmica 48(1), 67–90 (2007); (Preliminary version in: Distributed Computing. In: 15th International Conference, DISC 2001, Proceedings. LNCS, vol. 2180, pp. 166-179. Springer, Heidelberg (2001)
7. Dobrev, S., Flocchini, P., Prencipe, G., Santoro, N.: Searching for a black hole in arbitrary networks: Optimal mobile agents protocols. Distributed Computing 19(1), 1–18 (2006); (Preliminary version in: Proceedings of the 21st ACM Symposium on Principles of Distributed Computing, PODC 2002, pp. 153-161. ACM, New York (2002)
8. Dobrev, S., Kralovic, R., Santoro, N., Shi, W.: Black hole search in asynchronous rings using tokens. In: Calamoneri, T., Finocchi, I., Italiano, G.F. (eds.) CIAC 2006. LNCS, vol. 3998, pp. 139–150. Springer, Heidelberg (2006)
9. Flocchini, P., Santoro, N.: Distributed security algorithms by mobile agents. In: Chaudhuri, S., Das, S.R., Paul, H.S., Tirthapura, S. (eds.) ICDCN 2006. LNCS, vol. 4308, pp. 1–14. Springer, Heidelberg (2006)
10. Frederickson, G.N., Hecht, M.S., Kim, C.E.: Approximation algorithms for some routing problems. SIAM J. Comput. 7(2), 178–193 (1978)
11. Hohl, F.: Time limited blackbox security: Protecting mobile agents from malicious hosts. In: Vigna, G. (ed.) Mobile Agents and Security. LNCS, vol. 1419, pp. 92–113. Springer, Heidelberg (1998)
12. Hohl, F.: A framework to protect mobile agents by using reference states. In: Proceedings of the 20th International Conference on Distributed Computing Systems, ICDCS 2000, pp. 410–417. IEEE Computer Society, Los Alamitos (2000)
13. Klasing, R., Markou, E., Radzik, T., Sarracco, F.: Approximation bounds for black hole search problems. In: Anderson, J.H., Prencipe, G., Wattenhofer, R. (eds.) OPODIS 2005. LNCS, vol. 3974. Springer, Heidelberg (2006)

14. Klasing, R., Markou, E., Radzik, T., Sarracco, F.: Hardness and approximation results for black hole search in arbitrary networks. Theor. Comput. Sci. 384(2-3), 201–221 (2007); (Preliminary version in: Structural Information and Communication Complexity. In: 12th International Colloquium, SIROCCO 2005, Proceedings. LNCS, vol. 3499, pp. 200–215. Springer, Heidelberg (2005)
15. Ng, S.K., Cheung, K.W.: Protecting mobile agents against malicious hosts by intention spreading. In: Arabnia, H.R. (ed.) Proceedings of the International Conference on Parallel and Distributed Processing Techniques and Applications, PDPTA 1999. CSREA Press, pp. 725–729 (1999)
16. Sander, T., Tschudin, C.: Protecting mobile agents against malicious hosts. In: Vigna, G. (ed.) Mobile Agents and Security. LNCS, vol. 1419, pp. 44–60. Springer, Heidelberg (1998)

Remembering without Memory:
Tree Exploration
by Asynchronous Oblivious Robots

Paola Flocchini[1,*], David Ilcinkas[2,**], Andrzej Pelc[3,*], and Nicola Santoro[4,*]

[1] University of Ottawa, Canada
flocchin@site.uottawa.ca
[2] CNRS, Université Bordeaux I, France
david.ilcinkas@labri.fr
[3] Université du Québec en Outaouais, Canada
pelc@uqo.ca
[4] Carleton University, Canada
santoro@scs.carleton.ca

Abstract. In the effort to understand the algorithmic limitations of computing by a swarm of robots, the research has focused on the minimal capabilities that allow a problem to be solved. The weakest of the commonly used models is ASYNCH where the autonomous mobile robots, endowed with visibility sensors (but otherwise unable to communicate), operate in Look-Compute-Move cycles performed asynchronously for each robot. The robots are often assumed (or required to be) oblivious: they keep no memory of observations and computations made in previous cycles.

We consider the setting when the robots are dispersed in an anonymous and unlabeled graph, and they must perform the very basic task of *exploration*: within finite time every node must be visited by at least one robot and the robots must enter a quiescent state. The complexity measure of a solution is the number of robots used to perform the task.

We study the case when the graph is an arbitrary tree and establish some unexpected results. We first prove that there are n-node trees where $\Omega(n)$ robots are necessary; this holds even if the maximum degree is 4. On the other hand, we show that if the maximum degree is 3, it is possible to explore with only $O(\frac{\log n}{\log \log n})$ robots. The proof of the result is constructive. Finally, we prove that the size of the team is asymptotically *optimal*: we show that there are trees of degree 3 whose exploration requires $\Omega(\frac{\log n}{\log \log n})$ robots.

* Partially supported by NSERC. Andrzej Pelc is also partially supported by the Research Chair in Distributed Computing at the Université du Québec en Outaouais.
** This work was done during the stay of David Ilcinkas at the Research Chair in Distributed Computing of the Université du Québec en Outaouais and at the University of Ottawa, as a postdoctoral fellow.

A. Shvartsman and P. Felber (Eds.): SIROCCO 2008, LNCS 5058, pp. 33–47, 2008.
© Springer-Verlag Berlin Heidelberg 2008

1 Introduction

An important goal of theoretical research on computing by autonomous mobile robots has been to understand the algorithmic limitations of computing in such settings. The research has thus focused on the minimal capabilities that allow a problem to be solved by a swarm of robots. In the investigations, three models are commonly used: SYNCH, SSYNCH, and ASYNCH; the fully synchronous model SYNCH being the strongest, the asynchronous model ASYNCH being the weakest, the semi-synchronous model SSYNCH lying in between (see, e.g., [2,3,4,5,9,13,14,15,16]).

In ASYNCH the autonomous mobile robots are endowed with visibility sensors (but otherwise unable to communicate), are anonymous, are oblivious, and operate in asynchronous Look-Compute-Move cycles. In one cycle, a robot uses its sensors to obtain a snapshot of the current configuration (Look); then, based on the perceived configuration, it computes a destination (Compute), and moves there (Move); if the destination is the current position, the robot is said to perform a null move. Cycles are performed asynchronously for each robot; this means that, even if the operations are instantaneous, the time between Look, Compute, and Move operations is finite but unbounded, and is decided by the adversary for each action of each robot. The robots are oblivious: they keep no memory of observations and computations made in previous cycles. All robots are identical and execute the same algorithm. It is usually assumed that robots are capable of multiplicity detection: during a Look operation, a robot can determine if at some location there are no robots, there is one robot, or there are more than one robots; however, in the latter case, the robot might not be capable of determining the exact number of robots.

The asynchrony implies that a robot $\mathcal{R}$ may observe the position of the robots at some time t; based on that observation, it may compute the destination at some time $t' > t$, and Move to its destination at an even later time $t'' > t'$; thus it might be possible that at time t'' some robots are in different positions from those previously perceived by $\mathcal{R}$ at time t, because in the meantime they performed their Move operations (possibly several times). Since robots are oblivious, i.e., they do not have any memory of past observations, the destination is decided by a robot during a Compute operation solely on the basis of the location of other robots perceived in the previous Look operation.

In the literature, the ASYNCH model is used by researchers in the study of the coordination and control of autonomous mobile robots in the two-dimensional plane, which we shall term the *continuous* scenario. The computational capabilities of these robots when the spacial universe is a network or a graph, a scenario that we shall term *discrete*, has been recently investigated in [8,11], where the graph is a ring. In the discrete scenario, the computed destination in each cycle is either the node where the robot is currently situated or a node adjacent to it.

An important feature of the discrete scenario is the fact that the graph is totally anonymous: not only nodes but also edges are unlabeled, and there are no port numbers at nodes. This gives additional power to the adversary at the time when a robot must move. Indeed, it may happen that two or more edges incident to a node v currently occupied by the deciding robot look identical in the snapshot obtained during the last Look action , i.e., there is an automorphism of the tree which fixes v, carries empty nodes to empty nodes, occupied nodes to occupied nodes, and multiplicities to multiplicities, and carries one edge to the other. In this case, if the robot decides to take one of the ports corresponding to these edges, it may take any of the identically looking ports. We assume the worst-case decision in such cases, i.e., that the actual port among the identically looking ones is chosen by an adversary. This is a natural worst-case assumption and our algorithm is also resistant to such adversarial decisions.

We continue the study of computational capabilities of robots under the discrete scenario by considering the very basic task of *exploration*: within finite time every node must be visited by at least one robot and the robots must enter a quiescent state. The complexity measure of a solution is the number of robots used to perform the task. The problem of exploring a graph has been extensively studied in the literature under a variety of assumptions (e.g. see [1,6,7,10,12]) but not in the setting considered here. The only exception is [8] where we proved that the minimum number $\rho(n)$ of robots that can explore a ring of size n is $O(\log n)$ and that $\rho(n) = \Omega(\log n)$ for arbitrarily large n.

In this paper we consider the case when the graph is an arbitrary tree and establish some unexpected results. We first prove that, in general, exploration cannot be done efficiently. More precisely we prove that there are n-node trees where $\Omega(n)$ robots are necessary; this holds even if the maximum degree is 4. We then prove the existence of a *complexity gap*. We show that if the maximum degree of the tree is 3 then it is possible to explore it with only $O(\frac{\log n}{\log \log n})$ robots. The proof of the result is constructive. Finally, we show that the size of the team used in our solution is asymptotically *optimal*: there are trees of degree 3, whose exploration requires $\Omega(\frac{\log n}{\log \log n})$ robots.

Due to space limitations, the proofs are omitted.

2   Terminology and Preliminaries

We consider a n-node anonymous unoriented tree. Some nodes of the tree are occupied by robots. We will always assume that in an initial configuration of robots there is at most one robot in each node. The number of robots is denoted by k. A *complete d-ary tree* is a rooted tree, all of whose internal nodes have d children and all of whose leaves are at the same distance from the root. Nodes v and w are *similar* if there exists an automorphism of the tree T which carries v to w.

In order to formally define what a robot perceives during a Look action, we introduce the notion of the *view* of a rooted tree T occupied by robots, from its root v. This is defined by induction on the height of the tree T. If T consists only of v then $View(T, v) = (x, \emptyset)$, where $x = 0$, $x = 1$, or $x = *$, if there is 0, 1, or more than 1 robot in v, respectively. If T is of positive height, let $v_1, \ldots, v_m$ be children of the root v, and let $T_1, \ldots, T_m$ be subtrees rooted at $v_1, \ldots, v_m$, respectively . Then $View(T, v) = (x, \{View(T_1, v_1), \ldots, View(T_m, v_m)\})$, where x has the same meaning as before. Now, the snapshot taken by a robot located at v is simply $View(T, v)$. This formalism captures two essential assumptions about the perceptions of robots. First, a robot can distinguish between nodes occupied by 0, 1, or more than 1 robot, but cannot distinguish between numbers larger than 1 of robots located at the same node. Second, subtrees rooted at children of a node are not ordered: this is captured by considering the set of respective views, and not their sequence, in the recursive definition.

Two robots located at nodes v and w are called *equivalent*, if $View(T, v) = View(T, w)$. A node that is not occupied by any robot is called *empty*. When a node is occupied by more than one robot, we say that there is a *tower* in this node. A robot that is not a part of a tower is called *free*.

An exploration algorithm is a function whose arguments are views, and whose value for any given view $View(T, v)$ is either v or the equivalence class of one of its neighbors, with respect to the following equivalence relation $\sim$: $w_1 \sim w_2$ if there exists an automorphism f of the tree which fixes v, carries empty nodes to empty nodes, free robots to free robots, towers to towers, and such that $f(w_1) = w_2$. Note that $w_1 \sim w_2$ is equivalent to $View(T, w_1) = View(T, w_2)$. If the equivalence class returned by the algorithm for some view has more than one element then the choice of the neighbor in this class to which the robot will actually move, belongs to the adversary. If the value is v, we say that the move of the robot for the given view is the *null move*.

We say that exploration of a n-node tree is possible with k robots, if there exists an algorithm which, starting from any initial configuration of k robots without towers, and for any behavior of the adversary controling asynchrony and choices between equivalent neighbors, explores the entire tree and brings all robots to a configuration in which they all remain idle, i.e., there exists a time t after which all nodes are explored and all subsequent moves of robots are null moves. In fact, our negative results hold even for this weak (implicit) stopping condition, and our positive results (algorithms) are valid even with the following stronger (explicit) stopping condition: for any execution of the algorithm, there exists a time t after which all nodes are explored, and each robot knows that no non-null move of any robot (including itself) will ever occur. Obviously, if $k = n$, the exploration is already accomplished, hence we always assume that $k < n$.

3   Exploration of Trees

In this section we prove that, in general, exploration of n-node trees might require $\Omega(n)$ robots (even if the max degree is 4); we prove that, on the other hand, the minimum number of robots sufficient to explore all n-node trees of maximum degree 3 is $\Theta(\frac{\log n}{\log \log n})$.

3.1   Exploration of Arbitrary Trees

We show that there are arbitrarily large trees of maximum degree 4 whose exploration requires $\Omega(n)$ robots.

Theorem 1. *Exploration of a n-node complete 3-ary tree requires $\Omega(n)$ robots.*

3.2 Exploration of Trees of Maximum Degree 3: Upper Bound

We prove the following upper bound on the size of the team of robots capable to explore all n-node trees of maximum degree 3.

Main Theorem. *There exists a team of $O(\frac{\log n}{\log\log n})$ robots that can explore all n-node trees of maximum degree 3, starting from any initial configuration.*

This result is proved by showing an exploration algorithm using $O(\frac{\log n}{\log\log n})$ robots.

Overview of the Algorithm. The main idea of the algorithm is the following. The entire tree is partitioned into two or three subtrees, the number of parts depending on the shape of the tree. Parts are explored one after another by a team of three robots that sequentially visit leaves of this part. Since individual robots do not have memory, a specially constructed, dynamic configuration of robots, called the "brain", keeps track of what has been done so far. More precisely, the brain counts the number of already visited leaves and indicates the next leaf to be visited. It is also the brain that requires most of the robots used in the exploration process. The reason why $\Theta(\log n/\log\log n)$ robots are sufficient for exploration, is that the counting process is efficiently organized. The counting module of the brain consists of disjoint paths of logarithmic lengths, which are appropriately marked by groups of robots of bounded size. Paths are of logarithmic lengths because longer paths cannot be guaranteed to exist in all trees of maximum degree 3. Inside each of these paths a tower moves, indicating a numerical value by its position in the path. The combination of these values yields the current value of the number of visited leaves. Since the number of leaves may be $\Theta(n)$, we need a number x of paths, which can produce $\Theta(n)$ combinations of values, i.e., such that $(\Theta(\log n))^x = \Theta(n)$. This is the reason of constructing $\Theta(\log n/\log\log n)$ paths and thus using $\Theta(\log n/\log\log n)$ robots. We show how to construct these paths in any tree of maximum degree 3, and how to organize the counting process. The latter is complicated by the asynchronous behavior of the robots. During the switch of the counter from value i to $i+1$ robots move in the paths and a snapshot taken during the transition period shows a "blurred" picture: the old value is already destroyed while the new one is not yet created. This could confuse the robots and disorganize the process. Thus we use two counters acting together. They both indicate value i, then one of them keeps this value and the other transits to $i+1$. When this is completed, the first counter transits to $i+1$ and so on. This precaution permits to keep track of the current value during the process of incrementation. During the exploration of one part of the tree, the brain is located in another part and controls exploration remotely. After completing the exploration of one part, the brain is moved to the already explored part in order to let the exploring agents visit the rest of the tree.

There are two main difficulties in our algorithm. The first is to break symmetries that can exist in configurations of robots, in order to let them act independently and reach appropriate target nodes, in particular during the construction of the brain. The second challenge is the construction and relocation of the brain, as well as organizing its proper functioning by coordinating the two counters, regardless of the behavior of the adversary that controls asynchrony.

The algorithm is divided into the following phases. Phase 1 consists in moving all robots down the tree oriented in a specific way, without creating a tower, in order to create a large zone free of robots. When no robot can move further down, a tower is created to mark the end of Phase 1. In Phase 2, robots are moved from one part of the tree and create the brain in another part. If there are local symmetries, a leader is elected and breaks them by relocating to specific nodes of the tree. This is done to let the robots move independently from one part of the tree to another and occupy target positions. As a consequence, one part becomes almost empty, which facilitates its exploration. Phase 2 ends when the brain is at its place, properly initialized, and there remain only a tower and a free robot in the other part, that will explore this part. Phase 3 is the actual exploration of the part not containing the brain (or the larger of the two parts not containing the brain). This is done by visiting its leaves, one similarity class after another. Inside a similarity class, leaves are explored in a DFS manner, the brain keeping track of the progress of exploration. This phase ends when the brain indicates that the exploring robots are at the last leaf of the explored part. In Phase 4 the brain is relocated to the already explored part, and the exploring robots move to one of the unexplored parts. Again, Phase 4 ends when all robots are in their places and the brain is properly reinitialized (with the indication that one part is already explored). Finally, in the remaining phases the rest of the tree is explored, similarly as in Phase 3. There is a mechanism in the algorithm that enables robots to see what is the current phase, in order to avoid circular behavior. This is implemented by a special arrangement of robots, called signal, whose value increments from phase to phase.

Tools and Basic Properties. Before giving a detailed description of the algorithm we present some concepts that we will use in this description, and prove their basic properties. Let T be a n-node tree of maximum degree 3. Consider a team of k robots, where $c \log n / \log \log n \leq k \leq 2c \log n / \log \log n$, for an appropriately chosen constant c, and $k \equiv 5 \pmod 6$. The conditions on the constant c are explicitly given after the description of the algorithm.

Pieces

For each internal node v, consider the number of nodes in each of the subtrees rooted at neighbors of v, and let n_v be the maximum of these numbers. It is well known that either there exists exactly one node v for which $n_v \leq (n-1)/2$ (the centroid), or there is exactly one edge $\{v, w\}$, for which $n_v = n_w = n/2$ (the bicentroid). In each case we consider the oriented tree from the centroid or bicentroid down to the leaves. We will say that the tree is rooted in the centroid or bicentroid and use the usual notions of parent and children of a node.

Next we define the subtrees of T, called its *pieces*. If T has a centroid of degree 2 then there are two pieces T_1 and T_2 which are rooted at children of the centroid. If T has a centroid of degree 3 then there are three pieces T_1, T_2 and T_3 which are rooted at children of the centroid. Finally, if T has a bicentroid then there are two pieces T_1 and T_2 which are rooted at nodes of the bicentroid. Without loss of generality we assume that sizes of T_1, T_2 and T_3 are non-increasing. Hence $(n-1)/3 \leq |T_1| \leq n/2$ and $n/4 \leq |T_2| \leq n/2$. For every piece, we define its *weight* as the number of robots located in it. Thus we talk about the heaviest piece, a heavier piece, etc. A piece T_i is called *unique* if there is no other piece whose root has the same view as the root of T_i.

Core Zone
A node in a piece is *a core node*, if the size of the subtree rooted at this node is larger than the size k of the entire team of robots. The set of core nodes in a piece is called the *core zone* of the piece.

Lemma 1. *In any rooted tree of size x and such that every internal node has at most two children, the size of the core zone is at least $\frac{x+1}{k+1} - 1$.*

Since the size of any of the two largest pieces is at least $n/4$, Lemma 1 implies that the size of the core zone of any of these pieces is at least $\frac{n \log \log n}{10 c \log n}$.

Descending Paths
The basic component of the brain is a *descending path*. This is a simple path in a piece Q, whose one extremity is its node closest to the root of Q. It will be called the beginning of the path. The other extremity will be called its end. The size of such a path is the number of its nodes. We need sufficiently many pairwise disjoint descending paths, each sufficiently long, for all parts of the brain. The construction is a part of the proof of the following lemma.

Lemma 2. *For any sufficiently large m, every tree of maximum degree 3 and of size m contains at least $\log^2 m$ pairwise disjoint descending paths of size at least $\frac{1}{4} \log m$.*

The core zone is a tree of maximum degree 3, rooted in the root of the piece and has size $m \geq \frac{n \log \log n}{10 c \log n}$. Hence, for sufficiently large n, Lemma 2 guarantees the existence of at least $\log^2 m \geq \log n \geq 5 \log n / \log \log n$ pairwise disjoint descending paths of size at least $\frac{1}{4} \log m \geq \frac{1}{8} \log n$ in any of the two largest pieces.

Modules of the Brain
The brain consists of four parts: two *counters*, the *semaphore* and the *garbage*. Descending paths forming these parts will be situated in the core zone of a piece, each of the paths at distance at least 3 from the others, in order to allow correct pairing of beginnings and ends.

We now describe the structure of a counter. This is a collection of $q \in \Theta(\frac{\log n}{\log \log n})$ pairwise disjoint descending paths, of sizes $L+1, L+2, \ldots L+q$, where $L \in \Theta(\log n)$. We take paths of different lengths in order to easily distinguish them. Nodes of the ith path are numbered 1 to $L+i$ (where 1 corresponds to the beginning). Two towers will be placed in the first and third nodes of each path, thus marking its beginning. Similarly, three towers will be placed at the end of each path, separated by empty nodes, thus marking its end. Moreover there will be two or three robots moving from node 7 to node $L-8$ of each path. If these robots are located in the same node (thus forming a tower), their position codes a numerical value. By combining these values on all paths, we obtain the value of the counter. Since on each path there are $L - 14$ available positions, the value of the counter is the resulting integer written in base $L - 14$.

Let $q = \lceil 2 \log n / \log \log n \rceil$ and $L_1 = \frac{1}{10} \log n$. Take q of the descending paths described in the proof of Lemma 2 (chosen in an arbitrary deterministic way, identical for all robots, and excluding p_1), and in the ith path, where $1 \leq i \leq q$, take the lower part of size $L_1 + i$. These will be the descending paths of the first counter. Similarly, let $L_2 = L_1 + q + 1$. Take a set of q descending paths, other than those used for the first counter and other than p_1. In the ith path, where $1 \leq i \leq q$, take the lower part of size $L_2 + i$. These will be the descending paths of the second counter.

Another module of the brain is the semaphore consisting of two of the descending paths constructed in the proof of Lemma 2 (again excluding p_1). In each of these paths take the lower part of distinct constant sizes. The beginning and end of each path is marked similarly as in the counter. Likewise, there are two or three robots moving in each of these paths, their possible locations restricted to node 7 and 8 in each path. In each path, if these robots are located in the same node (thus forming a tower), they code one bit. Thus the semaphore has 4 possible values $00, 01, 11, 10$.

Finally, the garbage is the first descending path p_1 constructed in the proof of Lemma 2. This path has the property that its beginning is at the root of the piece. This path has length at least $\frac{1}{8} \log n$, and thus larger than the total number of robots, for sufficiently large n. The role of the garbage is to store all robots of the brain not used for the counters and the semaphore. The garbage is filled by putting a tower or a robot every 5 nodes in the path, until all robots are disposed. Therefore the end of the path is marked similarly as for paths in the counter and the semaphore, but the beginning is left unmarked.

Ordering of Robots

We first define a total order $\sqsubset$ on the set of all views. Let $V = View(T, v)$ and $V' = View(T', v')$. If the height of T is smaller than the height of T' then $V \sqsubset V'$. Otherwise, if the height of both trees is 0 then $(x, \emptyset) \sqsubset (x', \emptyset)$, if $x \leq x'$, where $0 < 1 < *$. If the height of both trees is positive, the order of views is the lexicographic order on the sequences $(x, View(T_1, v_1), \ldots, View(T_m, v_m))$, where views at children are ordered increasingly by induction.

We now define the following total preorder $\leq$ on the robots in the rooted tree T. Let $\mathcal{R}_1$ and $\mathcal{R}_2$ be two robots located at nodes v_1, v_2, at distances d_1 and d_2 from the root. (In the case of the bicentroid, we consider the distance to its closer extremity.) We let $\mathcal{R}_1 \leq \mathcal{R}_2$, if and only if, $d_1 < d_2$, or $d_1 = d_2 \wedge View(T, v_1) \sqsubset View(T, v_2)$. Note that the equivalence relation induced by this preorder is exactly the equivalence between robots defined previously. We say that a robot is larger (smaller) than another one meaning the preorder $\leq$. A robot not equivalent to any other is called *solitaire*.

Lemma 3. *The number of equivalent robots in any piece is either 1 or even.*

It follows from Lemma 3 that any unique piece with an odd weight must contain a solitaire.

Description of Algorithm Tree-exploration

Phase 1. There is no tower in the snapshot.

Goal: Empty the core zones of all pieces and create one tower in one piece.

We first free the core zones by moving every robot to an empty child, as long as such a child exists, except for up to two robots that may move from one piece to another. As described below, these exceptional robots are solitaires. The objective here is to have a unique heaviest piece with the additional property that it is either of odd weight or completely occupied by robots (i.e. every node of the piece is occupied by a robot). This is always possible because $k \equiv 5 \pmod 6$. Indeed, if there are two heaviest pieces, then there must exist a third piece of odd weight, and thus a solitaire of this piece (whose existence is guaranteed by Lemma 3) can move to one of the heaviest pieces, thus breaking the tie. If there is a unique heaviest piece, but of even weight and not completely occupied by robots, then there must exist another piece of odd weight, and thus a solitaire of this piece (whose existence, again, is guaranteed by Lemma 3) can move to the heaviest piece. Note that the case of three heaviest pieces is impossible because k is not divisible by 3.

As soon as the required properties hold in a piece P and the core zones are empty (except for possibly one robot in the core zone of P), a tower is created outside the core zone of P by moving a solitaire to an occupied node in such a way that at least half the robots in P, including a solitaire, are located outside the subtree rooted at the tower. The latter precaution is taken to have enough robots to form and subsequently move towers in Phase 2 using the solitaire. The way this is done will be described in the sequel.

Phase 1 has been clearly identified by the absence of towers in the snapshot. Such an easy characterization is not available in the subsequent phases, hence we use a gadget called *signal* to identify them. A signal is a largest set of at least 4 towers situated on a descending path inside a piece, such that consecutive towers are separated by two empty nodes. The value of a signal is $x - 1$, where x is the number of towers in it. This value will indicate the number of the current phase.

Phase 2. There is at least one tower and no signal in the snapshot.

In this phase piece P can be recognized as the unique piece where there is a tower outside the core zone and Q as the largest among pieces other than P (in the case of a tie Q is any piece with robots in the core zone.) Notice that, at the beginning of Phase 2, the core zone of Q does not contain any robots. Hence there is room in it for the brain.

Goal: Construct and initialize the brain in the core zone of piece Q, prepare the other pieces for exploration, and create the signal.

Stage 1. Goal: Move robots from P in order to construct the brain in Q and prepare P for exploration.

We now describe the way to form towers in P and move them to appropriate places in the descending paths forming the brain in Q. Robots migrate from piece P to piece Q, one or two robots at a time. The next robot or pair of robots starts its trip from P to Q only after the previous one is at its place. The aim is to occupy target nodes by towers. Nodes in descending paths are filled one path after another in a DFS post-order of beginnings of the paths. Thus a tower occupies a node v only after all robots in the subtree rooted at v are in their target positions. This rule applies to all descending paths of the brain, except the garbage. The latter is constructed at the end, after all other parts of the brain are completed. This is possible because the descending path containing the garbage starts at the root of the piece (path p_1 described in the proof of Lemma 2). The above migration of towers is done until there remains only a single tower and a solitaire in P. This prepares P for exploration.

There are two difficulties in performing this migration, both due to symmetries in configurations of robots. The first difficulty is to form towers consisting of only two robots in P and the other is to place such a tower in a specific target node in Q. (We want to restrict the size of towers in order to be able to create many of them using the available robots).

The essence of the first difficulty is that equivalence classes of robots can be large and thus it may be difficult to form a single small tower. (For example, if all robots in a piece are equivalent and occupy the same level, a single small tower cannot be formed without outside help.) We solve this problem by using a solitaire to break symmetry between two equivalent robots. More precisely, the solitaire moves to meet one of the equivalent robots thus creating a tower of two robots. At the same time the other equivalent robot becomes a solitaire.

The essence of the second difficulty is that if there are at least two equivalent target positions that a tower could occupy, the adversary could break the tower at the time when the tower tries to go down from the least common ancestor of these target nodes, sending each of the robots forming the tower to a different target node. We solve this problem by using a solitaire to first break the symmetry between these target positions. This solitaire, called the *guide* of the tower, is placed in one of these positions, thus indicating that the tower should go to the closest of the equivalent positions. As soon as the tower reaches its target, the solitaire is again available to break other symmetries, either those encountered when forming towers in P or when placing them in Q.

Stage 1 ends with the brain constructed in the core zone of piece Q. Moreover, in piece P there remain only a single tower and a robot without towers in its ancestors.

Stage 2. Goal: Empty the third piece P' (other than P and Q), if it exists.

This is done as follows. A largest robot of P', not in the root of P' (either a free robot or in a tower) goes to its parent. When there are no robots outside the root, the robots from the root of P' go to the garbage in Q. This way of merging all robots of P' at the root of this piece prevents accidental creation of a signal. Stage 2 ends when the ending condition of Stage 1 holds and piece P', if it exists, is empty.

Stage 3. Goal: Create the signal.

The signal is created at the bottom of the garbage (without considering towers marking its end). Towers descend in the garbage one at a time, until two sequences consisting of 4 towers, each at distance 3 from the preceding one, are created. These two sequences are separated by 5 empty nodes. Since there is no longer sequence of this type in the entire tree, the value of the newly created signal is 3. This completes Stage 3 and the entire Phase 2. (Note that we use two sequences forming a signal, rather than just one, in order to be able to move one of these sequences later on, without destroying the value of the signal. In fact we also need to leave two additional towers between these sequences, in order to update the value of signal from 3 to 4, when passing to Phase 4.)

From now on all towers in the entire tree are separated by at least one empty node. Hence if a tower moves and the adversary breaks it by holding back some of the robots of the tower, this can be recognized in subsequent snapshots and the moving tower can be reconstructed. Note that from now on we need not specify the existence of a tower in the snapshot, since the signal contains towers.

Phase 3. The value of signal is 3.
Goal: Explore P'': the largest of the pieces other than Q. (We explore this piece first to be able to relocate the brain into it in Phase 4: the other piece could be too small.)

At the beginning of Phase 3 both counters indicate value 0. Piece P'' will be explored by the free robot and the tower that are currently outside Q. They will be called exploring robot and exploring tower, respectively. These two entities explore leaves of P'' one similarity class after another in increasing order, induced by any total preorder of the nodes, with the following property: the equivalence classes induced by this preorder are the previously defined similarity classes. The entities move only if both counters indicate the same value i. Suppose that the jth class has size s_j. Let r be such that $i = s_1 + \cdots + s_d + r$, with $r \leq s_{d+1}$. Hence the brain indicates that the next leaf to be explored is the rth leaf in the $(d+1)$st class. If $r = 1$, the exploring robot goes to any leaf of the $(d+1)$st class. Otherwise, consider two cases. If r is even then let u be the leaf where the exploring robot is located. In this case the exploring tower goes to the (unique) closest leaf in the same similarity class. If r is odd then let v be the leaf where the exploring tower is located. In this case the exploring robot goes to the leaf w determined as follows. Let j be the length of the longest sequence of 1's counted from the right (least significant bit) of the binary expansion of the integer $(r-3)/2$. Order all leaves of the similarity class of v in any non-decreasing order of distances from v. The leaf w is the 2^{j+1}th node in this order. Notice that w is the closest leaf from v not yet explored.

Incrementing values of both counters from i to $i+1$ and moving the exploring robots according to those increments are complex actions involving relocation of many robots. Due to asynchrony, snapshots can be taken during these complex actions, potentially disorganizing the process. To ensure correct exploration, we artificially synchronize these actions using the semaphore. Its values change in the cycle $00, 01, 11, 10, 00, \ldots$. Note that the changes of values of the semaphore do not need additional synchronization, as each change involves a move of only one robot or tower. In the case of a move of a tower, the adversary can split the tower by delaying some of its robots and moving others, hence the value of the corresponding bit is unclear and robots must decide which value should be set. Nevertheless this is never ambiguous: for example, if the value of the first bit is 0 and the second is unclear, it must be set to 1 because, when the first bit is 0, the only possible change of the second bit is from 0 to 1. Other cases are similar.

At the beginning of Phase 3 the semaphore is at 00. This indicates that the first counter has to modify its value to $i + 1$, where i is the current value of the second counter. When this is done, the value of the semaphore changes to 01. This indicates that the second counter has to modify its value to the current value of the first counter. When this is done, the value of the semaphore changes to 11. This indicates that the exploring robot or the exploring tower (depending on the parity of the value shown by the counters) has to move to the neighbor of the leaf it occupies. When this is done, the value of the semaphore changes to 10. This indicates that the exploring entity which is in an internal node (i.e., the one that has just moved) has to move to the leaf indicated by the value of both counters, as explained above. When this is done, the value of the semaphore changes to 00.

Phase 3 is completed when the semaphore has value 11 and both counters have value $f + 1$, where f is the number of leaves in piece P''. At this time the value of signal is changed from 3 to 4 (by moving an additional tower down the garbage), thus marking the end of this phase. Note that, when both counters have value $f + 1$, all leaves of P'' are explored. There are two cases. If $P'' = P$ then at least one path between the root and a leaf of P'' has been explored when P was evacuated. Otherwise, at least one path between the root and a leaf of P'' has been explored when the exploring solitaire came from P to explore P''. Hence in both cases all leaves and at least one path between the root and a leaf have been explored. Since by the description of the exploration the explored part of P'' is connected, this implies that the entire piece has been explored.

Phase 4. The value of signal is 4.
Goal: Relocate the brain from Q to P'' (except when there are only two pieces and Q has few leaves, in which case exploration of Q is done immediately: see Subcase 2.2).

While the brain is relocated to P'', piece Q is emptied and thus ready to be explored. Piece Q is emptied in reverse order of its filling in Phase 2, i.e., robots that came last to Q leave it first. We will need the exploring solitaire and tower in piece Q in order to perform exploration during Phase 5. Hence while towers forming the old brain move from Q to P'', the solitaire and the exploring tower move in the opposite direction. This creates a problem when the tree has a long path of nodes of degree two, between the old brain and piece P'': there is no room to cross on this path. Hence for this class of trees we will use a particular technique. Consider two cases.

Case 1. There exist nodes v and w outside P'' such that the path from the root of P'' to each of them does not contain robots and there exists a path from a tower in Q to the root of P'' not containing robots and not containing v or w.

In this case there is no crossing problem. The solitaire and the exploring tower from Q can hide in v and w to let the towers from Q (that formed the old brain) move to P''.

Case 2. There are no nodes v and w as described in Case 1.

Let M be the largest integer such that $10c \log M / \log \log M \geq \log M$.

– Subcase 2.1. The number of leaves in piece Q is larger than M.

Since any tree of maximum degree 3 containing f leaves has height at least $\log f$, the condition on integer M implies that there exists a descending path in Q, with beginning u, satisfying the following properties:

(1) it is able to store all towers needed to explore Q, leaving distance 4 between consecutive towers. (We leave distance 4 not to confuse the sequence of towers with a signal.)

(2) there exist two leaves outside the tree rooted at u.

All towers from Q are moved to the above descending path leaving 3 empty nodes between consecutive towers, with the following exception. When moving the first five towers, the value of the signal is recreated using these towers. This is done before moving the second sequence of the signal created in Phase 2. After moving $2c \log M / \log \log M$ towers, all additional towers from this path are collapsed to one tower. After this compacting the condition of Case 1 holds because of property (2).

– Subcase 2.2. The number of leaves in piece Q is at most M.

In this case there are so few leaves that we can explore all of them without using a brain. First we recreate the signal in P with value 4, to record the phase number. Then all robots from Q go to the leaves. When all leaves are occupied, all robots go towards the root of Q forming a tower in this root, thus exploring the remaining nodes of Q. At this point the algorithm stops (explicit stopping condition).

Thus, after a finite number of moves in Case 2, either the exploration is completed (Subcase 2.2) or the algorithm transits to Case 1. From now on we suppose that the condition of Case 1 holds.

We continue Phase 4 by creating a signal with value 4 in piece P''. This is done by moving towers from the top of the garbage in Q and placing them outside the core zone in P''. The path forming the signal is of bounded length and thus there is enough space outside the core zone to place it. Moreover we place three additional towers in this path to be able to subsequently increase the value of the signal up to 7. After this is done we create the new brain in P'', similarly as in Phase 2. In particular, we use the solitaire as a guide to direct the towers coming from Q to their target positions. Note that all towers and robots in the core zone of Q are alone in their equivalence classes and thus there is no need to break symmetries using solitaires. When the counters and the semaphore of the new brain are created in P'', all robots from Q, except the exploring tower and solitaire are moved to the garbage of the new brain. Note that all the above actions are possible, since the solitaire and towers are able to move between pieces Q and P'' without crossing problems.

When there is only the exploring tower and solitaire in Q, the value of signal in P'' is incremented to 5. This ends Phase 4.

Phase 5. The value of signal is 5.
Goal: Explore piece Q and stop if there are only two pieces.

We proceed exactly as in Phase 3, this time exploring piece Q instead of P''. When the brain indicates that all leaves are explored, two situations are possible. If there are only two pieces in the tree, all nodes are already explored and the algorithm stops (explicit stopping condition). If there are three pieces, the value of signal is incremented to 6. This ends Phase 5.

Phase 6. The value of signal is 6.
Goal: Reinitialize the brain and relocate the exploring solitaire to the unexplored piece.

Both counters in the brain are reset to 0, the semaphore is reset to 00. The exploring solitaire moves to the root of the unexplored piece. The value of signal is incremented to 7. This ends Phase 6.

Phase 7. The value of signal is 7.
Goal: Explore the last piece and stop.

The piece containing only a solitaire is explored (using this solitaire and the tower from Q). This is done again as in Phase 3. When the brain indicates that all leaves are explored, exploration is completed and the algorithm stops (explicit stopping condition).

It remains to give the conditions on the constant c such that the number k of robots satisfies $c \log n / \log \log n \leq k \leq 2c \log n / \log \log n$. The constant c should be chosen so that there are sufficiently many robots to form the brain (including the markers of descending paths' extremities) and the exploring team. Note that if there are three pieces in the tree, and robots are initially equally divided among them, only $k/3$ robots will be used.

3.3   Exploration of Trees of Maximum Degree 3: Lower Bound

We now prove a lower bound on the number of robots necessary for exploration of complete binary trees, that matches the upper bound given by Algorithm Tree-exploration.

Theorem 2. $\Omega(\frac{\log n}{\log \log n})$ *robots are required to explore n-node complete binary trees.*

4   Concluding Remarks and Open Problems

A natural next research step would be the investigation of the exploration problem when the visibility of the robots is limited, e.g., to the immediate neighborhood. Notice that, in this case, exploration is not generally possible. Hence a limited visibility scenario could only work for some subset of initial configurations. Another line of research would be to equip robots with very small (e.g., constant) memory of past events and study how this additional power influences feasibility of exploration with limited or unlimited visibility. Finally, it would be interesting to extend our study to the case of arbitrary graphs, as well as to the stronger SSYNCH model.

References

1. Albers, S., Henzinger, M.R.: Exploring unknown environments. SIAM J. on Comput 29, 1164–1188 (2000)
2. Ando, H., Oasa, Y., Suzuki, I., Yamashita, M.: Distributed memoryless point convergence algorithm for mobile robots with limited visibility. IEEE Trans. on Robotics and Automation 15, 818–828 (1999)
3. Cieliebak, M., Flocchini, P., Prencipe, G., Santoro, N.: Solving the robots gathering problem. In: Baeten, J.C.M., Lenstra, J.K., Parrow, J., Woeginger, G.J. (eds.) ICALP 2003. LNCS, vol. 2719, pp. 1181–1196. Springer, Heidelberg (2003)
4. Cohen, R., Peleg, D.: Local algorithms for autonomous robot systems. In: Proc. 13th International Colloquium on Structural Information and Communication Complexity, (SIROCCO 2006). LNCS, vol. 3221, pp. 29–43. Springer, Heidelberg (2006)
5. Czyzowicz, J., Gasieniec, L., Pelc, A.: Gathering few fat mobile robots in the plane. In: Proc. 10th International Conference on Principles of Distributed Systems (OPODIS 2006). LNCS, vol. 4288, pp. 744–753. Springer, Heidelberg (2006)
6. Dessmark, A., Pelc, A.: Optimal graph exploration without good maps. Theoretical Computer Science 326, 343–362 (2004)
7. Fleischer, R., Trippen, G.: Exploring an unknown graph efficiently. In: Brodal, G.S., Leonardi, S. (eds.) ESA 2005. LNCS, vol. 3669, pp. 11–22. Springer, Heidelberg (2005)
8. Flocchini, P., Ilcinkas, D., Pelc, A., Santoro, N.: Computing without communicating: Ring exploration by asynchronous oblivious robots. In: Tovar, E., Tsigas, P., Fouchal, H. (eds.) OPODIS 2007. LNCS, vol. 4878, pp. 105–118. Springer, Heidelberg (2007)
9. Flocchini, P., Prencipe, G., Santoro, N., Widmayer, P.: Gathering of asynchronous robots with limited visibility. Theoretical Computer Science 337, 147–168 (2005)
10. Gasieniec, L., Pelc, A., Radzik, T., Zhang, X.: Tree exploration with logarithmic memory. In: Proc. 18th Annual ACM-SIAM Symposium on Discrete Algorithms (SODA 2007), pp. 585–594 (2007)
11. Klasing, R., Markou, E., Pelc, A.: Gathering asynchronous oblivious mobile robots in a ring. Theoretical Computer Science 390, 27–39 (2008)
12. Panaite, P., Pelc, A.: Exploring unknown undirected graphs. Journal of Algorithms 33, 281–295 (1999)
13. Prencipe, G.: Impossibility of gathering by a set of autonomous mobile robots. Theoretical Computer Science 384, 222–231 (2007)
14. Souissi, S., Défago, X., Yamashita, M.: Gathering asynchronous mobile robots with inaccurate compasses. In: Shvartsman, M.M.A.A. (ed.) OPODIS 2006. LNCS, vol. 4305, pp. 333–349. Springer, Heidelberg (2006)
15. Sugihara, K., Suzuki, I.: Distributed algorithms for formation of geometric patterns with many mobile robots. Journal of Robotic Systems 13(3), 127–139 (1996)
16. Suzuki, I., Yamashita, M.: Distributed anonymous mobile robots: formation of geometric patterns. SIAM J. Comput. 28, 1347–1363 (1999)

Average Binary Long-Lived Consensus: Quantifying the Stabilizing Role Played by Memory

Florent Becker[1], Sergio Rajsbaum[2], Ivan Rapaport[3], and Éric Rémila[1,*]

[1] Université de Lyon, LIP UMR 5668 CNRS - ÉNS Lyon - UCB Lyon 1, France
[2] Instituto de Matemáticas, Universidad Nacional Autónoma de México
[3] DIM and CMM, Universidad de Chile

Abstract. Consider a system composed of n sensors operating in synchronous rounds. In each round an *input vector* of sensor readings x is produced, where the i-th entry of x is a binary value produced by the i-th sensor. The sequence of input vectors is assumed to be *smooth*: exactly one entry of the vector changes from one round to the next one. The system implements a fault-tolerant averaging *consensus function* f. This function returns, in each round, a representative *output value* v of the sensor readings x. Assuming that at most t entries of the vector can be erroneous, f is required to return a value that appears at least $t + 1$ times in x. The *instability* of the system is the number of output changes over a random sequence of input vectors.

Our first result is to design optimal instability consensus systems with and without memory. Roughly, in the memoryless case, we show that an optimal system is D_0, that outputs 1 unless it is forced by the fault-tolerance requirement to output 0 (on vectors with t or less 1's). For the case of systems with memory, we show that an optimal system is D_1, that initially outputs the most common value in the input vector, and then stays with this output unless forced by the fault-tolerance requirement to change (i.e., a single bit of memory suffices).

Our second result is to quantify the gain factor due to memory by computing $c_n(t)$, the number of decision changes performed by D_0 per each decision change performed by D_1. If $t = \frac{n}{2}$ the system is always forced to decide the simple majority and, in that case, memory becomes useless. We show that the same type of phenomenon occurs when $\frac{n}{2} - t$ is constant. Nevertheless, as soon as $\frac{n}{2} - t \sim \sqrt{n}$, memory plays an important stabilizing role because the ratio $c_n(t)$ grows like $\Theta(\sqrt{n})$. We also show that this is an upper bound: $c_n(t) = O(\sqrt{n})$ for every t.

Our results are average case versions of previous works where the sequence of input vectors was assumed to be, in addition to smooth, *geodesic*: the i-th entry of the input vector was allowed to change *at most once* over the sequence. It thus eliminates some anomalies that occurred in the worst case, geodesic instability setting.

* Partially supported by Programs Conicyt "Anillo en Redes", Instituto Milenio de Dinámica Celular y Biotecnología and Ecos-Conicyt, and IXXI (Complex System Institute, Lyon).

A. Shvartsman and P. Felber (Eds.): SIROCCO 2008, LNCS 5058, pp. 48–60, 2008.

1   Introduction

Consider a system composed of n sensors sampled at synchronous rounds. In each round an *input vector* of sensor readings is produced, where the i-th entry of the vector is a value from some finite set V produced by the i-th sensor. To simplify the presentation, the sampling interval is assumed to be short enough, to guarantee that the sequence of input vectors is *smooth*: exactly one entry of a vector changes from one round to the next one.

There are situations where, for fault-tolerant purposes, all sensors are placed in the same location. Ideally, in such cases, all sensor readings should be equal. But this is not always the case; discrepancies may arise due to differences in sensor readings or to malfunction of some sensors. Thus, the system must implement some form of fault-tolerant averaging *consensus function* f, that returns a representative *output value* v of the sensor readings x. Assuming that at most t entries of a vector x can be erroneous, f is required to return a value that appears at least $t + 1$ times in x.

The same questions arise when consensus is done not between the values of sensors, but between the opinions of actors. Suppose for example that you have a server which can give several types of data to a bunch of clients. At a given time, each client has a favorite type of data it wants to receive, but the server can only broadcast one type of data to all the clients. If there is a cost to switching between requests (say, because one can no longer use cached data), then in order to serve as much clients as possible in the long-run, it might be wise to sometimes give a content that fewer of them want, but which we have already started serving.

In a social setting, the same kind of question arises whenever a group has to make a consensual decision. For example, consider a disc-jockey in a wedding party. There are both older people, who fancy dancing to a nice waltz, and younger ones, eager to get their kicks on techno music. Our disc-jockey has to make sure that the dance-floor is never too empty according to who is ready to dance at a given time. But if he changes the music too often, then nobody is going to be happy: stability matters. More seriously, in an election system, one might want to have a decision that is at the same time representative and stable, so that the policies which are decided have the time to be applied. In a setting where there is no term-mandate and decision-making is done live, we show that the stability can be enforced through election rules (i.e., the decision function).

In this context, the most natural function f is the one that returns the most common value of vector x. However, the *instability* of such function is high. In fact, as the next example shows ($n = 5$ and $t = 1$), the output value computed by this f could change from one round to the next one unnecessarily often:

inputs: $00011 \rightarrow 10011 \rightarrow 10010 \rightarrow 11010 \rightarrow \cdots$

outputs: $0 \rightarrow \quad 1 \rightarrow \quad 0 \rightarrow \quad 1 \rightarrow \cdots$

If instead of the previous f we consider the one that decides the smallest value in x that appears at least $t+1$ times, then no output changes would have occurred in the previous sequence (in the example $0 < 1$ and $t + 1 = 2$). Moreover, in order to reduce further the instability, we could consider a function that tries to stay with the ouput of previous rounds.

The worst case instability of consensus functions was studied in two previous papers [3,5]. The input sequence considered in those papers was assumed to be, in addition to smooth, *geodesic*: the i-th entry of the input vector was allowed to change *at most once* over the sequence. The instability of a consensus function was given by the largest number of output changes over any such sequence, called a *geodesic path*. Notice that a geodesic path must be finite, since the set V from which the input vectors draw their values is finite. The case $V = \{0, 1\}$ of binary input vectors was considered in [5]. The case of multi-valued input vectors, where the set V is arbitrary, turned out to be much more difficult and required higher-dimensional topological methods [3].

In this paper we initiate a study of the *average instability* of consensus functions. We tackle the case $V = \{0, 1\}$ of binary input vectors. We remove the geodesic requirement and therefore the smooth sequences of input vectors we consider here are random walks over the hypercube. If $P = X_0, X_1, \ldots$ is such a walk, then the average instability of a consensus function f is given by the fraction of time f changes its output over P. The first goal is –given n and t– to find out what function f minimizes the instability in the two possible scenarios: with memory and without memory. We obtain the following results.

For the memoryless case we show that a system D_0, that outputs 1 unless it is forced by the fault-tolerance requirement to output 0 (on vectors with t or less 1's), is optimal. For the case of systems with memory, we show that a system D_1, that initially outputs the most common value in the input vector, and then stays with its output unless forced by the fault-tolerance requirement to change, is optimal. Thus, a single bit of memory suffices to achieve optimal instability. We should point out that in order to compute the instability of D_1 we use a non-trivial result concerning the Ehrenfest Markov chain model (which gives a simple expression to the value of the expected time to go from state k to state $k + 1$ [14]).

Our second goal is to measure the stabilizing role played by memory. A natural way of doing this is by computing $c_n(t)$, the number of decision changes performed by D_0 per each decision change performed by D_1. We prove that $c_n(t) = O(\sqrt{n})$, and this upper bound is reached when $n/2 - t = \alpha\sqrt{n}$, with α constant. In contrast, if t or $n/2 - t$ are constant then $c_n(t)$ is also constant.

Our approach eliminates some anomalies that occured in the worst case geodesic instability setting. For instance, in the case of $t = 0$ (which is interesting because it leaves maximum freedom on the choice of f), it was proved in [5] that any optimal instability memoryless function must be *one-bit defined*, i.e., output the value of the i-th sensor. Intuitively, such a function has high instability. Indeed, in our average case setting, its instability is $1/n$, much higher than the optimal average instability of $1/2^n$ given by the function f_0 which is always 1 unless all sensors have a value 0.

As noted in [5], studying the instability of consensus functions may have applications in various areas of distributed computing, such as self-stabilization

[4] (indeed, see [10]), Byzantine agreement [1], real-time systems [11], complexity theory [8] (boolean functions), and VLSI energy saving [2,12,16] (minimizing number of transitions).

Also, it would be interesting to relate our results to natural phenomena that exhibit hysteresis (memory). It is known, for instance, that some biological phenomena exhibit hysteresis [9,15]. Did they appear in evolution as a way to minimize instability? The approach could also be applied in the social sciences. In fact, wouldn't it be possible to conceive an electoral system which instead of deciding by simple majority incorporates some memory in order to eliminate noise? Notice that system D_1, which minimizes instability, corresponds to an hysteretic switch.

This paper is organized as follows. In Section 2 we describe the model and define the instability measure. Section 3 considers memoryless consensus systems, while Section 4 considers the general case. In Section 5 we quantify the relevance of memory by analyzing the behavior of the gain factor for different values of t.

2   Instability

Let n, t be non-negative integers, $n \geq 2t + 1$. The *hypercube* of dimension n is a graph whose vertices $V_n = \{0,1\}^n$ are all binary n-dimensional vectors, called *input vectors*. The edges E_n are all pairs of vertices whose vectors differ in exactly one component. Notice that $|E_n| = n2^{n-1}$. The *distance* $d(x_1, x_2)$ between two vertices x_1, x_2 is equal to the number of entries in which they differ. Thus, $d(x_1, x_2) = d$ iff the shortest path between x_1 and x_2 in the hypercube is of length d. We denote by $\#_b(x)$ the number of entries in x that are equal to $b \in \{0,1\}$. The *corners* of the hypercube are the vertices 0^n and 1^n. The *d-neighborhood* of a vertex x of the hypercube, $\mathcal{N}^d(x)$, is the set of vertices at distance at most d from x. Thus, $\mathcal{N}^t(0^n) = \{x \mid \#_1(x) \leq t\}$, and similarly for 1^n. Since $n \geq 2t + 1$, $\mathcal{N}^t(0^n) \cap \mathcal{N}^t(1^n) = \emptyset$.

Let $x_0, x_1, \ldots$ be the vertices of a walk in the hypercube. We will consider functions f that assign, to each x_i, an *output value* d that satisfies the *fault-tolerance* requirement:

$$f(x_i) = d \Rightarrow \#_d(x_i) \geq t + 1.$$

In the *memoryless* case f is a function only of x_i. In general, f outputs d based on the previous vertices of the walk. Formally, a *system* is a tuple $D = (n, t, S, \tau, f)$, where S is a finite set of states that includes a special initial state $\perp \in S$, $\tau : V_n \times S \to S$ is the transition function, and $f : V_n \times S \to \{0,1\}$ is the consensus decision function. The fault-tolerance requirement implies, for all $x \in V_n$, $s \in S$:

$$f(x, s) = \begin{cases} 0 \text{ if } x \in \mathcal{N}^t(0^n) \\ 1 \text{ if } x \in \mathcal{N}^t(1^n) \end{cases}$$

An *execution* of the system is a sequence $(x_0, s_0, d_0) \to (x_1, s_1, d_1) \to \ldots$, where $s_0 = \perp$, $s_{i+1} = \tau(x_i, s_i)$, and $d_i = f(x_i, s_i)$. A triple (x_i, s_i, d_i) is a *configuration*.

We assume that if x is the current input vector, then the next input vector x' is taken in a random uniform way from the vectors at distance one from x in the hypercube. The initial input vector is chosen according to some distribution λ. Once the initial state x_0 is determined, so is the initial configuration, (x_0, s_0, d_0). The next configuration is produced by choosing at random a neighbor of x_0, say x_1, and we get the next configuration (x_1, s_1, d_1), where $s_1 = \tau(x_0, s_0)$ and $d_1 = f(x_1, s_1)$.

Formally, we define the following Markov process: the set of states is $V_n \times S$ and there is a transition from (x, s) to (x', s') if $\{x, x'\} \in E_n$ and $\tau(x, s) = s'$. Therefore, any random walk $X_0, X_1, X_2, \ldots$, where X_0 is chosen according to λ, defines an execution.

Each state (x, s) has an associated output value $f(x, s)$, so we may write $d_i = f(X_i)$ to be the output value associated to X_i. Let $c_{\lambda,l}(D)$ be the random variable defined by:

$$c_{\lambda,l}(D) = \frac{1}{l} \sum_{k=0}^{l-1} |d_{k+1} - d_k|.$$

Definition 1. The average instability of a consensus system D is $c(D) = \mathbb{E}(\lim_{l \to \infty} c_{\lambda,l}(D))$.

The Markov chain described above is finite and hence $c(D)$ exists by the ergodic theorem[1].

3   Average Instability of Memoryless Systems

In a memoryless system $|S| = 1$, and τ is irrelevant, so the system is defined by a triplet $D = (n, t, f)$, where $f : V_n \to \{0, 1\}$. In this case, the Markov chain is irreducible[2], and its set of states is V_n. That is, $X_0, X_1, X_2, \ldots$ is a λ-random walk on the hypercube, and its stationary distribution π is the uniform $\pi_x = 1/2^n$, for every $x \in V_n$ (for notation see [13]). The instability of f counts the number of times the function f changes its decision along a random walk. By the ergodic theorem, the fraction of time the random walk crosses bicolored edges (where changes in the decision take place) tends to the number of bicolored edges divided by $|E_n| = n2^{n-1}$.

Proposition 1. Let $D = (n, t, f)$ be a memoryless system. Then,

$$c(D) = \frac{\sum_{\{x,y\} \in E_n} |f(x) - f(y)|}{n2^{n-1}}$$

Proof. Since by definition $c(D) = \mathbb{E}(\lim_{l \to \infty} c_{\lambda,l}(D))$, it is sufficient to prove that

$$c_{\lambda,l}(D) \xrightarrow[l \to \infty]{} \frac{\sum_{\{x,y\} \in E_n} |f(x) - f(y)|}{n2^{n-1}} \qquad \text{a.s.}$$

---

[1] When S is not finite, $c(D)$ might not exist, but $c'(D) = \mathbb{E}(\liminf_{l \to \infty} c_{\lambda,l}(D))$ always exists, and can be considered as a lower bound for the cost.

[2] In an irreducible Markov chain, it is possible to get to any state from any state.

This follows directly from the ergodic theorem. Formally, we consider the Markov chain $(X_0, X_1), (X_1, X_2), \ldots$ in which each state is an arc $e = (X_i, X_{i+1}) \in \overrightarrow{E}_n$, an ordered couple of neighbor vertices of the hypercube. Therefore the function defined over the set of states (the arcs) is $f(X_i, X_{i+1}) = |f(X_{k+1}) - f(X_k)|$. In this standard approach one needs to verify that the chain is irreducible and that the unique invariant distribution is the uniform distribution in order to conclude that:

$$c_{\lambda,l}(D) = \frac{1}{l} \sum_{k=0}^{l-1} f(X_k, X_{k+1}) \xrightarrow[l \to \infty]{} \sum_{e \in \overrightarrow{E}_n} \pi_e f(e) = \frac{1}{2n2^{n-1}} \sum_{e \in \overrightarrow{E}_n} f(e)$$

$$= \frac{1}{n2^{n-1}} \sum_{e \in E_n} f(e) \qquad \text{a.s.}$$

3.1 Geodesic Worst Case vs. Average Instability

The worst case geodesic instability measure of [5,3] depends on the values f takes in a small part of the hypercube, given by a geodesic path (where the i-th entry of a vector changes at most once). In contrast, the average instability permits walks that traverse the whole hypercube. We are going to remark this difference by giving two examples for which the values behave in opposite ways.

Let us assume $t = 0$. In this case the only restrictions appear in the corners of the hypercube. More precisely, $f(d^n) = d$.

Suppose that the output of f depends exclusively on what happens in one particular processor (the i-th processor). In other words, consider the function $f^{(i)}(x) = x^{(i)}$, where $x = x^{(1)} \ldots x^{(n)}$. By common sense, this is clearly a bad strategy. But in terms of the geodesic analysis, this function appeared to be optimal [5]. Morover, it was proved that any optimal function must be of this form (when $t = 0$). The explanation comes from the fact that in a geodesic path, once the i-th coordinate changes, it can not change anymore. On the other hand, by Proposition 1, the average instability of $f^{(i)}$ is $c(f^{(i)}) = \frac{1}{n2^{n-1}} \frac{2^n}{2} = \frac{1}{n}$.

It is easy to see that the average instability of $f^{(i)}$ is far from being optimal. For example, assume n is odd. And let $f(x) = 1$ if and only if $x = 0^k 1^l$ with l odd. There is a geodesic path for which the function changes *all along the path*. Nevertheless, there is a small number of 1's in the hypercube. And, in fact, $c(f) = \frac{1}{n2^{n-1}} \frac{n}{2} = \frac{n}{2^n} << c(f^{(i)})$.

3.2 Optimal Memoryless Systems

Let Γ_i denote the set of vectors $x \in V_n$ satisfying $\#_1(x) = i$. In other words, Γ_i is the set of nodes of the hypercube at distance i from 0^n. Recall that $\mathcal{N}^k(x) = \{y \mid d(x,y) \leq k\}$. Let us define $D_0 = (n, t, f_0)$ with $f_0(x) = 0$ if and only if $x \in \mathcal{N}^t(0^n)$. In other words, f_0 is always 1 unless the fault-tolerance requirement forces the system to decide 0. We prove below that D_0 is an optimal memoryless system.

Theorem 1. *Let $D = (n, t, f)$ be a memoryless system. Then $c(D_0) \leq c(D)$, with $c(D_0) = \frac{\binom{n-1}{t}}{2^{n-1}}$.*

Proof. By Proposition 1, in order to compute $c(D_0)$, we need to count the number of bicolored edges $\{x, y\}$ induced by f_0. This number is $(n-t)\binom{n}{t}$ because $|\Gamma_t| = \binom{n}{t}$ and each vertex in Γ_t is connected to $n - t$ vertices in Γ_{t+1}. Thus, $c(D_0) = \frac{(n-t)\binom{n}{t}}{n2^{n-1}} = \frac{\binom{n-1}{t}}{2^{n-1}}$. It remains to show that there is no system $D = (n, t, f)$ of lower cost.

We show this using network flow theory. We orient the edges of he hypercube from Γ_i to Γ_{i+1}, and we assign a capacity 1 to each of these arcs. An output function f induces a cut, i.e., an (S, T)-partition of the vertices of the hypercube. In fact, S are the vertices that output 0 and T are the vertices that output 1. Since $\mathcal{N}^t(0^n) \subseteq S$ and $\mathcal{N}^t(1^n) \subseteq T$, we are in fact dealing with cuts separating Γ_t and Γ_{n-t}. The capacity of the cut is the number of edges starting in S and ending in T. This number (divided by $n2^{n-1}$) is precisely the average instability of D.

The cut induced by f_0 has capacity $(n - t)\binom{n}{t}$. If we prove that this is a minimum cut, we are done. We do this by describing a flow from Γ_t to Γ_{n-t} that saturates this cut, which proves the cut is minimum, by the min-cut/max-flow theorem. The flow on an arc linking a node of Γ_i to a node of Γ_{i+1} is uniformly $\frac{\binom{n-1}{t}}{\binom{n-1}{i}}$, for $t \leq i < n - t$. Notice that the law of flow conservation is satisfied since, at a vertex $x \in \Gamma_i$, $t < i < n - t$, one easily checks that the incoming flow $(n - i)\frac{\binom{n-1}{t}}{\binom{n-1}{i}}$ is equal to the outgoing flow $i\frac{\binom{n-1}{t}}{\binom{n-1}{i-1}}$. The flow of each arc e is at most 1, and the total transported flow is equal to $\binom{n}{i}(n - i)\frac{\binom{n-1}{t}}{\binom{n-1}{i}} = n\binom{n-1}{t}$. Finally $n\binom{n-1}{t} = (n - t)\binom{n}{t}$, which is equal to the capacity of the cut.

3.3   Symmetric Memoryless Systems: The Ehrenfest Urn Model

Random walks turn out to be particularily interesting when the function f is symmetric, i.e., when it depends only on the distribution of 0's and 1's in the input vector. These functions are of the follwing type: "if i sensors are measuring the value 1 then the output is d". In these cases, we can *project* the hypercube into a path, whose vertices are $\{0, \ldots, n\}$. We can therefore assume that f is defined over this set of vertices, instead of over the vertices of the hypercube. Instead of considering the state x we consider the state $i = \#_1(x)$. We get a new Markov chain with transitions:

$$\mathbb{P}\{i \to (i+1)\} = \frac{n - i}{n} \qquad \mathbb{P}\{i \to (i - 1)\} = \frac{i}{n}.$$

This process is known as the Ehrenfest urn model [7]. The probability of moving from i to $i + 1$ is simply the probability of moving from a vertex with i entries equal to 1 to a vertex with $i + 1$ entries equal to 1. There are $n - i$ such edges that can cause this to happen.

The invariant distribution of the Ehrenfest model (easily computable by projection of the uniform distribution of the hypercube) is known to be $\pi_i = \frac{\binom{n}{i}}{2^n}$. Therefore, the invariant distribution of the coupled states corresponding to the arcs $(i, i+1)$ and $(i+1, i)$ are $\pi_{(i,i+1)} = \frac{\binom{n}{i}}{2^n} \frac{n-i}{n}$, and $\pi_{(i+1,i)} = \frac{\binom{n}{i+1}}{2^n} \frac{i+1}{n}$. Thus, $\pi_{(i,i+1)} = \pi_{(i+1,i)} = \frac{\binom{n-1}{i}}{2^n}$.

Theorem 2. *Let $D = (n, t, f)$ be a symmetric memoryless system. Then,*

$$c(D) = \sum_{i=0}^{n-1} \binom{n-1}{i} \frac{|f(i+1) - f(i)|}{2^{n-1}}.$$

Proof. We know from the ergodic theorem that

$$c_{\lambda,l}(D) \xrightarrow[l \to \infty]{} \sum_{i=0}^{n-1} \pi_{(i,i+1)} |f(i+1) - f(i)| + \sum_{i=0}^{n-1} \pi_{(i+1,i)} |f(i+1) - f(i)| \quad \text{a.s.}$$

The result follows from the fact that $\pi_{(i,i+1)} = \pi_{(i+1,i)} = \frac{\binom{n-1}{i}}{2^n}$.

Remark 1. *We can recompute the instability of sytem D_0 introduced in previous section, because f_0 is symmetric: $f_0(i) = 1 \iff i > t$. Therefore, $c(D_0) = \binom{n-1}{t} \frac{1}{2^{n-1}}$.*

Remark 2. *D_0, in the context of the geodesic analysis in [5], appeared to be optimal only among the symmetric memoryless systems.*

4   Average Instability of Systems with Memory

Let us define the following system having only one bit of memory: $D_1 = (n, t, S_1, f_1, \tau_1)$, where $S_1 = \{\perp, 0, 1\}$ and

$$f_1(x, s) = \tau_1(x, s) = \begin{cases} 0 & \text{if } \#_1(x) \leq t \\ 1 & \text{if } \#_0(x) \leq t \\ maj(x) & \text{if } s = \perp \\ s & \text{otherwise} \end{cases}$$

Here, $maj(x)$ returns the most common bit value in $x \in \{0, 1\}^n$, and 1 in case of a tie.

Remark 3. *System D_1 is optimal in the geodesic cost model [5]. Nevertheless, the proof is much more complicated than in the memoryless case.*

Theorem 3. *For every system $D = (n, t, S, \tau, f)$, we have $c(D_1) \leq c(D)$. Moreover,*

$$c(D_1) = \left(2^{n-1} \sum_{k=t}^{n-t-1} \frac{1}{\binom{n-1}{k}} \right)^{-1}$$

and $\frac{1}{c(D_1)}$ is the average time necessary to pass from t to $n - t$ in the Ehrenfest model.

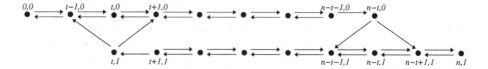

Fig. 1. The auxiliar Markov chain for D_1 (with $t = 2$)

Proof. Let $D = (n, t, S, \tau, f)$ be an arbitrary system. Let $x_0, x_1, \ldots, x_l$ be the first $l + 1$ vertices of a walk in the hypercube. We can associate to each of these vertices a symbol in $\{*, 0, 1\}$ depending on whether the system is not forced to decide ($t < \#_0(x_i), \#_1(x_i) < n - t$), it is forced to decide a 0 ($\#_1(x_i) \le t$) or it is forced to decide a 1 ($\#_0(x_i) \le t$). We therefore obtain a sequence $y \in \{*, 0, 1\}^{l+1}$. Let us delete the symbols $*$ in y in order to obtain a shorter binary string y'. Let $c'(y')$ be the number of changes in two consecutive symbols of y' (either 01 or 10). It is clear that the number of output changes of f over the same walk is at least $c'(y')$. Since $c'(y')$ is exactly the number of output changes of f_1 over the walk, we have that $c_{\lambda, l}(D_1) \le c_{\lambda, l}(D)$ over each random walk. Therefore, $c(D_1) \le c(D)$.

It remains to study the average instability of D_1. Since f_1 is symmetric, we can project the system into states of the form

$$(i, s) \in (\{0, \ldots, t-1\} \times \{0\}) \cup (\{t, \ldots, n-t\} \times \{0, 1\}) \cup (\{n-t+1, \ldots, n\} \times \{1\}),$$

where i denotes the number of 1's the processors are reading and s denotes the last output of f_1 (we are not considering the case $s = \bot$ because it appears only once, at the beginning). The transitions are the following:

$$\mathbb{P}\{(i, s) \to (i+1, s')\} = (\frac{n-i}{n}) \mathbf{1}_{\{f(i,s)=s'\}}$$

$$\mathbb{P}\{(i, s) \to (i-1, s')\} = (\frac{i}{n}) \mathbf{1}_{\{f(i,s)=s'\}}$$

The Previous Markov chain is irreducible and it has a stationary distribution π (to see that we only need to exhibit a positively recurrent state, for instance $(0, 0)$). We know that $c_{\lambda, l}(D_1)$ corresponds to the fraction of time the chain has been in states $(t, 1)$ or $(n-1, t)$ at time l. In fact, the system changes its decision each time it reads t 1's while its last decision was 1, or when it reads t 0's while its last decision was 0 (see Figure 1).

Then, by the ergodic theorem, $c_{\lambda, l}(D_1)$ converges almost surely to $\pi_{t,1} + \pi_{n-t,0}$. By symmetry, it follows that $c(D_1) = 2\pi_{t,1}$. We denote the expected time necessary to pass from a state (i, a) to a state (j, b) by $\mathbb{E}_{(i,a)}(T_{(j,b)})$. Since the inverses of the expected return times correspond to the stationary distribution, we have $c(D_1) = \frac{2}{\mathbb{E}_{(t,1)}(T_{(t,1)})}$. But, since each cycle passing through $(t, 1)$ also passes through $(n-t, 0)$,

$$\mathbb{E}_{(t,1)}(T_{(t,1)}) = \mathbb{E}_{(t,1)}(T_{(n-t,0)}) + \mathbb{E}_{(n-t,0)}(T_{(t,1)})$$

By symmetry, we have $\mathbb{E}_{(t,1)}(T_{(n-t,0)}) = \mathbb{E}_{(n-t,0)}(T_{(t,1)})$. On the other hand, for each $i \neq t$, we have $\mathbb{E}_{(t,1)}(T_{(i,0)}) = \mathbb{E}_{(t,0)}(T_{(i,0)})$. Thus we obtain:

$$c(D_1) = \frac{2}{2\mathbb{E}_{(t,0)}(T_{(n-t,0)})} = \frac{1}{\mathbb{E}_{(t,0)}(T_{(n-t,0)})}$$

The expected time to reach state $(n - t, 0)$ starting from $(t, 0)$ is equal to the expected time to reach state $n - t$ starting from state t in the classical Ehrenfest model. In fact, in this part of the walk the system will always be in state 0 and it only shifts to 1 when leaving the position $n - t$.

The last result follows from a theorem of [14], in which the authors prove that the expected time to go from state k to state $k + 1$ in the Ehrenfest model is equal to $\frac{1}{\binom{n-1}{k}} \sum_{j=0}^{k} \binom{n}{j}$. Thus,

$$\frac{1}{c(D_1)} = \sum_{k=t}^{n-t-1} \frac{1}{\binom{n-1}{k}} \sum_{j=0}^{k} \binom{n}{j}$$

Stating $k' = n - 1 - k$ we get:

$$\frac{1}{c(D_1)} = \sum_{k'=t}^{n-t-1} \frac{1}{\binom{n-1}{n-1-k'}} \sum_{j=0}^{n-k'-1} \binom{n}{j} = \sum_{k'=t}^{n-t-1} \frac{1}{\binom{n-1}{k'}} \sum_{j=k'+1}^{n} \binom{n}{j}$$

Thus, adding the two expressions of $\frac{1}{c(D_1)}$ we have:

$$\frac{2}{c(D_1)} = \sum_{k=t}^{n-t-1} \frac{2^n}{\binom{n-1}{k}}.$$

5   The Stabilizing Role Played by Memory

In order to quantify the stabilizing role played by memory we must compute the number of decision changes of the optimal memoryless system D_0 *per each* decision change performed by the optimal 1-bit of memory system D_1. More precisely, we must study the ratio $\frac{c(D_0)}{c(D_1)}$. Since we are interested in the asymptotic behavior $n \to \infty$, we note n instead of $n - 1$:

$$c_n(t) = \frac{c(D_0)}{c(D_1)} = \sum_{k=t}^{n-t} \frac{\binom{n}{t}}{\binom{n}{k}}$$

5.1   Upper Bounds

First we prove an upper bound that is independent of t.

Lemma 1. $c_n(t) = O(\sqrt{n})$

Proof. Let us assume (w.l.o.g.) that n is even.

$$c_n(t) = 2 \sum_{k=t}^{n/2} \frac{\binom{n}{t}}{\binom{n}{k}} = 2 \sum_{k=t}^{n/2} \frac{(t+1)(t+2)\ldots k}{(n-k+1)(n-k+2)\ldots(n-t)}.$$

Let $1 \le s_n \le n/2$ (s_n grows with n; later it becomes clear why we should choose $s_n = \sqrt{n}$). If $t + s_n > n/2$ then $c_n(t) \le 2 \sum_{k=n/2-s_n+1}^{n/2} 1 = 2s_n$. Let us consider the case $t + s_n \le n/2$. It follows:

$$c_n(t) = 2 \sum_{k=t}^{t+s_n} \frac{(t+1)(t+2)\ldots k}{(n-k+1)(n-k+2)\ldots(n-t)}$$

$$+2 \sum_{k=t+s_n+1}^{n/2} \frac{(t+1)(t+2)\ldots k}{(n-k+1)(n-k+2)\ldots(n-t)}$$

$$\le 2 + 2s_n +$$

$$2 \sum_{k=t+s_n+1}^{n/2} \frac{(t+1)\ldots(t+s_n)}{(n-t)\ldots(n-t-s_n+1)} \frac{(t+s_n+1)\ldots k}{(n-t-s_n)\ldots(n-k+1)}$$

$$\le 2 + 2s_n + 2(n/2 - t - s_n) \left(\frac{t+s_n}{n-t-s_n} \right)^{s_n}$$

Let $x = n/2 - t - s_n$. Let $g_n(x) = 2 + 2s_n + 2x \left(\frac{1-\frac{2x}{n}}{1+\frac{2x}{n}} \right)^{s_n}$. Since $c_n(t) \le g_n(x)$, our goal is to find the maximum of $g_n(x)$. By computing $g_n'(x_0) = 0$ we get

$$x_0 = \frac{ns_n}{2} \left(-1 + \sqrt{1 + \frac{1}{s_n^2}} \right) = \frac{ns_n}{2} \left(-1 + 1 + \frac{1}{2s_n^2} + o\left(\frac{1}{s_n^2} \right) \right) = \frac{n}{4s_n} + o\left(\frac{n}{s_n} \right).$$

Since $\frac{2x_0}{n} = \frac{1}{2s_n} + o\left(\frac{1}{s_n} \right)$, it follows:

$$c_n(t) \le 2 + 2s_n + 2 \left(\frac{n}{4s_n} + o\left(\frac{1}{s_n} \right) \right) \exp^{s_n (\log(1-\frac{1}{2s_n}+o(\frac{1}{s_n}))-\log(1+\frac{1}{2s_n}+o(\frac{1}{s_n})))}$$

$$\le 2 + 2s_n + 2 \left(\frac{n}{4s_n} + o\left(\frac{1}{s_n} \right) \right) \exp^{-1+o(1)} = O(s_n + \frac{n}{s_n})$$

Since either s_n or $\frac{n}{s_n}$ grows faster than $\sqrt{n}$, the best we can do is to choose $s_n = \sqrt{n}$ in order to conclude $c_n(t) = O(\sqrt{n})$.

5.2   Lower Bounds

The main question is whether there are values of t for which the previous upper bound is reached, i.e., whether $c_n(t) = \Theta(\sqrt{n})$ for some relation between n and t. We are going to see first that this does not happen for extremes values of t.

Lemma 2. *When t is constant, $\lim_{n \to \infty} c_n(t) = 2$.*

Proof. Notice first that $c_n(t) = 2 + \sum_{k=t+1}^{n-t-1} \frac{\binom{n}{t}}{\binom{n}{k}}$. When t is a constant, $\frac{\binom{n}{t}}{\binom{n}{t+1}} = \frac{\binom{n}{t}}{\binom{n}{n-t-1}} = O(1/n)$. For $t+2 \le k \le n-t-2$, $\frac{\binom{n}{t}}{\binom{n}{k}} = O(1/n^2)$ and there are $O(n)$ such k's. Thus, $\lim_{n \to \infty} c_n(t) = 2$.

The previous result can be explained as follows: for t constant, the Markov process will rarely enter a state where it needs to change its decision, either with D_0 or D_1. Thus, the chain, between two forced decisions, will be shuffled. Once shuffled, in half of the cases the chain will force the system to take the same decision it took before. On the other hand, when t is close to $n/2$, almost all the decisions are forced. In this case the ratio is also bounded by a constant. In fact,

Lemma 3. *When $t = n/2 - \beta$, $\lim_{n \to \infty} c_n(t) \le 2 + 2\beta$.*

Proof. We have $c_n(t) = 2 + \sum_{k=t+1}^{n-t-1} \frac{\binom{n}{t}}{\binom{n}{k}}$. The sum above contains (less than) 2β terms, which are all smaller than 1. Thus, $c_n(t) \le 2 + 2\beta$ when n tends to infinity.

The interesting case appears when t is close to $n/2$ but this distance grows with n. More precisely,

Lemma 4. *Let $t = n/2 - \alpha\sqrt{n}$ (with α constant). Then $c_n(t) = \Theta(\sqrt{n})$.*

Proof. It only remains to prove the lower bound (see Lemma 1).

$$
\begin{aligned}
c_n(n/2 - \alpha\sqrt{n}) &\ge \sum_{k=0}^{\alpha\sqrt{n}} \frac{\binom{n}{n/2-\alpha\sqrt{n}}}{\binom{n}{n/2-\alpha\sqrt{n}+k}} \\
&= \sum_{k=0}^{\alpha\sqrt{n}} \frac{(n/2 - \alpha\sqrt{n} + 1)(n/2 - \alpha\sqrt{n} + 2)\ldots(n/2 - \alpha\sqrt{n} + k)}{(n/2 + \alpha\sqrt{n})(n/2 + \alpha\sqrt{n} - 1)\ldots(n/2 + \alpha\sqrt{n} - k + 1)} \\
&\ge \sum_{k=0}^{\alpha\sqrt{n}} \left(\frac{n/2 - \alpha\sqrt{n}}{n/2 + \alpha\sqrt{n}}\right)^k \ge \alpha\sqrt{n} \left(\frac{1 - \frac{2\alpha}{\sqrt{n}}}{1 + \frac{2\alpha}{\sqrt{n}}}\right)^{\alpha\sqrt{n}} \\
&= \alpha\sqrt{n} \exp^{\alpha\sqrt{n}(\log(1 - \frac{2\alpha}{\sqrt{n}}) - \log(1 + \frac{2\alpha}{\sqrt{n}}))} \\
&= \alpha\sqrt{n} \exp^{-4\alpha + o(\frac{1}{\sqrt{n}})} = \Omega(\sqrt{n})
\end{aligned}
$$

References

1. Berman, P., Garay, J.: Cloture votes: $n/4$-resilient distributed consensus in $t+1$ rounds. Math. Sys. Theory 26(1), 3–19 (1993)
2. Chandrakasan, A.P., Brodersen, R.W.: Low power digital CMOS design. Kluwer Academic Publishers, Dordrecht (1995)

3. Davidovitch, L., Dolev, S., Rajsbaum, S.: Stability of Multi-Valued Continuous Consensus. SIAM J. on Computing 37(4), 1057–1076 (2007); Extended abstract appeared as Consensus Continue? Stability of Multi-Valued Continuous Consensus! In: 6th Workshop on Geometric and Topological Methods in Concurrency and Distributed Computing, GETCO 2004 (October 2004)
4. Dolev, S.: Self-Stabilization. The MIT Press, Cambridge (2000)
5. Dolev, S., Rajsbaum, S.: Stability of Long-lived Consensus. J. of Computer and System Sciences 67(1), 26–45 (2003); Preliminary version in Proc. of the 19th Annual ACM Symp. on Principles of Distributed Computing, (PODC 2000), pp. 309–318 (2000)
6. Diaconis, P., Shahshahani, M.: Generating a random permutation with random transpositions. Probability Theory and Related Fields 57(2), 159–179 (1981)
7. Ehrenfest, P., Ehrenfest, T.: Ueber zwei bekannte EingewÂd'nde gegen das Boltzmannsche H-Theorem. Zeitschrift für Physik 8, 311–314 (1907)
8. Kahn, J., Kalai, G., Linial, N.: The Influence of Variables on Boolean Functions. In: Proc. of the IEEE FOCS, pp. 68–80 (1988)
9. Kramer, B., Fussenegger, M.: Hysteresis in a synthetic mammalian gene network. Proc. Natl. Acad. Sci. USA 102(27), 9517–9522 (2005)
10. Kutten, S., Masuzawa, T.: Output Stability Versus Time Till Output. In: Pelc, A. (ed.) DISC 2007. LNCS, vol. 4731, pp. 343–357. Springer, Heidelberg (2007)
11. Kopetz, H., Veríssimo, P.: Real Time and Dependability Concepts. In: Mullender, S. (ed.) Distributed Systems, ch. 16, pp. 411–446. ACM Press, New York (1993)
12. Musoll, E., Lang, T., Cortadella, J.: Exploiting the locality of memory references to reduce the address bus energy. In: Proc. of the Int. Symp. on Low Power Electronics and Design, August 1997, pp. 202–207 (1997)
13. Norris, J.R.: Markov Chains. Cambridge Series in Statistical and Probabilistic Mathematics. Cambridge University Press, Cambridge (1998)
14. Palacios, J.L.: Another Look at the Ehrenfest Urn via Electric Networks. Advances in Applied Probability 26(3), 820–824 (1994)
15. Pomerening, J., Sontag, E., Ferrell, J.: Building a cell cycle oscillator: hysteresis and bistability in the activation of Cdc2. Nature Cell Biology 5, 346–351 (2003)
16. Su, C.-L., Tsui, C.-Y., Despain, A.M.: Saving power in the control path of embedded processors. IEEE Design & Test of Comp., 24–30 (1994)

Distributed Approximation Algorithm for Resource Clustering

Olivier Beaumont, Nicolas Bonichon, Philippe Duchon,
and Hubert Larchevêque

Université de Bordeaux, INRIA Bordeaux Sud-Ouest, Laboratoire Bordelais de
Recherche en Informatique

Abstract. In this paper, we consider the clustering of resources on large
scale platforms. More precisely, we target parallel applications consisting
of independant tasks, where each task is to be processed on a different
cluster. In this context, each cluster should be large enough so as to
hold and process a task, and the maximal distance between two hosts
belonging to the same cluster should be small in order to minimize latencies of intra-cluster communications. This corresponds to maximum
bin covering with an extra distance constraint. We describe a distributed
approximation algorithm that computes resource clustering with coordinates in $\mathbb{Q}$ in $O(\log^2 n)$ steps and $O(n \log n)$ messages, where n is the
overall number of hosts. We prove that this algorithm provides an approximation ratio of $\frac{1}{3}$.

1 Introduction

The past few years have seen the emergence of a new type of high performance
computing platform. These highly distributed platforms, such as BOINC [3]
or WCG [2] are characterized by their high aggregate computing power and
by the dynamism of their topology. Until now, all the applications running on
these platforms (seti@home [4], folding@home [1],...) consist in a huge number
of independent tasks, and all data necessary to process a task must be stored
locally in the processing node. The only data exchanges take place between
the master node and the slaves, which strongly restricts the set of applications
that can be performed on this platform. Two kind of applications fit in this
model. The first one consists in application, such as Seti@home, where a huge
set of data can be arbitrarily split into arbitrarily small amount of data that can
be processed independently on participating nodes. The other application that
are executed on these large scale distributed platforms correspond to Monte-
Carlo simulations. In this case, all slaves work on the same data, except a few
parameters that drive Monte Carlo simulation. This is for instance the model
corresponding to Folding@home. In this paper, our aim is to extend this last set
of applications. More precisely, we consider the case where the set of data needed
to perform a task is possibly too large to be stored at a single node. In this case,
both processing and storage must be distributed on a small set of nodes that will
collaborate to perform the task. The nodes involved in the cluster should have

A. Shvartsman and P. Felber (Eds.): SIROCCO 2008, LNCS 5058, pp. 61–73, 2008.

an aggregate memory larger than a given threshold, and they should be close enough (the latencies between those nodes should be small) in order to avoid high communication costs. In this paper, we focus on a preliminary subproblem, namely, that of efficiently forming groups of participating nodes that will be able to collaborate for solving these tasks.

This corresponds, given a set of weighted items (the weights are the storage capacity of each node), and a metric (based on latencies), to create a maximum of groups so that the maximal latency between two hosts inside each group is lower than a given threshold, and so that the total storage capacity of a group is greater than a given storage threshold. This problem turns out to be difficult, even if one node knows the whole topology (i.e. the available memory at each node and the latency between each pair of nodes). Indeed, even without the distance constraint, this problem is equivalent to the classical NP-complete bin covering problem [6]. Similarly, if we remove the constraint about storage capacity, but keep the distance constraint, the problem is equivalent to the NP-Complete disk cover problem [10].

Moreover, in a large scale dynamic environment such as BOINC, where nodes connect and disconnect at a high rate, it is unrealistic to assume that a node knows all platform characteristics. Therefore, in order to build the clusters, we need to rely on fully distributed schemes, where a node makes the decision to join a cluster based on its position, its weight, and the weights and positions of its neighbor nodes. In order to estimate the position of the nodes involved in the computation, we rely on mechanisms such as Vivaldi [7,9] that associate to each node a set of coordinates in a low dimension metric space, so that the distance between two points approximates the latency between corresponding hosts.

To the best of our knowledge, this paper is the first attempt to consider both criteria (memory and latency) simultaneously. Therefore, since our aim is also to design fully distributed algorithms, we limit our study to the case where the coordinates of the points lie in $\mathbb{Q}$ and we consider memory constraints only (one could imagine to add other constraints, such as a minimal aggregate computing power or a minimal aggregate disk storage).

In this paper, we present a fully distributed greedy algorithm that creates a number of bins approximating the optimal within a $\frac{1}{3}$ ratio. This result can be compared to some results on classical bin covering in centralized environment without the distance constraint. In this (easier) context, a $PTAAS$ (polynomial-time asymptotic approximation scheme) has been proposed for bin covering [8], i.e. algorithms A_ϵ such that for any $\epsilon > 0$, A_ϵ can perform, in a polynomial time, a $(1 - \epsilon)$-approximation of the optimal when the number of bins tends towards the infinite. Many other algorithms have been proposed for bin covering, such as [6], that provides algorithms with approximation ratio of $\frac{2}{3}$ or $\frac{3}{4}$, still in a centralized environment.

The $\frac{1}{3}$ approximation ratio algorithm we propose in this paper takes both memory and distance constraints into account and can be efficiently implemented in practice on a Peer To Peer (P2P for short) platform using a skip-graph [5] as overlay. The overall number of exchanged messages is of order $O(n \log n)$ and it takes at most $O(\log^2 n)$ rounds to converge.

The rest of the paper is organized as follow : in Section 2, we prove that any "reasonable" (i.e. satisfying two natural constraints) greedy algorithm leads to an approximation ratio of $\frac{1}{3}$. In Section 3, we present an efficient distributed algorithm to compute prefix sums using a skip graph [5]. This algorithm is later used in Section 4, where we detail the implementation and the theoretical properties of the distributed approximation algorithm we propose for solving bin covering problems with distance constraint.

2 Distance Constrained Bin Covering: Greedy Approximation

2.1 Bin Covering Subject to Distance Constraints

Since our aim is to build clusters whose aggregate power is large enough and such that any two nodes belonging to the same cluster are close enough, we introduce the "Distance Constrained Bin Covering" decision problem (DCBC for short).

Definition 1 (DCBC: Distance Constrained Bin Covering)

Input: − a set $S = \{s_1, ...s_n\}$ of elements,
  − a position function $pos : S \to E$ where (E, d) is a metric space,
  − a weight function $w : S \to \mathbb{Q}^+$,
  − a weight threshold W,
  − an integer K,
  − a distance bound d_{max}.
Output: Is there a collection of K pairwise disjoints subsets $S_1, ...S_K$ of S such that $\forall i \leq K$: $\sum_{s \in S_i} w(s) \geq W$ and $\forall (u, v) \in S_i, d(u, v) \leq d_{max}$?

Clearly, DCBC is NP-Complete, since the case where all elements are at the same location corresponds to the classical Bin Covering problem. In what follows, for the sake of clarity, we normalize the weights of the elements (divide them by W) and set $W = 1$ and we do not consider elements whose weight is larger than 1, since such an element can form a group by itself and can be removed. In the rest of the paper, we propose several approximation algorithms (both centralized and decentralized) for the corresponding optimization problem we call *max_DCBC*, in the restriction when the metric space E is taken to be $\mathbb{Q}$ with the usual distance.

2.2 Approximation Ratio of $\frac{1}{3}$

In this section, we propose an algorithm that provides an approximation ratio of $\frac{1}{3}$ for *max_DCBC*. We say that a bin is *interval-based* if the positions of all the elements of this bin belong to an interval and if any element whose position is in this interval belongs to the considered bin. On the other hand, we only deal with *greedy* bins, in the sense that the weight of a bin is strictly less than 2, considering that when the weight of a bin is 2 or more, any element of this bin can be removed without invalidating the bin.

The following Lemma states that any maximal solution consisting of minimal interval-based bins reaches an approximation ratio of $\frac{1}{3}$ with respect to the optimal (*i.e.* without the interval-based condition) solution.

Lemma 1. *A solution to* max_DCBC *problem that satisfies the following constraints*

P1: *Bins are interval-based and the weight of each bin is strictly smaller than 2,*
P2: *No valid interval-based bin can be created with the remaining elements, approaches the optimal solution within a ratio $\frac{1}{3}$.*

Note that these are the conditions satisfied by the solutions given by a greedy algorithm that repeatedly creates interval-based bins until none are possible any more.

Proof. Let us consider a solution $S = \{B_1, \dots B_k\}$ of *max_DCBC* satisfying the above conditions, and S^* an optimal solution (not necessarily satisfying the above conditions). Let us denote by $l(B_i)$ (resp. $r(B_i)$) the left (resp. right) bound of B_i. We assume that $\forall 1 \le i \le k-1, r(B_i) < l(B_{i+1})$, for the sake of simplicity.

We define the *extended area* of B_i as the interval $[l(B_i) - d_{max}, r(B_i) + d_{max}]$.

Clearly, because of P2, any bin in S^* intersects one of the $B_i's$ and is therefore included in its extended area.

Chains: We call chain a maximal sequence of consecutive bins $(B_j, \dots B_{j+k-1})$ such that, as in Figure 1, each extended area of each bin B_i of the sequence intersects the neighbouring bin $(r(B_i) + d_{max} \ge l(B_{i+1}))$. Consider the extended area of a chain, denoted as the interval I_c, where $I_c = [l(B_j) - d_{max}; r(B_{j+k-1}) + d_{max}]$ for each chain c of a solution.

Since any bin B in S^* is included in the extended area of a bin of our solution, each bin in S^* is included in the extended area of at least one chain. We arbitrarily affect each bin of S^* to a chain of our solution.

We prove in what follows that no more than $3k$ bins of S^* can be affected to a chain containing k bins, hence the approximation ratio of $\frac{1}{3}$.

We now claim that $w([l(B_j) - d_{max}; r(B_{j+k-1}) + d_{max}]) < 3k + 1$, for a chain $c = (B_j, \dots B_{j+k-1})$ containing k bins. This obviously implies that no more than $3k$ bins of S^* can be included in I_c. Indeed, the considered interval is the union of k bin intervals (each with weight $w(B_i) < 2$ as per P1), $k-1$ intervals of the form $(r(B_i), l(B_{i+1}))$ (each with weight strictly less than 1 as per P2), and of the two intervals $[l(B_j) - d_{max}, l(B_j))$ and $(r(B_{j+k-1}), r(B_{j+k-1}) + d_{max}]$ (also each with weight strictly less than 1). The total weight in the extended area of a chain containing k bins is $w(c) < 3k + 1$, thus no more than $3k$ bins of S^* can be affected to this chain. Summing over all chains, the number of bins in S^* cannot be larger than three times the number of bins in S.

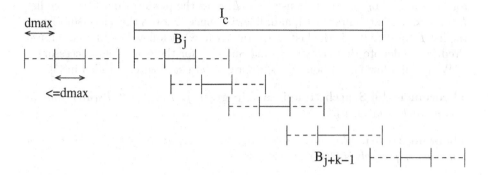

Fig. 1. Example of a chain of 4 bins

3 Preliminaries: Overlay and Prefix Sums Computation

Our approximation algorithm for max_DCBC, described in Section 4, runs on a special overlay network, and uses prefix sums computations as a preliminary step. In this section, we describe both elements.

3.1 The Skip-Graph Overlay

To each element of the list to group is associated a host of the network.

First we will present the overlay we use for the network, whose goal is to minimize the number of exchanged messages necessary to perform the bin covering of the elements, while being easy to maintain and construct. This overlay is a skip-graph [5], a structure inspired by skip-lists [12,11].

A skip-list is organized as a succession of ordered lists containing fewer and fewer nodes at each level i: for $i > 0$, each host at level $i - 1$ appears in the list of level i with probability p (we use $p = 1/2$). The hosts belonging to the level i list are linked to each other according to the coordinates' order of the basic list.

In the overlay network we use, hosts are organized as nodes of a skip-graph, and ordered by their positions, i.e their coordinates in $\mathbb{Q}$. To make notations simpler, we identify each host x with its position and write $x \leq y$ for $pos(x) \leq pos(y)$, where pos is the position function as defined in Section 2.

In a skip-graph, there are up to 2^i lists at level i, each indexed by a binary word of length i. Each host x generates a key $key(x)$, a sequence of random bits, using a sequence of random coin flips. At each level i, the level 0 list L_ϵ, containing all hosts ordered by coordinates, is partitioned into 2^i lists, each containing the hosts sharing the same first i key bits. Each host stops its key generation as soon as it is the only host in a level i list, i.e. as soon as no other host shares the exact first i key bits. Then it only keeps the first i bits of its key.

For an example, see Figure 2.

Notations. Let x be a host. We use $b_i(x)$ to denote the first i bits of $key(x)$, $h(x)$ to denote the number of bits of $key(x)$, and $L_i(x)$ the level i list x belongs to, *i.e.* $L_i(x) = L_{b_i(x)}$. We also use $\text{Pred}(L, x)$ as the predecessor of x in the list L (or $-\infty$ if x is the first in L), and, likewise, $\text{Succ}(L, x)$ as the successor of x in the list L (or $+\infty$ if x is the last in L). In fact in what follows we more often use $\text{Pred}_i(x)$ to denote $\text{Pred}(L_i(x), x)$ (and $\text{Succ}_i(x)$ for the equivalent successor).

We recall a few facts about the skip-graph data structure, detailed in [5]

Theorem 1 ([5] Search time in a skip-graph). *The expected search time in a skip graph is* $O(\log n)$.

Theorem 2 ([5]). *With high probability, the maximum height of an element in a n-node skip-graph is* $O(\log n)$.

We will use this overlay to compute the prefix sums for each host.

3.2   Prefix Sums Computation

We recall the fact that each host x has a weight $w(x) < 1$. We define the *prefix sum* $S(x)$ of a host x to be *the sum of the weights of every host in* L_ϵ *between the first host and* x, *included*. We also define a *level i partial prefix sum* $S_i(x)$:

$$S_i(x) = \sum_{\substack{z \in L_\epsilon \\ \text{Pred}_i(x) < z \leq x}} w(z). \tag{1}$$

Thus $S_i(x)$ is *the sum of the weights in the basic list, L_ϵ, of every host between its level i predecessor(excluded) and itself (included)*. It can also be defined as $S(x) - S(\text{Pred}_i(x))$.

To compute $S_i(x)$, we use the following recurrence: Thus we have:

$$S_i(x) = sum_{\substack{z \in L_{i-1}(x) \\ \text{Pred}_i(x) < z \leq x}} S_{i-1}(z). \tag{2}$$

(To prove equation 2, notice that the first z in the summation is $\text{Succ}_{i-1}(\text{Pred}_i(x))$, and the positive and negative terms cancel each other, leaving only $S(x)$ and $-S(\text{Pred}_{i-1}(\text{Succ}_{i-1}(\text{Pred}_i(x)))) = -S(\text{Pred}_i(x))$.)

Note that $S_0(x) = w(x)$ and $S(x) = S_{h(x)}(x)$.

Algorithm. Algorithm 1, running on each host, uses the skip-graph to compute the partial prefix sum for each host at each level. It runs upward in the levels of the skip-graph, ending at the highest level for each host. When the computation ends at a host x, it knows each of its partial prefix sums $S_i(x)$.

Time model. We consider that each message transmission is done in a unit time, and we do not consider computation time on each host.

Theorem 3. *Algorithm 1 computes all the prefix sums, using a skip-graph as overlay, in* $O(\log^2 n)$ *time steps and* $O(n \log n)$ *messages, with high probability.*

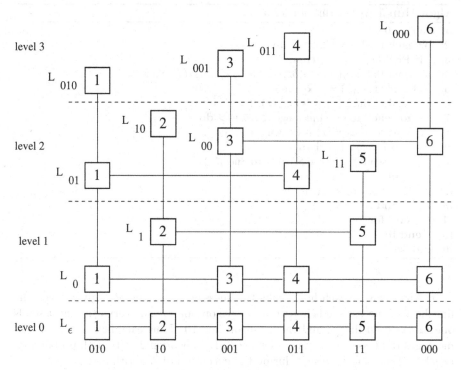

Fig. 2. Example of skip-graph, with 4 levels and 6 nodes in L_ϵ

Proof. **Number of messages**

- Each message has this form: $< i, b, w >$. We call i the message level, b the message key and w the message weight.
- Each message weight is null when it is first emitted (before the first retransmission).
- Each level i message is sent by a host to its level i successor.
- Whenever a level i message is received by a host x with a key not matching $b_i(x)$, the message sent in response (line 10 of Algorithm 1) is considered to be the same message, retransmitted with increased weight(this still counts in the total number of messages). Note that the first host x of each level i list sends automatically two messages, so that the first host to its right with a key not matching $b_i(x)$ also computes its S_{i+1} sum on line 12.
- As soon as a level i message is received by a host x with a key matching $b_i(x)$, it is not retransmitted anymore (line 12 of Algorithm 1), and its weight is used by the receiving host to compute its level $i + 1$ partial prefix sum.

Note that, given a list at level i, and two neighbouring hosts in this list, only two messages are exchanged between those two hosts at this level, sent by the leftmost one to the rightmost one.

Algorithm 1. Algorithm for a host x

1: $S_0(x) = 0$
2: **for each** level $i = 0, 1, 2, \ldots$ **do**
3:     **if** $\text{Pred}_i(x) == -\infty$ **then**
4:         send to $\text{Succ}_i(x)$: $< i, b_{i+1}(x), 0 >$, $< i, b_{i+1}(x), S_i(x) >$
5:     **else if** $\text{Succ}_i(x) \neq +\infty$ **then**
6:         send to $\text{Succ}_i(x)$: $< i, b_{i+1}(x), 0 >$
7:         **for each** received message $< i, b, w >$ **do**
8:             as soon as $S_i(x)$ is known:
9:             **if** $b \neq b_{i+1}(x)$ **then**
10:                 send $< i, b, w + S_i(x) >$ to $\text{Succ}_i(x)$
11:             **else**
12:                 $S_{i+1}(x) = S_i(x) + w$
13:             **end if**
14:         **end for**
15:     **end if**
16: **end for**

In Algorithm 1, each host sends two messages by level: at a given level, the first host of the list sends its two messages on line 4, and every other host sends one message on line 6, and the other on line 10. Thus each host x send $2h(x)$ messages (and receive as many messages). By Theorem 2, with high probability, $O(n \log n)$ messages are sent during the execution of Algorithm 1.

Time analysis

If $x_1 \ldots x_k$ are consecutive elements of a same level i list L, such that all belong to the same level $i + 1$ list L', and $\text{Pred}(L, x_1) \notin L'$, and $\text{Succ}(L, x_k) \notin L'$, we call $\{x_1, \ldots x_k\}$ a level i siblings set. Note that at each level, the largest siblings set is of length $O(\log n)$ with high probability. For more information on this result, see [12].

Each level i message is transmitted through a whole siblings set before it reaches its destination y. During these transmissions, it collects the partial prefix sums contributions to $S_i(y)$. When a host receives a message at a time T, it can retransmit it at a time $T + 1$.

If we write $T_i(x)$ for the instant where x learns $S_i(x)$, $T_i = \max_{x \in L_e} T_i(x)$, and M for the maximal size of all the siblings sets:

$$T_i(x) = \max_{j=0..m} (T_{i-1}(y_j) + m - j + 1) \tag{3}$$

where $y_0 = \text{Pred}_i(x)$, m is the size of the (potentially empty) level $i - 1$ siblings set between $\text{Pred}_i(x)$ and x, and $y_j, 1 \leq j \leq m$ is the jth host of this siblings set. Note that for all hosts x, $T_0(x) = 0$.

Thus

$$T_i(x) \leq \max_{j=0..m} (T_{i-1}(y_j) + m + 1) \leq T_{i-1} + M + 1. \tag{4}$$

Thus $T_i \leq T_{i-1} + M + 1$. Thus, if we write $H = \max_{x \in L_\epsilon} h(x)$, we have $T_H \leq H(1 + M)$. Both H and M are $O(\log n)$; the upper bound on the number of time steps follows.

Note that we suspect the $O(\log^2 n)$ bound is pessimistic, and that a tighter analysis would yield $\Theta(\log n)$.

4 A Distributed Approximation Algorithm

In this section, the n hosts to cluster are linked by a skip-graph as overlay network. Hosts are identified to their index in L_ϵ, *i.e.*. we write $i \leq j$ if the host i is before j in L_ϵ. Sending a message from a host i to the host j using the overlay network can be done with $O(\log |i - j|)$ messages with high probability.

First we present a distributed algorithm computing the clustering of a list of weighted hosts, without the distance restriction. Then this restriction is added to our algorithm in subsection 4.2. The final algorithm has an approximation ratio of $\frac{1}{3}$, as per Lemma 1.

Note that any greedy algorithm without distance restriction has an obvious approximation ratio of $\frac{1}{2}$, if it creates groups weighting strictly less than 2.

The following algorithms creates interval-based bins in L_ϵ. Each created bin corresponds to a cluster of the platform. The clusters are defined in the following way: each grouped host received the identifier of a host, called *leader*. A cluster is a set of hosts having the same leader. Here the leader of a cluster is always the leftmost host of the group.

4.1 Clustering Algorithm without Distance Constraints

The recursive treatment is based on a procedure that takes 3 parameters: li (left index), ri (right index) and rli (right leader index), with $li < rli \leq ri$. This routine is executed by host $j = \lfloor \frac{li+rli}{2} \rfloor$. This procedure creates clusters included in the intervals $[li, ri)$, having their leader in the intervals $[li, rli[$. To begin the clustering of the hosts, send $< 0, n, n >$ to host $\lfloor n/2 \rfloor$. The principle of this routine is quite simple. If host j can be leader of a cluster included in interval $[li, ri)$, it creates this cluster and recursively call the procedure on the two sublists respectively before and after the newly created cluster. If it cannot be leader, no other host after j can create such a cluster in interval (j, rli), hence it will only look for leaders in the sub-interval $[li, j - 1)$ for clusters included in $[li, ri)$.

The recursive routine is described in Algorithm 2. It uses two sub-routines: $\text{rb}(j)$ and $\text{buildCluster}(a, b)$. The function $\text{rb}(j)$ returns the minimum host k after j such that $p_k \geq 1 + p_j$, where p_j is the prefix sum of host j; *i.e.* the leftmost host that could possibly be the last host in an interval-based bin having j as its first host. By convention, $\text{rb}(j) = +\infty$ if no such host exists. The routine $\text{buildCluster}(a, b)$ defines a as leader for each host $a \leq j \leq b$.

Algorithm 2. Algorithm for a coordinator j

1: Receive $< li, rli, ri >$
2: **if** $j < ri$ **then**
3:    **if** $\text{rb}(j) < ri$ **then**
4:      buildCluster$(j, \text{rb}(j))$
5:      Send $< li, j, j >$ to host $\lfloor \frac{li+j}{2} \rfloor$
6:      **if** $\text{rb}(j) < rli$ **then**
7:         Send $< \text{rb}(j) + 1, rli, ri >$ to host $\lfloor \frac{\text{rb}(j)+rli+1}{2} \rfloor$
8:      **end if**
9:    **else**
10:      Send $< li, j, ri >$ to host $\lfloor \frac{li+j}{2} \rfloor$
11:    **end if**
12: **end if**

Theorem 4. *Consider n hosts linked by a skip-graph as overlay network. Algorithm 2 computes a clustering of those hosts such that: (i) each cluster is an interval of L_ϵ of total weight $w < 2$; (ii) each interval of L_ϵ of non-grouped hosts has a weight $w < 1$. Moreover, Algorithm 2 runs in $O(\log^2 n)$ steps and $O(n \log n)$ messages with high probability.*

Proof

Clustering: A host is said to be *coordinator* of an interval $[li; ri)$ when it receives a message $< li, rli, ri >$ and begins to execute Algorithm 2. It is coordinator of this interval until it begins to send messages on line 5 or on line 10. Thus, at each instant, for each host j, at most one host is the coordinator of an interval containing j.

Each coordinator creates, if it can, a cluster *included* in the interval of which it is coordinator. Then it designates at most two hosts as coordinators of sub-intervals not containing any host of the created cluster. So for two clusters having hosts $(j, k) \in [0; n)$ as leaders, $[j; \text{rb}(j)) \cap [k; \text{rb}(k)) = \emptyset$. Thus any host affected to a cluster can not be affected later to another cluster: a built cluster is definitively built. By definition of $\text{rb}(j)$, constructed clusters have a weight $w < 2$. (i) is thus proved.

To prove (ii), notice that there is a bijection between the set of coordinator hosts and the set of maximum intervals of non-grouped hosts. At the beginning, one host is coordinator of the whole interval of non-grouped hosts. Then at each interval subdivision, either a cluster is created (line 4), and two hosts are designated as coordinators of the two sub-intervals of non-grouped hosts at each side of the created cluster (these two sub-intervals are still maximal in terms of non-grouped hosts); or no cluster is created (line 9) and another host is designated coordinator of the same maximal interval of non-grouped hosts.

Thus at the end of the execution of the algorithm, each maximal interval $[li; ri)$ of non-grouped hosts has host li as coordinator. Since this coordinator does not create a cluster, $\text{rb}(li) \geq ri$. Hence the cumulated weight of this maximal interval is strictly less than 1. (ii) is thus proved.

Complexity: In the precomputation steps, the value $\mathrm{rb}(j)$ is computed by each host j. This can be done in two steps: first compute the prefix sum for each host (see Section 3.2). Then, based on the prefix sum, search for host $\mathrm{rb}(j)$. By Theorem 1, this search in a skip-graph takes for each host a $O(\log n)$ time w.h.p..

The routine $\mathrm{buildCluster}(a, b)$ takes $O(h \log h)$ messages and $O(\log h)$ steps with $h = b - a$. Since all built bins are pairwise disjoint intervals, all execution of buildCluster can be executed in parallel and will take $O(\log h_{max})$ time steps and $O(n \log n)$ messages total, w.h.p..

Each time a host executes Algorithm 2, it designates two hosts to be coordinators of two sub-intervals (lines 5 and 7), or one host to be coordinator of half the original one (line 10). Hence there are at most $\log n$ levels of recursion. As each call of Algorithm 2 takes $O(\log(rli - li))$ messages w.h.p., it takes $O(n \log n)$ messages and $O(\log^2 n)$ steps total w.h.p..

Corollary 1 (Approximation ratio). *The resulting clustering of Algorithm 2 has an approximation ratio of 1/3 with the optimal solution.*

Proof. To prove this corollary, consider Lemma 1, using $d_{max} = \max_{(u,v) \in L_\epsilon} (d(u, v))$. The first property is proved in Theorem 4, and the third property of Lemma 1 is obviously verified as all hosts are at distance less than d_{max} from each other.

4.2 Adding the Distance Restriction

We recall that each host j has a position $pos(j)$, and that hosts are ordered in L_ϵ by their position.

A host j is *eligible* if it can be the leader of a cluster, that is if $pos(\mathrm{rb}(j)) \leq 1 + pos(j)$. The function $\mathrm{neh}(j)$ return the *next eligible host* k after j. By convention, $\mathrm{neh}(j) = +\infty$ if no such host exists. As for the previous algorithm, $\mathrm{neh}(j)$ and $\mathrm{rb}(\mathrm{neh}(j))$ can be precomputed for each host j. Algorithm 3 is an adaptation of Algorithm 2 that produces clusters of bounded diameter.

Algorithm 3. Algorithm for a coordinator j

```
1:  Receive < li, rli, ri >
2:  if li < ri then
3:      if rb(neh(j)) < ri then
4:          buildCluster(neh(j), rb(neh(j)))
5:          Send < li, j, neh(j) > to host ⌊ li+j/2 ⌋
6:          if rb(neh(j)) + 1 < rli then
7:              Send < rb(neh(j)) + 1, rli, ri > to host ⌊ rb(neh(j))+rli+1/2 ⌋
8:          end if
9:      else
10:         Send < li, j, ri > to host ⌊ li+j/2 ⌋
11:     end if
12: end if
```

Theorem 5. *Consider n hosts with given positions, linked by a skip-graph as overlay network respecting the order of their positions. Algorithm 3 computes a clustering of those hosts with distance constrained such that: (i) each cluster is an interval of L_ϵ of diameter at most d_{max} and of weight $w < 2$; (ii) each interval of L_ϵ of non-grouped hosts of diameter $d \leq d_{max}$ has a total weight $w < 1$.*

Moreover, Algorithm 3 runs in $O(\log^2 n)$ steps and $O(n \log n)$ messages with high probability.

Proof. The complexity of this algorithm is clearly the same as Algorithm 2: at each execution of Algorithm 3, the interval $[li; rli]$ is split in at most two new sub-intervals, each of size at most half the size of the original interval $[li; rli]$.

(i): By definition of $\mathrm{rb}(j)$, and with a reasoning similar to the proof of Theorem 2, constructed clusters have a weight strictly lower than 2. Moreover, by definition of $\mathrm{neh}(j)$, created clusters have a diameter lower than $d_{max}(\leq d_{max})$.

(ii): The reasoning is still similar to the proof of Theorem 4. At the end of the algorithm the coordinator of $[li; ri]$ is li. As li does not create a cluster we deduce that $\mathrm{rb}(\mathrm{neh}(li)) \geq ri$, *i.e* the first host able to build a cluster at the right of li would build a cluster overtaking the right bound ri. By definition of rb and neh we deduce that no cluster of bounded diameter can be created in $[li; ri)$.

Corollary 2 (Approximation ratio). *The resulting clustering of Algorithm 2 has an approximation ratio of $1/3$ with the optimal solution.*

Proof. The proof of this corollary is a direct application of Lemma 1, of which the two necessary properties are verified, proved in Theorem 5.

5   Conclusions

In this paper we have presented a distributed approximation algorithm, running in $O(\log^2 n)$ steps and $O(n \log n)$ messages, that computes resource clustering for n hosts with coordinates in $\mathbb{Q}$. This algorithm provides an approximation ratio of $\frac{1}{3}$. We have restricted this work to a 1-dimensional case, but we are working on the extension of our results to higher dimensions or more general metric spaces.

As this work is meant to be used on large-scale platforms, it is necessary to make our algorithms able to handle a high degree of dynamicity on the hosts of the networks. Notably, they have to handle the dynamicity of the weights of the hosts, because each user of a large-scale platform is likely to want its resources back for its private use at any time.

It could also be interesting to work on a version in which each created cluster would have many criteria to satisfy. In fact in our work, we just considered that clusters had to offer a sufficient global storage capacity, but one may want clusters to additionally ensure, for example, sufficient computing power.

References

1. Folding@home, http://folding.stanford.edu/
2. World community grid, http://www.worldcommunitygrid.org
3. Anderson, D.P.: Boinc: A system for public-resource computing and storage. In: GRID 2004. Proceedings of the Fifth IEEE/ACM International Workshop on Grid Computing, Washington, DC, USA, pp. 4–10. IEEE Computer Society, Los Alamitos (2004)
4. Anderson, D.P., Cobb, J., Korpela, E., Lebofsky, M., Werthimer, D.: Seti@home: an experiment in public-resource computing. Commun. ACM 45(11), 56–61 (2002)
5. Aspnes, J., Shah, G.: Skip graphs. In: Proceedings of the fourteenth annual ACM-SIAM symposium on Discrete algorithms, pp. 384–393 (2003)
6. Assmann, S.F., Johnson, D.S., Kleitman, D.J., Leung, J.Y.T.: On a dual version of the one-dimensional bin packing problem. Journal of algorithms(Print) 5(4), 502–525 (1984)
7. Cox, R., Dabek, F., Kaashoek, F., Li, J., Morris, R.: Practical, distributed network coordinates. ACM SIGCOMM Computer Communication Review 34(1), 113–118 (2004)
8. Csirik, J., Johnson, D.S., Kenyon, C.: Better approximation algorithms for bin covering. In: Proceedings of the twelfth annual ACM-SIAM symposium on Discrete algorithms, pp. 557–566 (2001)
9. Dabek, F., Cox, R., Kaashoek, F., Morris, R.: Vivaldi: a decentralized network coordinate system. In: Proceedings of the 2004 conference on Applications, technologies, architectures, and protocols for computer communications, pp. 15–26 (2004)
10. Franceschetti, M., Cook, M., Bruck, J.: A geometric theorem for approximate disk covering algorithms (2001)
11. Munro, J.I., Papadakis, T., Sedgewick, R.: Deterministic skip lists. In: Proceedings of the third annual ACM-SIAM symposium on Discrete algorithms, pp. 367–375 (1992)
12. Pugh, W.: Skip lists: A probabilistic alternative to balanced trees. In: Workshop on Algorithms and Data Structures, pp. 437–449 (1989)

Sharpness:
A Tight Condition for Scalability

Augustin Chaintreau

Thomson
augustin.chaintreau@thomson.net

Abstract. A distributed system is scalable if the rate at which it completes its computation and communication tasks does not depend on its size. As an example, the scalability of a peer-to-peer application that transmits data among a large group depends on the topology and the synchronization implemented between the peers. This work describes a model designed to shed light on the conditions that enable scalability. Formally, we model here a collection of tasks, each requiring a random amount of time, which are related by precedence constraints. We assume that the tasks are organized along an euclidean lattice of dimension d. Our main assumption is that the precedence relation between these tasks is invariant by translation along any of these dimensions, so that the evolution of the system follows Uniform Recurrence Equations (UREs). Our main result is that scalability may be shown under two general conditions: (1) a criterion called "sharpness" satisfied by the precedence relation and (2) a condition on the distribution of each task completion time, which only depends on the dimension d. These conditions are shown to be tight. This result offers a universal technique to prove scalability which can be useful to design new systems deployed among an unlimited number of collaborative nodes.

1 Introduction

Scalability is usually regarded as an important if not critical issue for any distributed communication/computation system. However, it is in general difficult to describe formally in mathematical terms. This paper describes a model designed to shed light on the impact of synchronization among an ever-increasing number of nodes participating in a distributed system. The results we provide answer the following question: "Under which general conditions is a distributed system scalable, in the sense that the rate of tasks' completion remains the same independently of the number of participating nodes?" We characterize in particular the impact of three factors: 1- the organization of the local feedback and synchronization mechanisms deployed between the nodes (acknowledgment, etc.), 2- the global topology of the communication (in this paper, an euclidean lattice), and 3- the variations of local delay, or local computation time, which is captured in this model by random times to complete each step. This general problem is primarily motivated today to study the throughput of communication protocols in large scale data networks.

A. Shvartsman and P. Felber (Eds.): SIROCCO 2008, LNCS 5058, pp. 74–88, 2008.

Model. This paper analyzes the process of completion times **T** for a collection of tasks which are related together via precedence constraints. As an example of precedence constraint, a customer can be served in a server only once the previous customer has completed its service. As in a queueing system, each task, once it has started, takes a random amount of time to finish, which is called its *weight*. As an example, this random weight can represent either the time needed to exchange a message over a wireless link, a link shared with background data traffic, or the time needed to access some memory or processing unit. We assume that the weights of different tasks are independent random variables. The weight of a given task may occasionally take a large value (for instance, due to local congestion, server load, etc.); this is represented in the model by the weight distribution of this task. Hence, for a given initial condition of the system, the completion time of a task is a random variable that depends on the precedence relation with other tasks, and the (random) weights for all of them.

We focus here on the case where the collection of tasks is infinite and regular. Formally, we assume there exists a finite subset of indexes $\mathcal{H}$, which we call the *pattern*, and a dimension $d \geq 1$ such that the collection of tasks is

$$\{ \, a \times h \ \mid \ a \in \mathbb{Z}^d, \ h \in \mathcal{H} \, \} \, .$$

In other words, one can describe this system as a finite set $\mathcal{H}$ of tasks to be done locally, which is reproduced at every position of a d-dimensional lattice. Moreover, we assume that the precedence constraints that relate all the tasks of the system together are *invariant by translation*. Thus, a given task $(a \times h)$ depends on different tasks, some have the same position a in the lattice and a different index $h' \neq h$ chosen in $\mathcal{H}$, some have different positions in the lattice and any index h' in $\mathcal{H}$. The above assumption guarantees that, after translation, these precedence relations do not depend on the position a.

Main result. In this paper we provide sufficient and necessary conditions for the following property to hold: for any $a \in \mathbb{Z}^d$, and any $h \in \mathcal{H}$ we have

$$\exists M \in \mathbb{R} \quad \text{such that almost surely} \quad \limsup_{m \to \infty} \frac{1}{m} T_{(m \cdot a) \times h} < M \,, \qquad (1)$$

where $T_{a \times h}$ denotes the completion time for the task $(a \times h)$, and $(m \cdot a)$ denotes the vector $(m \cdot a_1, \ldots, m \cdot a_d) \in \mathbb{Z}^d$.

A system that satisfies the above property is called *scalable*, as this guarantees that the completion time grows linearly along any direction drawn in the lattice. Equivalently, the (random) set that contains all tasks completed before t grows with t according to a positive linear rate, in any given direction. As an example, if one dimension of $\mathbb{Z}^d$ denotes the sequence number of a packet to be received, this condition guarantees a positive throughput for each node in the system, even if the system itself is infinite.

We prove that scalability, as defined above, is characterized only by two conditions: one that deals with the organization of the precedence relation between tasks, which is called *sharpness*, and another condition that deals with the weight distributions, and which only depends on d. We also prove that these two conditions are tight, and in particular that a non-sharp system in dimension $d = 2$ is never scalable.

Implications. The result above, although it is stated in quite abstract terms, has some important consequences for the design of distributed systems. First, it proves that a large class of systems are scalable although they implement some closed feedback loop among an infinite number of nodes. In particular it shows that distributed reliable systems can be implemented using only *finite* buffers in nodes, while remaining truly scalable. A few examples of this counter-intuitive fact have already been shown in [1,2]. What is new in this paper is that we identify the ingredients of such scalability in a systematic way.

Second, under appropriate moment condition on weights, the scalability of a distributed system is shown to be equivalent to the sharpness condition, which itself corresponds to a finite number of linear inequalities. Proving scalability is therefore greatly simplified, and keeping this condition in mind can even help dimensioning new distributed systems (as happened for instance in [3]).

The assumption that the precedence relation among tasks should be invariant by translation might seem restrictive at first. It is to some extent necessary since designing a general model for irregular systems seems difficult, not to mention finding exact conditions for their scalability. However, we would like to point here that our model allows for patterns $\mathcal{H}$ of arbitrary finite size. It is therefore sufficient to model systems that are regular only at a certain scale. Moreover, it is often the case that irregular systems are included in a larger one that is regular; it is then sufficient to prove scalability by inclusion using stochastic ordering. Last, we have assumed that the systems is organized along an euclidean lattice, although in practice many distributed systems are organized along other hierarchical topologies. The sharpness condition and the scalability result presented here can be extended to this case, although it is far beyond the scope of this paper. The interested reader may found first results on general graphs reported in [4].

Relation with previous work. Scalability has been addressed in the past for distributed communication protocols and congestion control [2,5,3] , stream processing [6] and computational grids [7]. All these works addressed the issue of scale for a distributed system where a form of synchronization is implemented between a large number of nodes. In contrast, our work treats a general case, and identify for the first time a tight condition that characterizes scalability.

Our results extend recent advances in stochastic network theory on infinite tandem of queues with general service time distribution and blocking [1]. Rather than studying a certain feedback mechanism, our results determine which feedback systems make the throughput independent of the system size. In addition, most of the previous results deal only with networks of single server queues. In contrast, our model applies for any pattern $\mathcal{H}$ that contains an arbitrary finite number of tasks. As an example, it can be used to characterize tandem of any finite timed event graph. Such generalization is made possible, as in [1], through a formulation that reminds last-passage percolation time, which allows to use the powerful framework of subadditive ergodic theory [8]. However, no prior knowledge of last-passage percolation is needed to prove this result.

The systems we consider follow solutions of Uniform Recurrence Equations (UREs). UREs were introduced by Karp et al. in [9], where a general condition for the existence of a solution is presented. UREs have been used in the past to study synchronization in parallel computation, such as the discrete solution of differential equations. Our work points in a different direction: for the first time we study the solutions of general UREs when each step takes a random amount of time, and we characterize exactly when this solution grows linearly as its index grows. This is a stronger property as we prove that the sharpness criterion defined here is strictly stronger than the condition defined in [9] to classify UREs.

The organization of the paper is as follows. Section 2 presents the model sketched above in more details. Examples and relations with Uniform Recurrence Relations are explained. Section 3 defines the *sharpness* condition, and establishes the topological consequence of this criterion on the dependence paths between the tasks. Section 4 contains the main results of the paper. Section 5 concludes the paper with remarks on possible extensions.

NB: As for UREs, the boundary condition defining the system initial condition plays an important role in the analysis. In order to focus on the essential connection between sharpness and scalability, we choose to describe a single boundary case, corresponding to a system "initially empty". The result of this paper can be obtained under more general boundary condition (see [4] Chap.3).

2   Pattern Grid

2.1   Definition

Pattern grids are defined as directed graphs that follow an invariant property. Their vertices are indexed by both a multidimensional integer (*i.e.* the "position" of the task in the d-dimensional lattice $\mathbb{Z}^d$) as well as a local index chosen in a finite set $\mathcal{H}$. Formally, a graph $\mathcal{G}_{\text{patt}} = (\mathcal{V}, \mathcal{E})$ is called a *pattern grid*, with dimension d and pattern $\mathcal{H}$, if:

- The set of its vertices is $\mathcal{V} = \mathbb{Z}^d \times \mathcal{H}$.
- The set of its edges $\mathcal{E}$ is invariant by any translation in the lattice $\mathbb{Z}^d$;
    for all a, a', v in $\mathbb{Z}^d$ and h, h' in $\mathcal{H}$, we have
    $(a \times h) \rightarrow (a' \times h') \in \mathcal{E}$ if and only if $((a + v) \times h) \rightarrow ((a' + v) \times h') \in \mathcal{E}$.

In this work, we consider only the case of locally finite graphs (*i.e.* the number of edges that leave any vertex is finite). The invariant property implies that the degrees of vertices in this graph are uniformly bounded. We denote by H the cardinal of $\mathcal{H}$. When $H = 1$, the index h plays no particular role and the pattern grid follows a lattice.

To illustrate this definition, fix all coordinates in the left index (for instance, all of them null), consider the set of all the associated vertices (*e.g.* $(0, \ldots, 0) \times \mathcal{H}$)

and all the directed edges starting from any of those vertices. This defines a finite collection of local tasks to complete and relations between them and a few others "neighboring" tasks. The pattern grid is what is obtained when one reproduces this finite object at every site of an euclidean lattice.

2.2 Examples

Due to space constraints, we can only describe a few illustrating examples of the definitions above.

Infinite tandem of queues. The simplest case of a pattern grid with dimension 2 is an infinite line of single server queues (indexed by k) in tandem, serving customers (indexed by m). It works as follows: when a customer has completed its service in server k, he enters immediately the (infinite) buffer of server $(k+1)$, where it is scheduled according to a first-come-first-served discipline.

Let us consider all the tasks of the type "service of customer m in server k", which are naturally indexed by $(m, k) \in \mathbb{Z}^2$. The relation between them is essentially described by two precedence rules:

- $(m, k) \rightarrow (m, k - 1)$ (*i.e.* the service of m in k cannot start unless its service in server $(k - 1)$ is completed).
- $(m, k) \rightarrow (m - 1, k)$ (*i.e.* the service of m in k cannot start unless the previous customer has completed its service in k).

Hence this system is well described by a pattern grid with dimension 2, a pattern containing a single element, and the above edges.

Infinite tandem of queues with blocking. Let us now assume as in [1] that each server implements two queues (input and output), both with a finite size B_{IN} and B_{OUT}, such that buffers overflow are avoided by the appropriate blocking of service. It is not hard to see that such systems can be modeled by a pattern grid with dimension 2 and pattern $\mathcal{H} = \{i, o\}$. It contains the following edges:

- $(m, k) \times i \rightarrow (m, k - 1) \times o$ (*i.e.* serving a customer in k requires that he was forwarded from the previous server $(k - 1)$.)
- $(m, k) \times o \rightarrow (m, k) \times i$ (*i.e.* a customer cannot enter the output buffer before he has been served by this server.)
- $(m, k) \times i \rightarrow (m - 1, k) \times i$ (*i.e.* serving a customer requires that the previous customer has been served.)
- $(m, k) \times i \rightarrow (m - B_{OUT}, k) \times o$ (*i.e.* avoid B_{OUT} overflow.)
- $(m, k) \times o \rightarrow (m - B_{IN}, k + 1) \times i$ (*i.e.* avoid B_{IN} overflow.)

Fig.1 describes a portion of the graph defined in the two above examples. Other mechanisms of feedback fit in the same model. As an example, one may model TCP connections, with varying windows, organized in tandem and that implement back-pressure blocking [2].

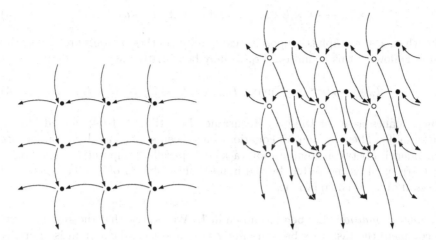

Fig. 1. Two examples of pattern grids: infinite tandem of single server queue (left), the same with input-output blocking (right) (i is represented by a white dot, o by a black dot, we set $B_{\text{IN}} = 1$, $B_{\text{IN}} = 2$)

2.3 Evolution Equation of a Pattern Grid

We now define how the relations between tasks, represented by edges, describe the evolution of the system: the edge $(a \times h) \to (a' \times h')$ in $\mathcal{E}$ represents that the task $(a \times h)$ cannot start unless $(a' \times h')$ has been completed. Vertex $(a' \times h')$ is then called an *immediate predecessor* of $(a \times h)$.

We associate with each vertex $(a \times h)$ of the pattern grid a *weight* denoted by $\mathtt{W}(a \times h)$, representing the time needed to complete this task, once it has started. For instance in the two examples shown above, this weight is the service time of customer m in server k. It is generally random; sometimes it can be taken equal to zero (like for $(m, k) \times o$ in the second example above, if the delay between two servers is neglected).

We consider the process of completion time for every task

$$\mathbf{T} = \left\{ \ T_{a \times h} \in \mathbb{R} \cup \{-\infty\} \ \big| \ a \in \mathbb{Z}^d, \ h \in \mathcal{H} \ \right\} .$$

Assuming that a task begins as soon as all its immediate predecessors have been completed, we have for all a in $\mathbb{Z}^d$ and h in $\mathcal{H}$:

$$T_{a \times h} = \mathtt{W}(a \times h) + \max \left\{ T_{a' \times h'} \big| (a \times h) \to (a' \times h') \in \mathcal{E} \right\} . \qquad (2)$$

As an example, for the infinite tandem of queues described above it becomes Lindley's equation:

$$T_{(m,k)} = \mathtt{W}(m, k) + \max \left(T_{(m,k-1)}, T_{(m-1,k)} \right) .$$

Relation with UREs. It is easy to see that another way to characterize the set $\mathcal{E}$ of edges in a pattern grid is via a collection of *dependence sets*: a finite collection $(\Delta_{h,h'})_{h,h' \in \mathcal{H}}$ of subsets of $\mathbb{Z}^d$ indexed by $\mathcal{H}^2$, such that

$$(a \times h) \rightarrow (a' \times h') \in \mathcal{E} \text{ if and only if } (a' - a) \in \Delta_{h,h'} . \tag{3}$$

Note that, as the graph is supposed locally finite, all these subsets are necessarily finite. Following this definition, Eq.(2) may be rewritten as

$$T_{a \times h} = W(a \times h) + \max \left\{ T_{(a+r) \times h'} \, \middle| \, r \in \Delta_{h,h'}, \ h' \in \mathcal{H} \right\} . \tag{4}$$

which defines a set of Uniform Recurrence Equations (UREs), already introduced in [9]. These systems of equations have been studied as they characterize dependencies between computation tasks in a parallel computation. The article by Karp et al. is motivated by the numerical resolution of discrete version of classical differential equations.

Boundary condition. Let beg be chosen in $\mathcal{H}$. We assume that the system starts to complete the task $(0 \times \text{beg})$ at time $t = 0$, and that the system is "initially empty" (*i.e.* all tasks $(a \times h)$ such that a has at least one negative coordinate are supposed to be initially complete). In other words, we introduce $\mathcal{G}_{\text{patt}}^{[0]}$ the pattern grid where the weight of any task $(a \times h)$ is replaced by $-\infty$ whenever a has at least one negative coordinate. The process $\mathbf{T}$ is then a solution of the following system of equations:

$$\begin{cases} T_{0 \times \text{beg}} = W(0 \times \text{beg}), \\ T_{a \times h} = W(a \times h) + \max_{(a \times h) \rightarrow (a' \times h') \in \mathcal{E}} T_{a' \times h'} \ , \text{ for } (a \times h) \neq (0 \times \text{beg}). \end{cases} \tag{5}$$

One can immediately check that the following defines a solution of Eq.(5):

$$\forall a \in \mathbb{Z}^d, h \in \mathcal{H}, \ T_{a \times h} = \sup \left\{ W(\pi) \, \middle| \, \begin{matrix} \pi \text{ a path in } \mathcal{G}_{\text{patt}}^{[0]} \\ \pi : a \times h \rightsquigarrow 0 \times \text{beg} \end{matrix} \right\} , \tag{6}$$

where a *path* π is defined following the natural definition of paths in directed graph, and its weight $W(\pi)$ is the sum of the weights of all its vertices. The supremum is taken over all possible paths, where we include paths that contain a single vertex and no edge.

Note that whenever a contains a negative coordinate, the supremum is equal to $-\infty$. When the precedence relation between tasks is acyclic (*i.e.* when the system has no deadlock, see the next section) one can show by induction that this solution is unique for every task $(a \times h)$ for which a path exists in the above supremum.

One may rephrase Eq.(6) as "The completion time of task $(a \times h)$ is the maximum sum of weights along a dependence path leading from $(a \times h)$ back to the origin task $(0 \times \text{beg})$." By an analogy with models from statistical physics, this may be called the *last-passage percolation time* in $(a \times h)$. It is important to note that, as in percolation model, these variables exhibit super-additive property, such that one can benefit from the subadditive ergodic theorem which generalizes the law of large number [8].

3   Sharpness

In this section, we characterize the properties of the paths in the pattern grid
with a single condition: the sharpness criterion. We first prove under this condition that the combinatorial properties of these paths follow the connected subsets
of a lattice (also called *lattice animals* [10]). When this condition is not verified,
we show in dimension 2 that the combinatorial properties of these paths are
radically different.

3.1   Definitions

Dependence graph, simple cycle. For any pattern grid, we define the associated *dependence graph* as the following directed multi-graph, where all edges
are labeled with a vector in $\mathbb{Z}^d$:

$$\begin{cases} \text{Its set of vertices is } \mathcal{H}. \\ \text{Its set of edges is } \{h \rightarrow h' \text{ with label } r \mid r \in \Delta_{h,h'}, \, h, h' \in \mathcal{H}\}. \end{cases}$$

Note that according to this definition, $h \rightarrow h'$ is an edge of the dependence
graph with label r if and only if for all $a \in \mathbb{Z}^d$, $(a \times h) \rightarrow ((a + r) \times h')$ is in $\mathcal{E}$.
We represent the dependence graph associated with the two examples of 2.2 in
Figure 2.

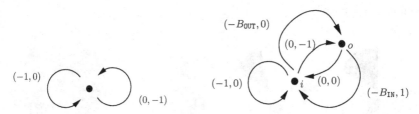

Fig. 2. Two examples of dependence graphs: for an infinite tandem (left), same with
input and output blocking (right)

A *path* in the dependence graph follows the natural definition of a path in a
graph. Its *size* is given by the number of vertices it contain (the same vertex can
be included multiple times); its *associated vector* is the sum of the label for the
edges that it contains.

A path drawn in the dependence graph which begins and ends in the same
vertex h of $\mathcal{H}$ is called a *cycle*. It is a *simple cycle* if it does not contain any
other cycle. In other words, all vertices visited by this cycle are distinct except
the first and last one, which are necessarily the same. As a consequence, a simple
cycle contains at most $H + 1$ vertices (including multiplicity), and the collection
of simple cycles is finite.

Sharpness condition. We denote by $\mathcal{C}$ the set of all vectors associated with a
simple cycle in the dependence graph. For two vectors u and v in $\mathbb{Z}^d$, $<u, v>$
denotes their scalar product, $<u, v> = \sum_{i=1}^{d}(u_i \cdot v_i)$.

Condition 1. *The following conditions are equivalent*

(i) *There exists $s \in \mathbb{Z}^d$, such that $\forall r \in \mathcal{C}$, $<r, s> \; < 0$.*

(ii) *There exists $s \in \mathbb{Z}^d$, such that $\forall r \in \mathcal{C}$, $<r, s> \; \leq -1$.*

(iii) *There exists a hyperplane of $\mathbb{R}^d$ such that all vectors in $\mathcal{C}$ are contained in an open half-space defined by this hyperplane.*

A pattern grid is then called sharp. A vector s satisfying (ii) is called a sharp vector.

Condition (iii) may be seen as a rewriting of (i) in geometric terms, (i) implies (ii) since the family $\mathcal{C}$ contains a finite number of vectors. In practice, to determine whether a sharp vector exists, one has to extract all the simple cycles in the dependence graph, and then to solve a finite system of linear inequalities.

As an example $s = (1,1)$ is a sharp vector for the infinite tandem in Section 2.2. For the second example, $s = (2,1)$ is a sharp vector since we obviously assume $B_{\text{IN}} + B_{\text{OUT}} \geq 1$.

Geometric interpretation. Let us define the *cone* $\mathtt{Cone}(\mathcal{C})$ containing all linear combinations of elements in $\mathcal{C}$ with non-negative coefficients. Assuming that the pattern grid is sharp, this cone intersects one hyperplane only in 0 and is otherwise contained in one of the open half space defined by this hyperplane. In other words, the angle of this cone should be acute. By analogy, the pattern grid is then called "sharp".

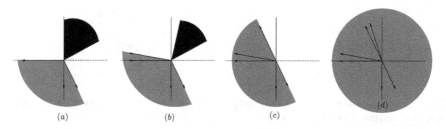

Fig. 3. Geometric representation: (a) and (b) represent families that admit a sharp vector, (c) and (d) families that do not admit such a vector

Some examples are shown in Figure 3 for the case of dimension 2. Different families $\mathcal{C}$ have been represented, containing from 3 to 6 vectors. The cone generated by positive linear combination of this family is shown in gray. We have shown in black the directions that can define a sharp vector, for the cases (a) and (b), where such a vector may be found. Case (c) shows an example of family containing opposite vectors, making it impossible to find a sharp vector. In the case (d), the cone created by positive linear combination of vectors is the whole space $\mathbb{R}^2$, such that, again, no sharp vector may be found. We prove in §3.3, that these cases depict all possible situations for dimension 2.

Relation with deadlock avoidance. A loop in the graph defining the pattern grid corresponds to a deadlock of the system, since it denotes that a task indirectly depends on its own completion to start. Karp et. al proved necessary and sufficient conditions to avoid such deadlock, and showed that they characterize system of Uniform Recurrence Equations where an explicit solution can be constructed [9]. Note that a deadlock corresponds to a cycle drawn in the dependence graph whose associated vector is null. In other words, the system has no deadlock if and only if $0 \notin \mathcal{C}$.

One may immediately observe that sharpness implies deadlock avoidance (*e.g.* as a direct consequence of (i)). It is however less obvious that the sharpness condition is indeed strictly stronger than deadlock avoidance (see Section 4.2).

3.2   Why Is a Sharp Vector Useful ?

The main consequence of sharpness is that one can define a direction in the grid so that dependence paths between tasks remain close to that direction. It comes from the following fact: a path in the dependence graph is more or less a concatenation of a large number of cycles. We can then limit the size of a dependence path based only on its direction in the grid.

Lemma 1. *Suppose that a pattern grid admits a sharp vector s. There exist two constants B, C such that, for any path $\pi : (a \times h) \rightsquigarrow (a' \times h')$*

$$|\pi| \leq B + C \cdot <a - a', s> \ .$$

Proof. Let us introduce the *residue* of a pattern grid, for a sharp vector s.

$$\mathbf{Res} = \max \left\{ (\ <r, s>\ )^+ \,|\ r \text{ assoc. with } \pi \text{ and } |\pi| \leq H \ \right\} \ .$$

It is a finite maximum by definition, because the dependence graph contains a finite number of vertices and edges. Note that for the case where the motif set $\mathcal{H}$ contains a single element, this residue is null, because every path is a cycle. The above result is implied by the following result on paths in the dependence graph: for any path π associated with r,   we have $|\pi| \leq H\,(1 + \mathbf{Res} - <r, s>\ )$.

We will prove this fact by induction on the size of π. First, from the definition of the residue, this result holds trivially for any path π whose size is less than or equal to H.

If π has a size strictly larger than H, then it contains a cycle, and hence a simple cycle, σ. We can write $\pi = \pi_1 \circ \sigma \circ \pi_2$. The path $\pi_1 \circ \pi_2$ is well defined, as the vertex ending π_1 is also the one starting π_2. The path $\pi_1 \circ \pi_2$ has necessarily a smaller size than π. If we assume by induction that it satisfies the result of the theorem, we can deduce:

$$\begin{aligned}
|\pi| &\leq |\pi_1 \circ \pi_2| + |\sigma| - 1 \leq |\pi_1 \circ \pi_2| + H + 1 - 1 \\
&\leq H(1 + \mathbf{Res} - <r_{\pi_1} + r_{\pi_2}, s>\ ) + H \\
&\leq H(1 + \mathbf{Res} - <r_{\pi_1} + r_{\pi_2}, s>\ ) - H <r_\sigma, s> , (\ \text{since } <r_\sigma, s> \ \leq -1\ ) \\
&\leq H(1 + \mathbf{Res} - <r_\pi, s>\ ).
\end{aligned}$$

This bound only depends on the positions in the grid of the two extreme nodes of this path. Hence this result provides an upper bound on the size of any dependence path between two given tasks in the pattern grid. We need slightly more than that: we aim at bounding the maximum weight of any path between two given nodes. Hence we have to capture in addition the combinatorial property of the sum of weights found in such paths.

We can address this second problem as follows: Since $\mathcal{H}$ is a finite set, we can construct a one-to-one correspondence between $\mathbb{Z}^d \times \mathcal{H}$ and $\mathbb{Z}^d$. Let us introduce the following definition, we say that a subset of $\mathbb{Z}^d$ is lattice-connected if it is connected according to the neighbor relation of an undirected lattice. A subset of $\mathbb{Z}^d \times \mathcal{H}$ is called lattice-connected if its associated subset in $\mathbb{Z}^d$ (by the above correspondence) is lattice connected.

Lemma 2. *Suppose that a pattern grid admits a sharp vector* s. *There exist two constants* B, C *such that, for any path* $\pi : (a \times h) \rightsquigarrow (a' \times h')$

$\exists \xi$ *a lattice-connected set, such that* $\pi \subseteq \xi$ *and* $|\xi| \leq B + C \cdot {<}a - a', s{>}$.

Proof. The proof is an application of Lemma 1. We introduce the *radius* of a pattern grid as the finite maximum

$$\texttt{Rad} = \max\left\{ \|r\|_\infty \text{ for } r \in \Delta_{h,h'}, h, h' \in \mathcal{H} \right\} .$$

It is the maximum difference on one coordinate between vertices $a \times h$ and $a' \times h'$ that are connected by an edge in the pattern grid. One can show that any path π may be augmented into a lattice-connected subset which contains at most $|\pi| \cdot (H + d \cdot \texttt{Rad})$. A formal version of this argument, based on a one-to-one correspondence between $\mathbb{Z}^d \times \mathcal{H}$ and $\mathbb{Z}^d$ is detailed in [4].

This turns out to be a very powerful tool, because the class of lattice-connected subset, also known as lattice animals, have been well characterized from a combinatorial and probabilistic standpoint [10],[11].

3.3   Why Is a Sharp Vector Necessary ?

Focusing on dimension $d = 2$, we describe properties of non-sharp pattern grids. These results are only used later to prove that sharpness is a tight condition of scalability. We start by a result showing that a non-sharp pattern grid always exhibits some pathological case: the proof of this lemma may be found in [12].

Lemma 3. *If we consider a family of vectors* $\mathcal{C}$ *in* $\mathbb{Z}^2$ *that does not admit a sharp vector, then one of the following statements is true:*

   (*i*) *It contains the vector* 0.

   (*ii*) *It contains two opposite vectors: there exist* e *and* f *in* $\mathcal{C}$ *such that*

$$\exists \alpha \in \mathbb{R}, \alpha > 0, \text{ such that } e = -\alpha \cdot f .$$

   (*iii*) *It contains a generating triple: there exist* e, f, g *in* $\mathcal{C}$ *such that*

$$\begin{cases} {<}e, f{>} < 0 \ {<}e, g{>} < 0 \\ {<}\tilde{e}, f{>} > 0 \ {<}\tilde{e}, g{>} < 0 \end{cases}, \text{ with } {<}\tilde{e}, e{>} = 0.$$

Let us define a pattern grid as *irreducible* if its dependence graph is strongly connected (*i.e.* there always exists a path leading from h to h', for any h and h'). A non-irreducible pattern grid can be decomposed using the strongly connected components of the dependence graph, and studied separately. The next result, a consequence of Lemma 3 proves that dependence paths in non-sharp pattern grid cannot be bounded as in Lemma 1. Due to space constraint, we omit the proof which may be found in Appendix B.2 of [12].

Corollary 1. *We consider a pattern grid, irreducible, with dimension $d = 2$ that does not admit a sharp vector. We pick an arbitrary vertex of this graph as an origin. There exists a vertex $v \times h$, such that we can build a path of size arbitrary large from $v \times h$ to the origin.*

4   Scalability

In this section, we establish the main result of this paper: under a moment condition, a distributed system represented by a sharp pattern grid is scalable. Moreover, we prove that the rate of completion along any direction converges to a deterministic constant. We then prove that the sharp condition is necessary for scalability, at least for dimension 2, when one avoids degenerate cases. A simple example is provided to illustrate how sharpness is a stronger condition than the one defined in [9].

4.1   The Sharp Case

Moment condition. The weight of $a \times h$ is supposed to follow that depends only on h and is upper bounded by $\bar{s}$, for the stochastic ordering,
$$\forall u \in \mathbb{R}, \quad \text{we have} \quad \mathbb{P}\left[W(a \times h) \geq u\right] \leq \mathbb{P}\left[\bar{s} \geq u\right].$$

Condition 2. *We assume* $\displaystyle \int_0^{+\infty} P(\bar{s} \geq u)^{1/d} du < \infty.$

Condition 2 implies $\mathbb{E}[(\bar{s})^d] < +\infty$. It is implied by $\mathbb{E}[(\bar{s})^{d+\epsilon}] < +\infty$ for any positive ϵ, but it is slightly more general.

Theorem 1. *Let $\mathcal{G}_{\text{patt}}$ be a pattern grid satisfying Condition 1 and 2, then*

(i) *The system is scalable.*

$$\exists M \in \mathbb{R} \quad \text{such that almost surely} \quad \limsup_{m \to \infty} \frac{1}{m} T_{(m \cdot a) \times h} < M.$$

(ii) *If there exists a path $\pi : a \times \text{beg} \rightsquigarrow 0 \times \text{beg}$ with non-negative coordinates (i.e. such that $W(\pi) \neq -\infty$ and $T_{a \times \text{beg}} \neq -\infty$), then*

$$\lim_{m \to \infty} \frac{T_{(m \cdot a) \times \text{beg}}}{m} = l \in \mathbb{R} \quad \text{almost surely and in } \mathbb{L}^1.$$

As a consequence, the throughput of an infinite number of queues organized in tandem is positive, with or without blocking, whenever Condition 2 is verified by the service time. Condition 2 is almost tight since one can build a counter-example when $\mathbb{E}[(\bar{s})^d] = \infty$ [1].

Proof. We prove a slightly more general result, that for any a there exists $M \in \mathbb{R}$ such that, almost surely

$$\max_{h,h' \in \mathcal{H}} \limsup_{m \to \infty} \frac{1}{m} \left(\sup_{\pi: ((m \cdot a) \times h) \rightsquigarrow (0 \times h')} W(\pi) \right) \leq M < +\infty$$

Theorem 1.1 in [11] tells us that in a lattice $\mathbb{Z}^d$, where weights satisfy Condition 2, the maximum weight of a lattice-connected set ξ grows linearly: There exists $N \in \mathbb{R}$ such that, when $n \to \infty$,

$$\frac{1}{n} \left(\max_{\xi \text{ latt. conn., } |\xi|=n, \, 0 \in \xi} W(\xi) \right) \to N < \infty \text{ almost surely and in } \mathbb{L}^1.$$

For any fixed h and h', as a consequence of Lemma 2, a path π leading from $((m \cdot a) \times h)$ to $(0 \times h')$ is contained in a lattice connected subset ξ with size smaller than

$$|\xi| \leq (H^2 + d \cdot \text{Rad} \cdot H)(1 + \text{Res} + m \cdot \text{<}a, s\text{>}),$$

where s is a sharp vector for this grid. All these connected subset contain in particular $0 \times h'$. The weight of a path is then upper bounded by the maximum weight of a connected subset that contains this fixed point. We deduce

$$\limsup_{m \to \infty} \frac{1}{m} \sup_{\pi: ((m \cdot a) \times h) \to (0 \times h')} W(\pi) \leq (H^2 + d \cdot \text{Rad} \cdot H) \cdot \text{<}a, s\text{>} \cdot N < \infty.$$

The proof of (ii) relies on the sub-additive ergodic theorem. It is omitted due to space constraint and may be found in the Appendix A of [12].

4.2   The Non-sharp Case

Let us first illustrate with an example the case of a non-sharp pattern grid. The pattern grid represented in Figure 4 avoids deadlock since one cannot build a cycle in the dependence graph associated with a null vector. It is not sharp as two opposite vectors are associated with cycles in the dependence graph.

As shown in Figure 4 on the right, one can construct a path from any vertex to the origin by following first the top left direction (remaining only over black vertices), following an edge towards a white vertex before crossing the y-axis, then following a right bottom line, remaining on white vertices. It can be seen that starting from coordinate (m, k), the length of this path for large m and k is of the order of $(m + k)^2$. This proves that the $T_{(m,0) \times b} \approx m^2$, and thus that the growth rate associated with this direction is more than linear (*i.e.* its associated completion rate in this direction is zero).

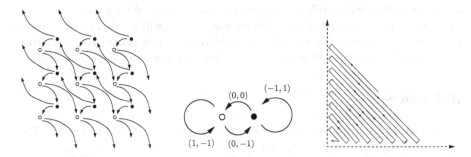

Fig. 4. Example of a pattern grid that avoids deadlock but does not satisfy the sharpness condition: represented via pattern grid (left), dependence graph (middle), shape of a path with a "super-linear" size (right)

We now prove that the phenomenon found above is not an exceptional case but that it always occurs when the sharp condition is not verified. Just like for the study made in Section 3.3, we consider irreducible pattern grid with dimension 2. The proof of the following result is in Appendix B.3 in [12].

Theorem 2. *Let $\mathcal{G}_{\mathtt{patt}}$ be an irreducible pattern grid with dimension 2 that does not admit a sharp vector and* $\mathbf{T}$ *be any solution of Eq.(5).*

We assume that there exists a, with only positive coordinates, and $h \in \mathcal{H}$, such that a path $a \times h \rightsquigarrow 0 \times h$ exists and always has non-negative coordinates. We also assume that all weights are non negative and not identically null, then

$$\exists a' \in \mathbb{Z}^d, \quad \lim_{m \to \infty} \frac{T_{(m \cdot a') \times h}}{m} = +\infty \quad \text{almost surely and in expectation.}$$

5   Concluding Remarks

In this paper we have proved a general result on the scalability of distributed systems. It was obtained in two steps. First, we have proved that the dependence paths in a general precedence relation can be bounded by a scalar product whenever a *sharp vector* exists. Second, the completion of tasks associated with the system have been analyzed taking advantage of subadditive ergodic theory. The scalability result implies that the random set that contains all completed tasks grows linearly with time, in any direction. This work refines the classification of distributed computing systems introduced in [9] via Uniform Recurrence Equations (UREs).

Some aspects of this method have not been included in the paper, due to space constraint. Let us now review them briefly. All the results presented here can be shown for the case of a random pattern grid (*i.e.* where the collection of edges is random, with a distribution invariant per translation). Similarly, the same result can be obtained with different boundary conditions, allowing to construct stationary regimes and analyze stability of large scale system. Another important extension that we could not describe in this paper is when the topology does not follow a lattice but different types of infinite graphs, such as trees. We refer to [4] for a first account of these extensions, which will be described more in future work. In a more longer term, we wish to study how other synchronization between nodes (like the presence of conflicting services) may impact scalability.

References

1. Martin, J.: Large tandem queuing networks with blocking. Queuing Systems, Theory and Applications 41, 45–72 (2002)
2. Baccelli, F., Chaintreau, A., Liu, Z., Riabov, A.: The one-to-many TCP overlay: A scalable and reliable multicast architecture. In: Proceedings of IEEE INFOCOM, vol. 3, pp. 1629–1640 (2005)
3. He, J., Chaintreau, A.: BRADO: scalable streaming through reconfigurable trees (extended abstract). In: Proceedings of ACM Sigmetrics, pp. 377–378 (2007)
4. Chaintreau, A.: Processes of Interaction in Data Networks. PhD thesis, INRIA-ENS (2006),
   http://www.di.ens.fr/~chaintre/research/AugustinChaintreauPhD.pdf
5. Jelenkovic, P., Momcilovic, P., Squillante, M.S.: Buffer scalability of wireless networks. In: Proceedings of IEEE INFOCOM, pp. 1–12 (2006)
6. Xia, C., Liu, Z., Towsley, D., Lelarge, M.: Scalability of fork/join queueing networks with blocking. In: Proceedings of ACM Sigmetrics, pp. 133–144 (2007)
7. Chen, L., Reddy, K., Agrawal, G.: Gates: A grid-based middleware for processing distributed data streams. In: HPDC 2004. Proceedings of the 13th IEEE International Symposium on High Performance Distributed Computing, pp. 192–201 (2004)
8. Kingman, J.: Subadditive ergodic theory. Annals of Probability 1(6), 883–909 (1973)
9. Karp, R.M., Miller, R.E., Winograd, S.: The organization of computations for uniform recurrence equations. J. ACM 14(3), 563–590 (1967)
10. Gandolfi, A., Kesten, H.: Greedy lattice animals II: linear growth. Annals Appl. Prob. 1(4), 76–107 (1994)
11. Martin, J.: Linear growth for greedy lattice animals. Stochastic Processes and their Applications 98(1), 43–66 (2002)
12. Chaintreau, A.: Sharpness: a tight condition for scalability. Technical Report CR-PRL-2008-04-0001, Thomson (2008)

Discovery of Network Properties with All-Shortest-Paths Queries

Davide Bilò[1], Thomas Erlebach[2], Matúš Mihalák[1], and Peter Widmayer[1]

[1] Institute of Theoretical Computer Science, ETH Zurich, Switzerland
{dbilo,mmihalak,widmayer}@inf.ethz.ch
[2] Department of Computer Science, University of Leicester, United Kingdom
te17@mcs.le.ac.uk

Abstract. We consider the problem of discovering properties (such as the diameter) of an unknown network $G(V, E)$ with a minimum number of queries. Initially, only the vertex set V of the network is known. Information about the edges and non-edges of the network can be obtained by querying nodes of the network. A query at a node $q \in V$ returns the union of all shortest paths from q to all other nodes in V. We study the problem as an online problem – an algorithm does not initially know the edge set of the network, and has to decide where to make the next query based on the information that was gathered by previous queries. We study how many queries are needed to discover the diameter, a minimal dominating set, a maximal independent set, the minimum degree, and the maximum degree of the network. We also study the problem of deciding with a minimum number of queries whether the network is 2-edge or 2-vertex connected. We use the usual competitive analysis to evaluate the quality of online algorithms, i.e., we compare online algorithms with optimum offline algorithms. For all properties except maximal independent set and 2-vertex connectivity we present and analyze online algorithms. Furthermore we show, for all the aforementioned properties, that "many" queries are needed in the worst case. As our query model delivers more information about the network than the measurement heuristics that are currently used in practise, these negative results suggest that a similar behavior can be expected in realistic settings, or in more realistic models derived from the all-shortest-paths query model.

1 The Problem and the Model

Dynamic large-scale networks arise in our everyday life naturally, and it is no surprise that they are the subject of current research interest. Both the natural sciences and the humanities have their own stance on that topic. A basic prerequisite is the network itself, and thus, before any study can even begin, the actual representation (a map) of a network has to be obtained. This can be a very difficult task, as the network is typically dynamic, large, and the access to it may be limited. For example, a map of the Internet is difficult to obtain, as the network consists of many autonomous nodes, who organize the

A. Shvartsman and P. Felber (Eds.): SIROCCO 2008, LNCS 5058, pp. 89–103, 2008.
© Springer-Verlag Berlin Heidelberg 2008

physical connections locally, and thus the network lacks any central authority or access point.

There are several attempts to obtain an (approximate) map of the Internet. A common approach, on the level of Autonomous Systems (ASs), is to inspect routing tables and paths stored in each router (passive measurement) or directly ask the network with a traffic-sending probe (active measurement). Data obtained by such measurements are used in heuristics to obtain (approximate) maps of the Internet, see e.g. [1,2,3,4].

As performing such measurements at a node is usually very costly (in terms of time, energy consumption or money), the question of minimizing the number of such measurements arises naturally. This problem was formalized as a combinatorial optimization problem and studied in [5]. The map of a network (and the network itself) is modeled as an undirected graph $G = (V, E)$. The nodes V represent the communication entities (such as ASs in the Internet) and the edges represent physical communication links. A measurement at a node $v \in V$ of the network is called a *query at* v, or simply a *query* v. Each query q gives some information about the network. The *network discovery* problem asks for the minimum number of queries that discover the whole network. In [5] the *layered-graph query model* (LG for short) is defined: a query q returns the union of all shortest paths from q to every other node. In this paper we refer to the LG query model as the *all-shortest-paths* query model. Network discovery is an online problem, where the edges and non-edges (a pair $\{u, v\}$ is a non-edge, if it is not an edge) are initially not known and an algorithm queries vertices of V one by one, until all edges and non-edges are discovered.

Having a map of a network G at our disposal, various aspects of G can be studied. For example, the routing aspects of G are influenced by the diameter, average degree, or connectivity of G. Other graph properties that are studied in the networking community include, for example, a maximal/maximum independent set, minimal/minimum dominating set, shell index, the decision whether the graph is bipartite, power-law, etc. All these properties can be computed from the map of G.

If only a single parameter of a network is desired to be known, obtaining the whole map of the network may be too costly. In this work we address the problem of computing (an approximation of) network properties (such as the diameter of G) in an online way: given an unknown network (only the nodes are known in the beginning), discover a *property* (or an approximation of a property) of the network (graph) with a minimum number of queries. The properties that we address in this paper are the diameter of the graph, a minimal dominating set, a maximal independent set, minimum degree, maximum degree, edge connectivity and vertex connectivity. We use standard graph-theoretic terminology and notation, as it is described for example in [6].

We assume the all-shortest-paths query model, i.e., a query q returns the union of all shortest paths from v to every other node. The result of the query q can be viewed as a layered graph: all the vertices at distance i from q form a layer $L_i(q)$, and the query returns all information between any two layers,

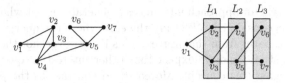

Fig. 1. A graph G (left) and the result of a query at node v_1 as a layered graph (right)

i.e., if u and v are from different layers, then the query returns whether $\{u, v\}$ is an edge or a non-edge. We depict the result of a query graphically as in Fig. 1. For simplicity we sometimes write L_i instead of $L_i(q)$, if it is clear from the context which node is queried. We denote by E_q and $\overline{E}_q$ the set of edges and non-edges, respectively, that are discovered by query q. In the all-shortest-paths query model, E_q is the set of edges whose endpoints have different distance from q, and $\overline{E}_q$ is the set of non-edges whose endpoints have different distance from q. By E_Q and $\overline{E}_Q$ we denote the set of edges and non-edges that are discovered by queries Q, i.e., $E_Q = \bigcup_{q \in Q} E_q$ and $\overline{E}_Q = \bigcup_{q \in Q} \overline{E}_q$. The graph G_Q is the graph on V with the edge set E_Q. Finally, we denote by $\mathrm{comp}(G, Q)$, the set of all graphs G' with vertex set V containing all the edges in E_Q and all non-edges in $\overline{E}_Q$.

It is easy to observe that querying all vertices of G discovers all the edges and non-edges of G and thus any property of the graph can be derived from this information. We are interested in algorithms that deliver minimum-sized query sets that reveal the necessary information about the sought network property. An online algorithm for the (approximate) discovery of a network property is called *c-competitive*, if the algorithm delivers, for any input graph G, a query set Q of size at most $c \cdot \mathrm{Opt}$, where Opt is the optimum number of queries that discover the (approximation of the) property. By an approximate discovery of a property we understand a computation of a value A which is "close" to the actual value O of the property. We require $A \geq O$, if we want to approach O from above (we call the property a *minimization property*), or $A \leq O$, if we want to approach O from below (we call the property a *maximization property*). We will treat the diameter as a minimization property. We call an online algorithm a ρ-*approximation* algorithm for the problem of discovering a minimization property if for any input graph G it discovers a ρ-approximation of the property, i.e., if for the numerical value A returned by the algorithm, and the actual value O of the property, we have $O \leq A \leq \rho \cdot O$. For example, a ρ-approximation, c-competitive algorithm for the diameter discovery problem is an algorithm that discovers a graph G_Q for which the diameter diam_{G_Q} is at most $\rho \cdot \mathrm{diam}_G$, and queries at most c times more queries than an optimal offline ρ-approximation algorithm.

Related Work. Deciding exactly (and deterministically) a graph-theoretic property of a given graph where the measure of quality is the number of accessed entries in the adjacency matrix of the graph is a well understood area. Rivest and Vuillemin [7] show that any deterministic procedure for deciding any non-trivial monotonous n-vertex graph property must examine $\Omega(n^2)$ entries in the adjacency matrix representing the graph. Each such examination of an entry can be seen as a query. Our approach introduces a general concept where other types of queries can be considered. We study the case where the query at a vertex returns all shortest paths from that vertex. This is, however, not the only possible query model to study, and we expect that other interesting query models will be studied following this concept. Moreover, in contrast to the previous work, we study the problem as an online problem, and thus evaluate the quality of algorithms using the competitive ratio.

An active and related field of research is the well-established area of *property testing*, in which a graph property is asked to be probabilistically examined with possibly few edge-queries on the edges of the graph. The aim of such property-testing algorithms is to spend time that is sub-linear or even independent of the size of the graph. In property testing, a graph possessing an examined property $\mathcal{P}$ shall be declared by the algorithm to have property $\mathcal{P}$ with probability at least $3/4$, and a graph that is "far" from having property $\mathcal{P}$ should be declared by the algorithm not to have property $\mathcal{P}$ with probability at least $3/4$. A survey on property testing can be found for example in [8]. Our work differs from property testing in the type of query we make, and in that we consider deterministic strategies.

The all-shortest-paths query model was introduced by Beerliová et al. for studying the mapping process of large-scale networks [5]. The authors studied the problem of discovering all edges and all non-edges of an unknown network with possibly few queries. They presented, among other results, a randomized $O(\sqrt{n}\log n)$-competitive algorithm, and lower bounds 3 and $4/3$ on the competitive ratio of any deterministic and randomized algorithm, respectively. A query set that discovers the edges and non-edges of the network is also called a *resolving set* and the minimum-size resolving set is called a *basis* of the underlying graph, and the size of the basis is the *dimension* of the graph. A graph-theoretic and algorithmic overview of this topic can be found in [9] and [10], respectively.

Our Contribution. We consider several graph properties in the property discovery setting with the all-shortest-paths query model. We first study the discovery of the diameter of an unknown graph G. We present and use a new technique of querying an "interface" between two parts of a graph G. Using k "interfaces" leads to a $(1+\frac{1}{k+1})$-approximation algorithm for the discovery of the diameter of G. The "interface" is in our case a layer of vertices which are at the same distance from an initial query q_0. Considering the competitive ratio as well, and setting $k = 1$, we can present a $(\frac{3}{2} + \frac{2p-1}{\ell})$-approximation, $(\frac{n}{2p})$-competitive algorithm, where ℓ is the maximum distance from q_0 (which is at least half of the diameter of G), and p is a parameter, $p < \ell/4$. We present a lower bound $\sqrt{n} - 1/2$ for the competitive ratio of any algorithm computing a minimal dominating set. We also present an algorithm which queries at most $O(\sqrt{d \cdot n})$ vertices, where d is the size of a minimum dominating set of G. For the problem of finding a maximal independent set we show a lower bound $\sqrt{n}$ on the competitive ratio of any algorithm. We further study the discovery of 2-edge and 2-vertex connectivity of G, and show a lower bound $n/2$ on the competitive ratio of any algorithm for discovering a bridge or an articulation vertex of G. We also present an $n/2$-competitive algorithm which discovers whether G is 2-edge connected. For the problem of discovering the maximum and the minimum degree of G, we present lower bounds $n/2$ and $n/2$, respectively, for the competitive ratios of any algorithm.

2   Discovering the Properties

In the following we use a common approach to the (approximate) discovery of a graph property of a given graph G: select a query set Q such that the resulting graph G_Q has the same (or approximately similar) graph property.

2.1   Discovering the Diameter

Following the general approach, we want to find a (possibly) small query set Q, such that the resulting graph $G_Q = (V, E_Q)$ has a diameter which is a good approximation of the diameter of G.

It has been previously observed [11] that a single query $q \in V$ yields a 2-approximation of the diameter of G. To see this, let q be a vertex of G. Let v be the vertex with the maximum distance from q. Let ℓ denote this distance, i.e., $d(q, v) = \ell$. Clearly, diam $\geq \ell$. Also, for any two nodes $u, v \in V$, $d(u, v) \leq d(u, q) + d(v, q) \leq 2\ell$. Thus, the diameter of G_q is at most 2ℓ, and therefore it is at most twice the diameter of G.

The following example shows that in general, unless we discover the whole network, we cannot hope for a better approximation than 2. Consider two graphs: $G_1 = K_n$, the complete graph, and $G_2 = K_n \setminus \{u, v\}$, the complete graph minus one edge $\{u, v\}$. The diameter of G_1 is 1, and the diameter of G_2 is 2. For any query q, but u or v, the result looks all the same, a star graph centered at q. Thus, we know that the diameter is at most 2, but cannot obtain a better approximation until all the vertices (but one) are queried. As any deterministic algorithm can be forced to query $V \setminus \{u, v\}$ first, the example shows that there is no deterministic $(2 - \epsilon)$-approximation algorithm with less than $n - 1$ queries.

If the diameter of the graph is larger than two (e.g. a growing function in n, such as $\log n$), the following strategy guarantees a better approximation ratio. We first make an arbitrary query $q \in V$. This splits the vertices of V into layers L_i, where L_i contains the vertices at distance i from q. As a next step we query all vertices at layer L_k (we will show that $k = \frac{3}{4}\ell$ is a good choice). See Fig. 2 for an illustration of the upcoming discussion. From the information that we gain after querying all vertices in L_k we want to improve the upper bound or the lower bound for the diameter, and thus the approximation ratio of our algorithm. Thus, the algorithm computes the diameter of $G' := G_{\{q\} \cup L_k}$ (the discovered part of G), and reports it as the approximate solution. In the following we discuss the quality of such an approximation. Let u and v be the vertices whose distance is the diameter of G'.

If a shortest path between u and v goes via vertices of the queried layer L_k, the actual distance between u and v will be discovered in G' (and the approximation

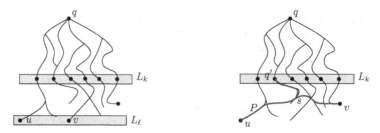

Fig. 2. The initial query q splits the vertices of G into ℓ layers $L_1, L_2, \ldots L_\ell$. The distance $d(u, v)$ between any two nodes $u, v \in V$ is at most $d(u, q) + d(q, v) \leq 2\ell$, but can be shorter if edges within the same layer are present.

ratio will be 1). Thus, we concetrate on the cases where the shortest path between u and v does not go via L_k.

Case 1. If u and v lie both within layers $L_1, \ldots, L_{k-1}$, then clearly $d_{G'}(u, v) \leq 2(k-1)$. This type of nodes guarantees an approximation ratio of $2(k-1)/\ell$ (as the diameter of G is at least ℓ).

Case 2. If both u and v lie within layers $L_{k+1}, \ldots, L_\ell$, and every shortest path in G between u and v goes via vertices of layers $L_{k+1}, \ldots, L_\ell$, we can obtain the following bounds on $d_G(u, v)$. Trivially, $d_G(u, v) \leq d_G(u, q') + d_G(q', v) = d_{G'}(u, q') + d_{G'}(q', v)$, for any $q' \in L_k \cup \{q\}$. Let P be a shortest path in G between u and v. Let $s \in V$ be a vertex on P that is closest to L_k and let q' be a vertex in L_k which is closest to s. We obtain $d_G(q', u) \leq d_G(q', s) + d_G(s, u) \leq (\ell-k) + d_G(s, u)$, and similarly $d_G(q', v) \leq d_G(q', s) + d_G(s, v) \leq (\ell-k) + d_G(s, v)$. Thus, $d_G(q', u) + d_G(q', v) \leq 2(\ell-k) + d_G(s, u) + d_G(s, v) = 2(\ell-k) + d_G(u, v)$. As $d_G(q', u) = d_{G'}(q', u)$ and $d_G(q', v) = d_{G'}(q', v)$, we obtain $d_{G'}(q', u) + d_{G'}(q', v) - 2(\ell-k) \leq d_G(u, v) \leq d_{G'}(q', u) + d_{G'}(q', v)$, and the approximation ratio obtained for this type of vertices is at most $\frac{d_{G'}(q', u) + d_{G'}(q', v)}{\max\{\ell, d_{G'}(q', u) + d_{G'}(q', v) - 2(\ell-k)\}}$. We now distinguish two cases. First, if $d_{G'}(q', u) + d_{G'}(q', v) - 2(\ell-k) \leq \ell$, then the approximation ratio is at most $\frac{\ell + 2(\ell-k)}{\ell} = \frac{3\ell - 2k}{\ell}$. Second, if $d_{G'}(q', u) + d_{G'}(q', v) - 2(\ell-k) > \ell$, then the approximation ratio is of the form $\frac{x}{x - 2(\ell-k)}$, which is maximized (under the condition that $x - 2(\ell - k) \geq \ell$) for $x = \ell + 2(\ell - k)$. Thus the approximation ratio is at most $\frac{3\ell - 2k}{\ell}$.

Hence, taking all cases into account, the approximation ratio of the algorithm is $\max\{1, \frac{2(k-1)}{\ell}, \frac{3\ell - 2k}{\ell}\}$. To minimize the approximation ratio, we need to set $2(k-1) = 3\ell - 2k$, i.e., $k = \frac{3\ell + 2}{4}$, which leads into $\text{diam}_{G_Q}/\text{diam}_G \leq \frac{3}{2} - \frac{1}{\ell}$. We assume, for simplicity of presentation, that every fractional computation results in an integral number (such as the query level $k = \frac{3\ell + 2}{4}$). In reality one has to round the numbers, which can "shift" the queried layer by half, i.e., $|[k] - k| \leq 0.5$ (by $[k]$ we denote the rounding of k). This results in a small additive error of order $\frac{1}{\ell}$ in the approximation ratio of the diameter. Observe that this error approaches zero, as ℓ (and the diameter) grows with n. For simplicity we sometimes omit these small rounding errors in the statements about approximation ratios.

It is not difficult to imagine that querying more layers leads to a better approximation of the diameter. This is indeed the case. We only state the theorem here. The proof of this more general statement can be found in [12].

Theorem 1. *Let ℓ be the maximum distance from an initial query q to a vertex of G. Let $Q = \{q\} \cup L_{k_1} \cup L_{k_2} \cup \ldots L_{k_s}$, $s \geq 1$, $k_i < k_{i+1}$, $i = 1, \ldots, s-1$, where $k_i = \ell/2 + i \cdot \frac{\ell}{2(s+1)}$. Then the query set Q leads to a graph G_Q for which the diameter diam_{G_Q} is a $1 + \frac{1}{s+1}$ approximation of the diameter of G.*

So far we have been mainly concerned with the quality of the approximation but we did not consider the number of queries we make. A problem of the previous algorithm is that the right choice of layer L_k where we make the queries may result in many queries (say, $n - \ell$ in the worst case, if the layer L_k contains almost all vertices of G). If we want to maintain a bounded competitive ratio, we have to be careful about the choice of L_k, which leads to a bi-criteria optimization problem.

Bi-criteria Optimization. To keep some control over the number of queries, a natural choice is to allow some freedom in the choice of the layer L_k. Thus, we do not set $k = \frac{3}{4}\ell + 0.5$, but parametrize the choice of k and allow k to be in the range $\{\frac{3}{4}\ell + 0.5 - p, \ldots, \frac{3}{4}\ell + 0.5 + p\}$, where p is a parameter. The algorithm now picks the layer L_k with the minimum number of vertices among all layers L_i, $i \in \{\frac{3}{4}\ell + 0.5 - p, \ldots, \frac{3}{4}\ell + 0.5 + p\}$. Thus, the size of L_k is at most $n/2p$, which is also the upper bound on the competitive ratio of the algorithm. Relaxing p allows to keep the number of queries small, but can harm the approximation quality, while setting p very small improves the approximation but leaves no control over the number of queries. Clearly, a meaningful choice of p is in the range $\{0, 1, 2, \ldots, \frac{1}{4}\ell - 0.5\}$.

Repeating the previous case analysis, the upper bounds on the approximation ratio for the different cases are 1, $2(k-1)/\ell$, and $\frac{3\ell - 2k}{\ell}$. As $3\ell - 2k \leq 3\ell - 2(\frac{3}{4}\ell + 0.5 - p) = \frac{3}{2}\ell - 1 + 2p$ and $2(k-1) \leq 2(\frac{3}{4}\ell + 0.5 + p - 1) = \frac{3}{2}\ell + 2p - 1$ we obtain that the approximation ratio is $\max\{1, 3/2 + \frac{2p-1}{\ell}, 3/2 + \frac{2p-1}{\ell}\} = 3/2 + \frac{2p-1}{\ell}$. The parameter p can be used to tweak the approximation ratio and the competitive ratio of the algorithm, which are $3/2 + \frac{2p-1}{\ell}$ and $n/2p$, respectively.

Theorem 2. *Let G be any graph and q a query which results in ℓ layers. Then there is an algorithm, parametrized by $p \in \{0, 1, 2, \ldots, \lfloor\frac{1}{4}\ell - 0.5\rfloor\}$, which delivers a $(3/2 + \frac{2p-1}{\ell})$ approximation of the diameter of an unknown graph, and is $n/2p$ competitive.*

2.2   Discovering a Minimal Dominating Set

In this section we consider the problem of discovering a minimal dominating set in G. We provide an algorithm that discovers a minimal dominating set of G with $O(\sqrt{d \cdot n})$ queries, where d is the size of a minimum dominating set of G. The algorithm, which we simply call *Alg*, works as follows. It starts from an empty set D and grows it by adding vertices step by step so that D will eventually be a minimal dominating set. At each step, Alg queries two vertices x and y (an x-vertex and a y-vertex, respectively). The first vertex x is chosen arbitrarily among the vertices that are not yet dominated by D. The algorithm queries x and the information of the query decides the next choice of vertex y; y is chosen among the set of neighbors of x in such a way that it maximizes the set of *newly* dominated nodes by y (i.e., the subset of neighbors $N(y)$ of y which are at distance 2 from x and which are not neighbors of any vertex belonging to our partial solution D). Both x and y are put into D. It can happen that the query x has only one layer, and hence y does not dominate any new vertex, and thus D is not minimal (y can be removed from D). Similarly, if y dominates all neighbors of x and some vertices from $L_2(x)$, x is obsolete, and D is not minimal. Thus, at the end, we modify D to make it minimal.

Theorem 3. *The set D returned by Alg is a minimal dominating set in G. Moreover, in order to discover D, the algorithm makes $O(\sqrt{d \cdot n})$ queries, where d denotes the size of a minimum dominating set in G.*

Proof. It is clear that the returned set D is a minimal dominating set. It remains to show the bound on the number of queries. Let $\{z_1, \ldots, z_d\} \subseteq V$ be a minimum dominating set in G. We partition the set V into subsets C_i, $i = 1, \ldots, d$: The set $C_i \subseteq V$ contains z_i and all the neighbors of z_i that are not in $\{z_1, \ldots, z_d\}$ and that are not in any of the previous sets C_j, $j < i$.

Let X and Y denote the x-vertices and y-vertices, respectively, produced by the algorithm. Every x-vertex belongs to a single set C_i. Let X_i, $i = 1, \ldots, d$, denote the vertices of X that belong to C_i. We consider the vertices of X_i in the reverse order in which they have been queried by the algorithm. Let k_i denote the size of X_i and let $x_1^i, \ldots, x_{k_i}^i$ denote the reverse order. For each vertex x_j^i we denote by y_j^i the corresponding y-vertex (which was chosen in the same step as x_j^i). Now observe that (i) there are at least ℓ uncovered vertices in X_i (and thus in C_i, too) before querying x_ℓ^i, (i.e., at least the vertices $x_1^i, \ldots, x_\ell^i$); and (ii) at least ℓ uncovered vertices are covered during the while loop in which x_ℓ^i and y_ℓ^i are queried (as z_i, a neighbor of x_ℓ^i, has at least ℓ undominated neighbors in C_i at that time, and y_ℓ^i is chosen to maximize the number of newly dominated vertices).

Consequently, we have that all vertices are covered when $\sum_{i=1}^{d} \sum_{\ell=1}^{k_i} \ell = n$, i.e., when $\sum_{i=1}^{d} k_i(k_i + 1) = 2n$. The algorithm queries at most $|X| + |Y| = 2|X| = 2\sum_{i=1}^{d} k_i$ vertices. We are thus interested in how big the sum $\sum_{i=1}^{d} k_i$ can be. We consider a maximization linear program (LP) $\max \sum_{i=1}^{d} k_i$ with the constraints $k_i \geq 0$, $\forall i = 1, \ldots, d$, and $\sum_{i=1}^{d} k_i(k_i + 1) = 2n$.

By the Cauchy-Schwartz inequality we have that an optimal solution of the above LP is $k_1, \ldots, k_d = \frac{\sqrt{\frac{8n}{d} + 1} - 1}{2}$ which is at most $\sqrt{\frac{2n}{d}}$. This implies that $|X| = \sum_{i=1}^{d} k_i \leq d\sqrt{\frac{2n}{d}} = \sqrt{2dn}$. $\square$

Now we construct an example in which it is possible to compute a minimal dominating set of size $d \geq \sqrt{n} - 1/2$ after querying one specific vertex, but

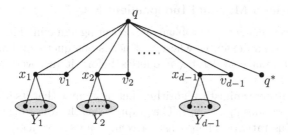

Fig. 3. The lower bound construction for a minimal dominating set

any algorithm needs at least d queries before being able to compute a minimal dominating set.

The graph G has the following structure (see Fig. 3 for an illustration). The vertices in V are partitioned into three sets $L_0 = \{q\}, L_1 = \{q^*, x_1, \ldots, x_{d-1}\} \cup \{v_1, \ldots, v_{d-1}\}$ and $L_2 = Y_1 \cup \cdots \cup Y_{d-1}$, where all the sets $Y_1, \ldots, Y_{d-1}$ have cardinality d. All vertices but those in L_2 are connected to q. Moreover, for all $i = 1, \ldots, d-1$, vertex x_i is also connected to the vertex v_i and all vertices in Y_i. It is easy to see that both $\{q, x_1, \ldots, x_{d-1}\}$ and $\{q^*, x_1, \ldots, x_{d-1}\}$ are minimum dominating sets of G.

First we prove it is enough to query q^* to find a minimal dominating set of G. Indeed, after querying q^*, we discover all edges of G except the ones linking x_i with the vertex v_i. The layers of q^* are $\{q^*\}, \{q\}, L_1 \setminus \{q^*\}, L_2$ (ordered according to the distance from q^*). The query q^* also discovers that q^* is connected to q only, and that, considering only the edges between the layers, vertices of Y_i are adjacent with x_i only. It is now an easy observation that from the information of query q^* the algorithm can infer that both $\{q, x_1 \ldots, x_{d-1}\}$ and $\{q^*, x_1, \ldots, x_{d-1}\}$ are minimal dominating sets in G.

Now let Alg be any deterministic algorithm and let us assume that it has queried any set $Q \subseteq V \setminus \{q^*\}$ with $|Q| < d$ and such that Q contains q (notice that we can always ensure that q is the first vertex queried by the algorithm). One can show that the algorithm cannot guarantee the minimality of any dominating set of G_Q; moreover, it can be proved that the set of vertices that are indistinguishable to the algorithm and that contains q^* has size at least $d - |Q| + 1$. Finally, one can prove that there are at least $d - |Q| + 1$ indistinguishable vertices in every Y_i. As a consequence, we can claim that Alg needs at least d queries for discovering a minimal dominating set of G, as we can force the algorithm to make the next query not equal to q^*. Expressing d in terms of n, we obtain a lower bound of $d = \sqrt{n + \frac{1}{4}} - \frac{1}{2} \geq \sqrt{n} - 1/2$. The missing details of the proof can be found in [12].

Theorem 4. *There is a graph for which any algorithm needs to query at least $\sqrt{n} - 1/2$ vertices before it discovers a minimal dominating set, while an optimum offline algorithm needs only one query. Thus no algorithm can achieve a better competitive ratio than $\sqrt{n}$ for the problem of discovering a minimal dominating set.*

2.3   Discovering a Maximal Independent Set

In this section we consider the problem of discovering a maximal independent set in G. We construct an example where Opt needs one query, and any algorithm can be forced to make at least $\sqrt{n}$ queries before it discovers any maximal independent set.

Let Alg be any deterministic algorithm. Let us assume that its first query is at node q_1 (out of n nodes $v_1, \ldots, v_n$). The graph G has the following structure (see Fig. 4 for an illustration). There exists a *central* node c which is connected to

Fig. 4. Construction of a graph G for which any algorithm needs $\sqrt{n}$ queries to discover a maximal independent set

every node in V, and forms a maximal independent set on its own. Thus, Opt can make a query at this node and discover that c is a maximal independent set. We add other edges to G to make it impossible for any algorithm to find a maximal independent set with less than $\sqrt{n}$ queries. First, we split the vertices of V into three groups: $L_0 = \{q_1\}$, L_1, and L_2. Vertex q_1 is in L_0, $\sqrt{n}$ vertices are in L_2, and the rest of the vertices is in L_1. The central vertex c is in L_1. Vertex q_1 is connected to every vertex in L_1, and all vertices in L_1 are also connected to L_2, and c is connected to every vertex in L_1 (hence, c is indeed connected to every vertex). There is no edge within vertices in L_2. The query at q_1 splits the vertices into two layers L_1 and L_2. Observe first that $X_1 := \{q_1\} \cup L_2$ is a maximal independent set and there is no other one containing a vertex from X_1. Any algorithm discovering X_1 as an independent set needs to query all but one nodes in L_2, which is $\sqrt{n} - 1$ (no query in L_1 can discover any information on non-edges within L_2). Observe that any such query does not discover any information about edges and non-edges within L_1. If Alg does not query only in L_2 (and thus cannot discover X_1 with less than $\sqrt{n}$ queries), let q_2 be the first node that is queried in L_1. Because all the nodes in this layer look the same to the algorithm, the algorithm can be forced to query q_2 at any node of L_2. The edge construction within L_2 is a recursive construction: the query q_2 splits L_1 into two layers: $L_{1,1}$ and $L_{1,2}$, where, $L_{1,2}$ has $\sqrt{n} - 1$ nodes, c is in $L_{1,1}$, q_2 is connected to every node in $L_{1,1}$, and $L_{1,1}$ is connected to every node $L_{1,2}$. There is no edge in $L_{1,2}$. Again, $X_2 := \{q_2\} \cup L_{1,2}$ is the only maximal independent set containing a vertex from X_2, and any algorithm needs $|L_{1,2}| - 1 = \sqrt{n} - 2$ nodes to discover X_2. If Alg queries also in $L_{1,1}$, the nodes within $L_{1,1}$ are split recursively into three parts $\{q_3\}$, $L_{1,1,1}$, and $L_{1,1,2}$, with the obvious size and edge-set. This recursive splitting can obviously run for at least $\sqrt{n}$ times, which shows that no deterministic algorithm can guarantee to find a maximal independent set of a graph with less than $\sqrt{n}$ queries.

Theorem 5. *There is a graph for which any algorithm needs to query at least $\sqrt{n}$ vertices before it discovers a maximal independent set, while an optimum offline algorithm needs only one query. Thus there is no $o(\sqrt{n})$-competitive algorithm for the problem of discovering a maximal independent set.*

2.4   Discovering a Bridge or an Articulation Node of G

In this section we discuss two related properties of G. We want to discover whether the graph G has an articulation node or a bridge. An articulation node of G is a vertex v such that the induced graph on $V \setminus \{v\}$ is not connected. A bridge is an edge e for which the graph $G \setminus e$ is not connected. We show that if the graph contains an articulation node, no algorithm is better than $n/2$-competitive, and if the graph contains a bridge, similarly, no algorithm can achieve a competitive ratio better than $n/2$. We also present an $n/2$-competitive algorithm for the bridge discovery problem.

We begin with the bridge discovery problem. Consider the graph G from Fig. 5. G has an even number of vertices, and consists of one node q_0 connected to all remaining $n-1$ vertices $v_1, \ldots, v_{n-1}$. The vertices v_{2i-1} and v_{2i}, $i = 1, \ldots, (n-2)/2$, form an edge. The graph contains exactly one bridge – the edge $\{q_0, v_{n-1}\}$. Any algorithm can be forced to make the first query at q_0. Thus, all the remaining vertices lie within the same layer L_1, and look indistinguishable to the algorithm. We can force the next query to be at v_1. This query keeps the vertices $v_3, v_4, \ldots, v_{n-1}$ indistinguishable to the algorithm, and does not give any information on the bridge $\{q_0, v_{n-1}\}$. Hence, next time the algorithm queries a vertex in this group of vertices, we can force it to query v_3. Thus, using the recursive approach, any algorithm can be forced to query at least vertices $v_1, v_3, v_5, \ldots, v_{n-3}$, which then together discover the bridge $\{q_0, v_{n-1}\}$. Observe that an optimum algorithm can query v_{n-1} to discover the bridge. This shows the lower bound $n/2$.

For the problem of discovering an articulation node we prove a lower bound of $n/2$ by modifying the input graph G according to the vertices queried by the algorithm (i.e., we assume that the adversary is adaptive to the algorithm). The graph G will be a super-graph of a star centered at q such that a node $q^* \neq q$ is incident with q only. In this case, by querying q^* we can claim that q is an articulation node as we discover that q^* has degree 1. Before explaining

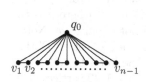

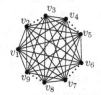

Fig. 5. Bridge discovery   **Fig. 6.** Assigning vertex w' **Fig. 7.** A clique minus some
　　　　　　　　　　　　　　to the query q_i 　　　　　edges (dotted lines)

how the idea behind the proof of the lower bound works, we provide some new definitions. First, given a set of queries Q, we define a Q-*block* as a maximal set of vertices in $V \setminus \{q\}$ that are connected in the graph $G_Q \setminus \{q\}$. Clearly, if $Q = V$, we discover the whole graph, and thus G has an articulation node iff there are at least two Q-blocks in the original graph G. The idea of the lower bound is to prevent any algorithm to learn this information soon. In every Q-block B of G_Q we consider a special vertex – an *anchor*. An anchor is a vertex for which the query set Q does not reveal whether the anchor is connected to another anchor in the original graph, i.e., Q is not enough to distinguish G from another $G' \in \mathtt{comp}(G, Q)$ (recall that $\mathtt{comp}(G, Q)$ is the set of all graphs G' which give the same query results for queries in Q), i.e., we do not know whether all Q-blocks are connected to one another after querying Q, hence we cannot claim that G is (is not) 2-vertex connected. Clearly, in order to claim that G is 2-vertex connected, the algorithm has to prove that $V \setminus \{q\}$ is a Q-block, i.e., all the graphs in $\mathtt{comp}(G, Q)$ are 2-vertex connected. Conversely, in order to claim that G is not 2-vertex connected, then the algorithm has to prove that all the graphs in $\mathtt{comp}(G, Q)$ are not 2-vertex connected.

Now, let us consider any deterministic algorithm. As all vertices are indistinguishable, we may assume that the algorithm starts by querying $Q = \{q_0 = q\}$. Clearly, for each vertex x in $V \setminus \{q\}$, we have that $\{x\}$ is a Q-block whose anchor vertex is x. As all vertices $V \setminus \{q\}$ are indistinguishable, we can assume that the algorithm queries $q_1 \neq q^*, q$. In this case we grow the Q-block $B = \{q_1\}$ by merging it with two other Q-blocks $B' = \{x'\}$ and $B'' = \{x''\}$, with $x', x'' \neq q^*$. Basically, we add the edges $\{q_1, x'\}$ and $\{x', x''\}$ to G. Notice that there are 2-vertex connected graphs in $\mathtt{comp}(G, \{q_0, q_1\})$ as we do not know whether there are edges connecting two anchor vertices to each other. Finally we let x'' be the new anchor vertex of the Q'-block B, where $Q' = Q \cup \{q_1\}$. At a generic step, let us assume that the algorithm queried all the vertices in Q, and let us assume that $\mathtt{comp}(G, Q)$ contains a 2-vertex connected graph and a graph with an articulation node. The algorithm can either choose to query a vertex q' in the Q-block B we grew so far or not. In the first case, notice that the new information discovered is maximized when q' is exactly the anchor vertex of the Q-block B. In the case where q' is from B and q is the anchor vertex a of B, we merge B with two other blocks $B' = \{x'\}$ and $B'' = \{x''\}$, where $x', x'' \neq q^*$ (it is worth noticing that all vertices but q and those in B are indistinguishable in G_Q) by simply adding edges $\{a, x'\}$, and $\{x', x''\}$ to G. Let $Q' = Q \cup \{q'\}$. In the new graph, x'' is the new anchor of the enlarged Q'-block B' containing the old block B. In the case where query q' is outside B, we merge two singleton Q-blocks $\{q'\}$ and $\{x'\}$ to B by adding edges $\{q', x'\}$, and $\{x', a\}$ to G, where a is the anchor vertex of B, and x' is any vertex outside B and not equal to q'. Notice that in this new construction, a remains the anchor of the new Q'-block B' that contains the original Q-block B (where $Q' = Q \cup \{q'\}$). The lower bound of $n/2$ follows from the fact that the algorithm queries at least $(|B| - 1)/2$ vertices of B.

Theorem 6. *For the problem of discovering a bridge or an articulation node there is no better deterministic algorithm than $n/2$-competitive.*

We now present a simple algorithm for determining whether a graph G is 2-edge connected. The algorithm needs at most $\lceil \frac{n}{2} \rceil$ queries. The algorithm makes an arbitrary initial query q_0. The resulting layered graph $G_{\{q_0\}}$ is used by the algorithm to choose the next queries. We denote by q_i the query that is made by the algorithm in the i-th step, and by Q_i all the queries (including q_i) made so far. Observe that if there is i such that there is only one edge e between L_i and L_{i+1}, the edge e is a bridge of G. Observe also that if G has a bridge $e \in E$, it has to appear as an edge in the result of the query q_0. Thus, choosing query q_{i+1}, we can concentrate on those edges of G_{q_0}, which are not part of any cycle of G_{Q_i}. While there are such edges (and thus candidates for a bridge), the algorithm picks among all such edges the farthest endpoint from q_0, and queries it. We claim that this algorithm terminates, and that the algorithm knows at the end whether the graph has a bridge or not, and that it makes at most $\lfloor (n-1)/2 \rfloor$ queries on top of q_0 (and is thus $\lceil n/2 \rceil$-competitive).

Let q_i be the query of the algorithm in step i, and let $e_i = \{u_i, q_i\}$ be the bridge of $G_{Q_{i-1}}$ with q_i the farthest endpoint from q_0 among all bridges of $G_{Q_{i-1}}$. Let ℓ_i denote the distance of q_i from q_0. Let $R(q_i)$ be the set of vertices from layers L_j, $j \geq \ell_i$, which can be reached from q_i by a path which uses at most one vertex from each L_j, $j \geq \ell_i$. (i.e., if we orient the edges according to the increasing distance from q_0, the set $R(q_i)$ is the set of all vertices for which there exists a directed path from q_i). Thus, $R(q_i)$ forms a component of $G_{Q_{i-1}} \setminus \{e_i\}$, as there cannot be any edge with endpoints in the same layer leaving $R(q_i)$ (otherwise e_i would no longer be a bridge in $G_{Q_{i-1}}$). Let us assume that e_i is not a bridge in G. Then there exists a cycle C in G which contains the edge e_i. The cycle C has to contain a not yet discovered edge $e_c = \{w, w'\}$ which is adjacent to a vertex w in $R(q_i)$, and to a vertex $w' \notin R(q_i)$. The vertices w and w' have to be from the same layer L_j, $j \geq \ell_i$ (as it was not discovered by q_0). Clearly, q_i discovers this edge $\{w, w'\}$, as the distance from q_i to w is $j - \ell_i$ (as $w \in R(q_i)$), and the distance from q_i to w' is bigger than $j - \ell_i$ (as $w \notin R(q_i)$). As $\{w, w'\}$ is a newly discovered edge, it follows that w' was not queried before. To show that at most $\lfloor (n-1)/2 \rfloor$ queries are made by the algorithm after the query q_0, we want to assign one unqueried vertex to one queried vertex. In our case we assign w' to q_i (notice that w could possibly be equal to q_i, and thus cannot be assigned to q_i). We now show that w' is not already assigned to a previously queried vertex q_k, $k < i$, with $\ell_k \geq \ell_i$. Figure 6 depicts the situation. If this is the case, w' is assigned to query q_k because w' is an endpoint of an edge $\{w', w''\}$ which was discovered by query q_k, and which is a part of a cycle that shows that q_k is not an endpoint of a bridge in G. Thus, $w'' \in R(q_k)$ and $w' \notin R(q_k)$. Clearly, the distance between q_k and w' is $j - \ell_k + 1$. The distance between q_k and w has to be $j - \ell_k + 1$ as well, as the edge $\{w, w'\}$ is not discovered by q_k. But this is not possible. The shortest path from q_k to w cannot go via a vertex from layer L_s, $s < \ell_k$ (the distance would be bigger than $j - \ell_k + 1$). Thus, the shortest path between q_k and w goes only via vertices of layers L_s, $s \geq \ell_k$. But then e_i cannot be a bridge in $G_{Q_{i-1}}$: The shortest path from q_k to w, the shortest path from w to q_i, and the path from q_i to q_k via q_0 induce a cycle with e_i, using edges known after query q_k. This is a contradiction, and thus w' is not assigned to q_k and can be assigned to q_i.

Thus, if e_i is not a bridge, we will discover at least one new edge e_c that includes e_i into a cycle of G, and one of the endpoints of e_c can be assigned to q_i. If we do not discover any such edge, the edge e_i is a bridge of G. The assignment argument shows that after q_0 we query at most $\lfloor (n-1)/2 \rfloor$ vertices. The termination of the algorithm follows from the fact that we can query at most n vertices, and from the fact that if G_{Q_i} contains a bridge, then its endpoint further from q_0 was not queried yet, and we still have a vertex to query in step $i+1$.

Theorem 7. *There is an $\lceil n/2 \rceil$-competitive algorithm for the problem of discovering a bridge of a graph.*

2.5 Discovering the Min/Max Degree of G

We investigate how many queries are needed in order to discover the minimum degree of G, and the maximum degree of G. The lower bound construction for the problem of finding an articulation node (Section 2.4) shows an example where any deterministic algorithm needs at least $n/2$ queries to discover the minimum degree of G, whereas an optimum algorithm needs only one query, yielding a lower bound $n/2$ on the competitive ratio of deterministic algorithms. For the problem of discovering the maximum degree we similarly present a lower bound $n/2$ on the competitive ratio of deterministic algorithms. Consider a graph G with $n = 2k+1$ vertices, which is constructed from a complete graph K_n by deleting the "even" edges $\{v_{2i}, v_{2i+1}\}$, $i = 1, \ldots, k$ from the cycle $v_1, v_2, v_3, \ldots, v_n$. An example of such a graph for $n = 9$ is in Fig. 7. Observe that v_1 is the only vertex of the graph which has degree $n-1$, and thus the maximum degree of G can be discovered by one query at v_1. On the other hand, any other vertex v_i has exactly $n-2$ neighbors, which are indistinguishable with the query. Thus, every deterministic algorithm can be forced to query k vertices before it can distinguish v_1 from other vertices, and therefore the algorithm makes at least $k+1$ queries before it reveals the maximum degree of G.

3 Conclusions

We have introduced the online problem of discovering graph properties with all-shortest-paths queries, and considered in more detail the discovery of the diameter, a minimal dominating set, a maximal independent set, the 2-edge connectivity, the 2-vertex connectivity, the maximum degree, and the minimum degree of an unknown graph. We have presented lower bounds for the problems, and also an $O(\sqrt{d \cdot n})$-competitive algorithm for the minimal dominating set discovery, and an optimal $\frac{n}{2}$-competitive algorithm for the bridge discovery problem. We have also introduced a technique of querying an interface of a graph G_Q, which may prove to be helpful in other discovery settings. Furthermore we have shown an adaptive-adversary lower bound construction, which is the first adaptive construction in the discovery setting as introduced in [5].

Our work was motivated by the current intensive activities in the area of mapping the Internet. The all-shortest-path queries model the information that is obtained from routing tables of BGP routers. Of course, our assumption of getting *all* shortest paths is not reflected fully in reality – it certainly is a simplification which helps to analyze the problem. In reality, we would assume to get much less information. The lower bounds presented in this paper suggest, however, that in any realistic situation we cannot hope for better results.

References

1. Cheswick, B.: Internet mapping project, http://www.cheswick.com/ches/map/
2. Govindan, R., Tangmunarunkit, H.: Heuristics for Internet map discovery. In: Proceedings of the 19th Conference on Computer Communications (IEEE INFOCOM), Tel Aviv, Israel, March 2000, pp. 1371–1380 (2000)
3. DIMES: Mapping the internet, http://www.netdimes.org/
4. Route Views Project: University of Oregon, http://www.routeviews.org
5. Beerliová, Z., Eberhard, F., Erlebach, T., Hall, A., Hoffmann, M., Mihalák, M.: Network discovery and verification. IEEE Journal on Selected Areas in Communications (JSAC) 24(12), 2168–2181 (2006)
6. Diestel, R.: Graph Theory, 3rd edn. Graduate Texts in Mathematics, vol. 173. Springer, Heidelberg (2005)
7. Rivest, R.L., Vuillemin, J.: On recognizing graph properties from adjacency metrices. Theoretical Computer Science 3, 371–384 (1976)
8. Ron, D.: Property testing. In: Handbook of Randomized Computing, vol. II, pp. 597–649. Kluwer Academic Publishers, Dordrecht (2001)
9. Chartrand, G., Zhang, P.: The theory and applications of resolvability in graphs: A survey. Congressus Numerantium 160, 47–68 (2003)
10. Barrat, A., Erlebach, T., Mihalák, M., Vespignani, A.: A (short) survey on network discovery. Technical report, DELIS – Dynamically Evolving, Large-Scale Information Systems (2008)
11. Ram, L.S.: Tree-based graph reconstruction. Research Report of the European Graduate Program Combinatorics-Computation-Geometry (CGC) (March 2003)
12. Bilò, D., Erlebach, T., Mihalák, M., Widmayer, P.: Discovery of network properties with all-shortest-paths queries. Technical Report 591, Department of Computer Science, ETH Zurich (April 2008)

Recovering the Long-Range Links
in Augmented Graphs

Pierre Fraigniaud[1,*], Emmanuelle Lebhar[1,**], and Zvi Lotker[2]

[1] CNRS and University Paris Diderot
Pierre.Fraigniaud@liafa.jussieu.fr
[2] Ben Gurion University
Emmanuelle.Lebhar@liafa.jussieu.fr

Abstract. The *augmented graph* model, as introduced by Kleinberg (STOC 2000), is an appealing model for analyzing navigability in social networks. Informally, this model is defined by a pair (H, φ), where H is a graph in which inter-node distances are supposed to be easy to compute or at least easy to estimate. This graph is "augmented" by links, called *long-range* links, which are selected according to the probability distribution φ. The augmented graph model enables the analysis of *greedy routing* in augmented graphs $G \in (H, \varphi)$. In greedy routing, each intermediate node handling a message for a target t selects among all its neighbors in G the one that is the closest to t in H and forwards the message to it.

This paper addresses the problem of checking whether a given graph G is an augmented graph. It answers part of the questions raised by Kleinberg in his Problem 9 (Int. Congress of Math. 2006). More precisely, given $G \in (H, \varphi)$, we aim at extracting the base graph H and the long-range links R out of G. We prove that if H has high clustering coefficient and H has bounded doubling dimension, then a simple local maximum likelihood algorithm enables to partition the edges of G into two sets H' and R' such that $E(H) \subseteq H'$ and the edges in $H' \setminus E(H)$ are of small stretch, i.e., the map H is not perturbed too greatly by undetected long-range links remaining in H'. The perturbation is actually so small that we can prove that the expected performances of greedy routing in G using the distances in H' are close to the expected performances of greedy routing using the distances in H. Although this latter result may appear intuitively straightforward, since $H' \supseteq E(H)$, it is not, as we also show that routing with a map more precise than H may actually damage greedy routing significantly. Finally, we show that in absence of a hypothesis regarding the high clustering coefficient, any local maximum likelihood algorithm extracting the long-range links can miss the detection of at least $\Omega(n^{5\varepsilon}/\log n)$ long-range links of stretch at least $\Omega(n^{1/5-\varepsilon})$ for any $0 < \varepsilon < 1/5$, and thus the map H cannot be recovered with good accuracy.

* Additional supports from the ANR projects ALADDIN and ALPAGE, and from the COST Action 295 DYNAMO.
** Additional supports from the ANR project ALADDIN, and from the COST Action 295 DYNAMO.

1   Introduction

Numerous papers that appeared during the last decade tend to demonstrate that several types of interaction networks share common statistical properties, encompassed under the broad terminology of *small worlds* [35,36,37]. These networks include the Internet (at the router level as well as at the AS level) and the World Wide Web. Actually, networks defined in frameworks as various as biology (metabolic and protein networks), sociology (movies actors collaboration network), and linguistic (pairs of words in english texts that appear at most one word apart) also share these statistical properties [20]. Specifically, a network is said small world [39] if it has low density (i.e., the total number of edges is small, typically linear in the number of nodes), the average distance between nodes is small (typically polylogarithmic as a function of the number of nodes), and the so-called *clustering* coefficient, measuring the local edge density, is high (i.e., significantly higher than the clustering coefficient of Erdös-Rényi random graphs $\mathcal{G}_{n,p}$). Other properties often shared by the aforementioned networks include *scale free* properties [6] (i.e., fat tailed shapes in the distributions of parameters such as node degree), limited growth of the ball sizes [2,21], or low doubling dimension [38].

A lot remains to be done to understand why the properties listed above appear so frequently, and to design and analyze models capturing these properties. Nevertheless, there is now a common agreement on their presence in interaction networks. The reason for this agreement is that, although the statistical validity of some measurements is still under discussion [4], many tools (including the controversial Internet Traceroute) have been designed to check whether a network satisfies the aforementioned properties.

This paper addresses the problem of checking another important property shared by social networks: the *navigability* property.

It was indeed empirically observed that social networks not only possess small average inter-node distance, but also that short routes between any pair of nodes can be found by simple decentralized processes [9,34]. One of the first papers aiming at designing a model capturing this property is due to Kleinberg [23], where the notion of *augmented graphs* is introduced. Informally, an augmented graph is aiming at modeling two kinds of knowledge of distances available to the nodes: a global knowledge given by a base graph, and a local knowledge given by one extra random link added to each node. The idea is to mimic the available knowledge in social networks, where individuals share some global distance comparison tool (e.g., geographical or professional), but have also private connections (e.g., friendship) that are unknown to the other individuals. We define an *augmented graph model* as a pair (H, φ) where H is a graph, called *base graph*, and φ is a probability distribution, referred to as an *augmenting distribution* for H. This augmenting distribution is defined as a collection of probability distributions $\{\varphi_u, u \in V(H)\}$. Every node $u \in V(H)$ is given one extra link[1], called *long-range link*, pointing to some node, called the *long-range contact* of u. The destination v of such a link is chosen at random with probability $\Pr\{u \rightarrow v\} = \varphi_u(v)$. (If $v = u$ or v is a neighbor of u, then no link is added). In this paper, a graph $G \in (H, \varphi)$ will often be denoted by $H + R$ where H is the base graph and R is the set of long-range links resulting from the trial of φ yielding G.

An important feature of this model is that it enables to define simple but efficient decentralized routing protocols modeling the search procedure applied by social entities in Milgram's [34] and Dodd's et al [9] experiments. In particular, *greedy routing* in (H, φ) is the oblivious routing process in which every intermediate node along a route from a source $s \in V(H)$ to a target $t \in V(H)$ chooses among all its neighbors (including its long-range contact) the one that is the closest to t *according to the distance measured in H*, and forwards to it. For this process to apply, the only "knowledge" that is supposed to be available at every node is its distances to the other nodes in the base graph H. This assumption is motivated by the fact that, if the base graph offers some nice properties (e.g., embeddable in a low dimensional metric with small distorsion) then the distance function dist_H is expected to be easy to compute, or at least to approximate, locally.

Lots of efforts have been done to better understand the augmented graph model (see, e.g., [1,5,7,11,12,13,14,15,29,30,31,32], and the survey [25]). Most of these works tackle the following problem: given a family of graph $\mathcal{H}$, find a family of augmenting distributions $\{\varphi_H, H \in \mathcal{H}\}$ such that, for any $H \in \mathcal{H}$, greedy routing in (H, φ_H) performs efficiently, typically in polylog(n) expected number of steps, where $n = |V(H)|$. Kleinberg first showed that greedy routing performs in $O(\log^2 n)$ expected number of steps on any square mesh augmented with an appropriate harmonic distribution [23]. Among the works that followed Kleinberg's seminal results, an informative result due to Duchon et al [10] states that any graph of bounded growth can be augmented so that greedy routing performs in polylog(n) expected number of steps. Slivkins [38] extended this result to graphs of bounded doubling dimension, and even doubling dimension at most $O(\log \log n)$. This bound on the doubling dimension is tight since [16] proved that, for any function $d(n) \in \omega(\text{polylog}(n))$, there is a family of graphs of doubling dimension $d(n)$ for which any augmentation yields greedy routing performing in $\omega(\text{polylog}(n))$ expected number of steps[2].

Despite these progresses in analyzing the augmented graph model for small worlds, the key question of its validity is still under discussion. In [25], Kleinberg raised the question of how to check that a given network is an augmented graph (Problem 9). This is a critical issue since, if long-range links are the keystone

---

[1] By adding $k_u \geq 1$ long-range links to node u, for every $u \in V(H)$, instead of just one, with $\Pr(k_u = k) \sim 1/k^\alpha$ for some $\alpha > 1$, the model can also capture the scale-free property. For the sake of simplicity however, we will just assume $k_u = 1$ for every $u \in V(H)$.

[2] The notation $d(n) \in \omega(f(n))$ for some functions f and d means that $d(n)/f(n)$ tends to infinity when n goes to the infinity.

of the small world phenomenon, they should be present in social networks, and their detection should be greatly informative. This paper aims at answering part of this detection problem.

1.1 The Reconstruction Problem

This paper addresses the following reconstruction problem: given an n-node graph $G = H + R \in (H, \varphi)$, for some unknown graph H and unknown distribution φ, extract a good approximation H' of H such that greedy routing in G using distances in H' performs approximately as well as when using the distances in the "true" base graph H. More precisely, the expected number of steps of greedy routing in H' has to be the one in H up to a polylogarithmic factor. Note that, for every edge in R one extremity is the long-range contact of the other. Nevertheless, there is no a priori orientation of these edges when G is given.

To measure the quality of the approximation H' of H, we define the *stretch* of a long-range link between u and v as $\text{dist}_H(u, v)$. Then, the extracted base graph H' is considered to be of good quality if it contains H and does not contain too many long-range links of large stretch. Indeed, we want to approximate H by H' as close as possible not only for the purpose of efficient routing using the metric of H', but also because the augmented graph model assumes that distances in H are easy to compute or approximate. Therefore, the map of distances of H' is wished to be close to the one of H.

In addition to its fundamental interest, the reconstruction problem may find important applications in network routing. In particular, if the base graph H offers enough regularity to enable distance computation using node names (or labels) of small size, then critical issues of storage and quick access to routing information (such as the ones currently faced for Internet [26,33]) can be addressed. Indeed, applying greedy routing in the network using solely the distances in H may be sufficient to achieve fast routing (i.e., performing in expected polylogarithmic number of steps).

1.2 Methodology

In statistics, one of the most used techniques is the maximum likelihood method [22]. Applied to our problem, this would lead to extract the long-range links based on their probability of existence. Precisely, the method would select S as the set of the n long-range links such that

$$\Pr(G \,|\, S \text{ is the set of long-range links})$$

is maximum. This brute force approach however requires testing an exponential number of sets, and it requires some knowledge about the distribution φ. For instance, in [3,8], the authors assume that R is a random power law graph added on top of the base graph H. Motivated by the experimental results in [28], and the analytical results in [10,23,27,38], we consider augmenting distributions where $\varphi_u(v)$ is inversely proportional to the size of the ball of radius $\text{dist}_H(u, v)$ centered at u. We call such kind of augmenting distributions *density based* distributions. They are the ones enabling an efficient augmentation of graphs with bounded ball growth, and, up to modifying the underlying metric by weighting nodes, of graphs with bounded doubling dimension.

Fixing a class of augmenting distributions still does not suffice for applying the maximum likelihood method because of the large number of sets. One way to overcome this difficulty it to consider every edge separately. More precisely, we consider *local* maximum likelihood methods defined as follows.

Definition 1. *An algorithm $\mathcal{A}$ for recovering the base graph H from $G \in (H, \varphi)$ is a* local maximum likelihood algorithm *if and only if $\mathcal{A}$ decides whether or not an edge $e \in E(G)$ is a long-range link solely based on the value of*

$$\Pr(G \mid e \in E(H)).$$

Applying a local maximum likelihood algorithm however requires some information about the local structure of the base graph H. For instance, in [8], the base graph H is assumed to possess a clustering property characterized by a large number of edge-disjoint paths of bounded length connecting the two extremities of any edge. In [3], the clustering property is characterized by a large amount of flow that can be pushed from one extremity of an edge to the other extremity, along routes of bounded length. Motivated by the statistical evidences demonstrating that social networks are locally dense, we consider a clustering property stating that every edge participates in at least $c \cdot \log n / \log \log n$ triangles for some positive constant c. Note that this function grows very slowly, and that its output for practical values of n is essentially constant: for a network with one billion nodes, our assumption states that every edge participates to at least $6c$ triangles. (For a network with one billion billions nodes, this bound becomes $10c$). Note also that even though we focus on the number of triangles, our approach could easily be adapted to apply on many other types of local structures whose characteristics would enable distinguishing local connections from remote connections.

1.3 Our Results

First, we present a simple local maximum likelihood algorithm, called EXTRACT, that, given an n-node graph $G = H + R \in (H, \varphi)$, where H has a clustering coefficient such that every edge participates in $\Omega(\log n / \log \log n)$ triangles, and φ is a density-based augmenting distribution, computes a partition (H', R') of $E(G)$. This partition satisfies $E(H) \subseteq H'$ and, for any $\beta \geq 1$, if X is the random variable counting the number of links in $R \setminus R'$ of stretch at least $\log^{\beta+1} n$, then $\Pr\{X > \log^{2\beta+1}(n)\} \leq 1/n$ whenever the maximum degree Δ of H satisfies $\Delta \leq O(\log^\beta n)$. That is, Algorithm EXTRACT is able to almost perfectly reconstruct the map H of G, up to long-range links of polylogarithmic stretch. It is worth mentioning that Algorithm EXTRACT runs in time close to linear in $|E(G)|$, and thus is applicable to large graphs with few edges, which is typically the case of small world networks.

Our main positive result (Theorem 1) is that if in addition H has bounded growth, then greedy routing in G using the distances in H' performs in polylog(n) expected number of steps between any pair. This result is crucial in the sense that Algorithm EXTRACT is able to approximate the base graph H and the set R of long-range links accurately enough so that greedy routing performs efficiently. In fact, we prove that the expected slow down of greedy routing in G using the distances in H' compared to greedy routing in (H, φ) is only polylog(n). Although this latter result may appear intuitively straightforward since $H' \supseteq E(H)$, we prove that routing with a map more precise than H may actually damage greedy routing performances significantly.

We also show how these results can be generalized to the case of graphs with bounded doubling dimension.

Finally, Theorem 2 proves that the clustering coefficient plays a crucial role for extracting the long-range links of an augmented graph using local maximum likelihood algorithms. We prove that *any* local maximum likelihood algorithm extracting the long-range links in some augmented graph with low clustering coefficient fails. In fact, this is true even in the case of cycles augmented using the harmonic distribution, that is even in the case of basic graphs at the kernel of the theory of augmented graphs [23]. We prove that any local maximum likelihood algorithm applied to the harmonically augmented cycle fails to detect at least $\Omega(n^{5\varepsilon}/\log n)$ of the long-range links of length $\Omega(n^{1/5-\varepsilon})$ for any $0 < \varepsilon < 1/5$.

2   Extracting the Long-Range Links

In this section, we first focus on the task of extracting the long-range links from an augmented graph $G = H + R \in (H, \varphi)$ without knowing H. The efficiency of our extraction algorithm in terms of greedy routing performances will be analyzed in the next section. As will be shown in Section 4, extracting the long-range links from an augmented graph is difficult to achieve in absence of a priori assumptions on the base graph H and on the augmenting distribution φ. Before presenting the main result of the section we thus present the assumptions made on H and φ.

The clustering coefficient of a graph H is aiming at measuring the probability that two distinct neighboring nodes u, v of a node w are neighbors. Several similar formal definitions of the clustering coefficient appear in the literature. In this paper, we use the following definition. For any node u of a graph H, let $N_H(u)$ denotes the neighborhood of u, i.e., the set of all neighbors of u in H.

Definition 2. *An n-node graph H has clustering $c \in [0, 1]$ if and only if c is the smallest real such that, for any edge $\{u, v\} \in E(H)$,*

$$\frac{|N_H(u) \cap N_H(v)|}{n} \geq c.$$

For instance, according to Definition 2, each edge of a random graph $G \in \mathcal{G}_{n,p}$ with $p \simeq \frac{\log n}{n}$ has expected clustering $1/n^2$ up to polylogarithmic factors. In our results, motivated by the fact that interaction networks have a clustering coefficient much larger than uniform random graphs, we consider graphs in (H, φ) for which the clustering coefficient of H is slightly more that $1/n$, that is every edge participates in $\Omega(\log n / \log \log n)$ triangles.

We also focus on augmenting distributions that are known to be efficient ways to augment graphs of bounded growth (or bounded doubling dimension) [10,23,38]. For any node u of a graph H, and any $r > 0$, let $B_H(u, r)$ denote the ball centered at u of radius r in H, i.e., $B_H(u, r) = \{v \in V(G) \mid \text{dist}_H(u, v) \leq r\}$.

Definition 3. *An augmenting distribution φ of a graph H is density-based if and only if $\varphi_u(u) = 0$, and for every two distinct nodes u and v of H,*

$$\varphi_u(v) = \frac{1}{Z_u} \frac{1}{|B_H(u, \text{dist}_H(u, v))|}$$

where $Z_u = \sum_{w \neq u} 1/|B_H(u, \text{dist}_H(u, w))|$ is the normalizing coefficient.

Density-based distributions are motivated by their kernel place in the theory of augmented graphs, as well as by experimental studies in social networks. Indeed, density-based distributions applied to graphs of bounded growth roughly give a probability $1/k$ for a node u to have its long-range contact at distance k, which distributes the long-range links equivalently over all scales of distances, and thus yields efficient greedy routing. In addition, Liben-Nowell et al. [28] showed that in some social networks, two-third of the friendships are actually geographically distributed this way: the probability of befriending a particular person is inversely proportional to the number of closer people.

Notation. According to the previous discussion, for any $\beta \geq 1$, we consider the family $\mathcal{M}(n, \beta)$ of n-node density-based augmented graph models (H, φ) where H has clustering $c \geq \Omega(\frac{\log n}{n \log \log n})$ and maximum degree $\Delta \leq O(\log^\beta n)$.

We describe below a simple algorithm, called EXTRACT, that, given an n-node graph G and a real $c \in [0, 1]$, computes a partition (H', R') of the edges of G. This simple algorithm will be proved quite efficient for reconstructing a good approximation of the base graph H and a good approximation the long-range links of a graph $G \in (H, \varphi)$ when H has high clustering and φ is density-based.

> **Algorithm EXTRACT:**
> **Input:** a graph G, $c \in [0, 1]$;
> $R' \leftarrow \emptyset$;
>   **For** every $\{u, v\} \in E(G)$ **do**
>   **If** $\frac{1}{n} |N_G(u) \cap N_G(v)| < c$ **then** $R' \leftarrow R' \cup \{u, v\}$;
> $H' \leftarrow E(G) \setminus R'$;
> **Output:** (H', R').

Note that the time complexity of Algorithm EXTRACT is $O(\sum_{u \in V(G)} (\deg_G(u))^2)$, i.e., close to $|E(G)|$ for graphs of constant average degree. More accurate outputs could be obtained by iterating the algorithm using the test $\frac{1}{n} |N_{H'}(u) \cap N_{H'}(v)| < c$ until H' stabilizes. However, this would significantly increase the time complexity of the algorithm without significantly improving the quality of the computed decomposition (H', R'). The main quantifiable gain of iterating Algorithm EXTRACT would only be that H' would be of clustering c, and would be maximal for this property. Finally, note also that Algorithm EXTRACT involves local computations, and therefore could be implemented in a distributed manner.

The result hereafter summarizes the main features of Algorithm EXTRACT.

Lemma 1. *Let $(H, \varphi) \in \mathcal{M}(n, \beta)$, and $G \in (H, \varphi)$. Let c be the clustering coefficient of H. Assume $G = H + R$. Then Algorithm EXTRACT with input (G, c) returns a partition (H', R') of $E(G)$ such that $E(H) \subseteq H'$, and:*

$$\Pr(X > \log^{2\beta+1} n) \leq O\left(\frac{1}{n}\right),$$

where X is the random variable counting the number of links in $R \setminus R'$ of stretch at least $\log^{\beta+1} n$.

Proof. Since H has clustering c, for any edge $\{u, v\}$ in $E(H)$, $\frac{1}{n}|N_H(u) \cup N_H(v)| \geq c$, and therefore $\{u, v\}$ is not included in R' in Algorithm EXTRACT. Hence, $E(H) \subseteq H'$. For the purpose of upper bounding $\Pr(X > \log^{2\beta+1} n)$, we first lower bound Z_u, for any $u \in G$. We have for any $u \in G$, $Z_u \geq \deg_H(u)/(\deg_H(u) + 1) \geq 1/2$.

Let $S \subseteq R$ be the set of long-range links that are of stretch at least $\log^{\beta+1} n$. We say that an edge $\{u, v\} \in R$ *survives* if and only if it belongs to H'. For each edge $e \in S$, let X_e be the random variable equal to one if e survives and 0 otherwise, when R is the set of random links chosen according to φ.

Let $e = \{u, v\} \in S$. For e to be surviving in H', it requires that u and v have at least $c \cdot n$ neighbors in common in G. If w is a common neighbor of u and v in G, then, since $\text{dist}_H(u, v) \geq \log n > 2$, at least one of the two edges $\{w, u\}$ or $\{w, v\}$ has to belong to R. Note that u and v can only have one common neighbor w such that both of these edges are in R because we add at most one long-range link to every node, and $u, v \in S$. Thus, there must be at least $c \cdot n - 1$ common neighbors w for which only one of the edges $\{w, u\}$ or $\{w, v\}$ is in R. The following claim upper bound the probability of this event. (Due to lack of space, the proofs of all claims are omitted).

Claim 1. $\Pr\{X_e = 1\} \leq 1/n$ *where $e = \{u, v\} \in S$.*

To compute the probability that at most $\log^{2\beta+1} n$ edges of S survive in total, we use virtual random variables that dominate the variables X_e, $e \in R$, in order to bypass the dependencies between the X_e. Let us associate to each $e \in S$ a random variable Y_e equals to 1 with probability $1/n$ and 0 otherwise. By definition, Y_e dominates X_e for each $e \in S$ and the Y_e are independently and identically distributed. Note that, the fact that some long-range link e survives affects the survival at most Δ^2 other long-range links of R, namely, all the potential long-range links between $N_H(u)$ and $N_H(v)$. Therefore the probability that k links of S survive is at most the probability that k/Δ^2 of the variables Y_e are equal to one. In particular we have: $\Pr\{\sum_{e \in S} X_e > \log^{2\beta+1} n\} \leq \Pr\{\sum_{e \in S} Y_e > \log^{2\beta+1} n/\Delta^2\}$. Using Chernoff' inequality, we have the following claim.

Claim 2. $\Pr\{\sum_{e \in S} Y_e > \log^{2\beta+1} n/\Delta^2\} \leq 1/n$.

From Claim 2, we directly conclude that $\Pr\{\sum_{e \in S} X_e > \log^{2\beta+1} n\} \leq \frac{1}{n}$.   □

3   Navigability

In the previous section, we have shown that we can almost recover the base graph H of an augmented graph $G \in (H, \varphi)$: very few long-range links of large stretch remain undetected with high probability. In this section, we prove that our approximation H' of H is good enough to preserve the efficiency of greedy routing. Indeed, although it may appear counterintuitive, being aware of more links does not necessarily speed up greedy routing. In other words, using a map $H' \supseteq H$ may not yield better performances than using the map H, and actually it may even significantly damage the performances. This phenomenon occurs because the augmenting distribution φ is generally chosen to fit well with H, and this fit can be destroyed by the presence of a few more links in the map. This is illustrated by the following property.

Property 1. *There exists an n-node augmented graph model (H, φ) and a long-range link e such that, for $\Omega(n)$ source-destination pairs, the expected number of steps of greedy routing in (H, φ) is $O(\log^2 n)$, while greedy routing using distances in $H \cup \{e\}$ takes $\omega(\text{polylog}(n))$ expected number of steps.*

Proof. Let H be the $2n$-node graph consisting in a path P of n nodes $u_1, \dots, u_n$ connected to a d-dimensional ℓ_∞-mesh M of n nodes. Precisely, M is the n-node graph consisting of k^d nodes labeled $(x_1, \dots, x_d)$, $x_i \in \mathbf{Z}_k$ for $1 \leq i \leq d$, where $k = n^{1/d}$. Node $(x_1, \dots, x_d)$ of M is connected to all nodes $(x_1 + a_1, \dots, x_d + a_d)$ where $a_i \in \{-1, 0, 1\}$ for $1 \leq i \leq d$, and all operations are taken modulo k. Note that, by construction of M, the distance between any two nodes $x = (x_1, \dots, x_d)$ and $y = (y_1, \dots, y_d)$ is $\max_{1 \leq i \leq d} \min\{|y_i - x_i|, k - |y_i - x_i|\}$. Hence, the diameter of M is $\lfloor n^{1/d}/2 \rfloor$. Assume that P is augmented using the harmonic augmenting distribution h, and M is augmented using some augmenting distribution ψ. It is proved in [16] that, for any augmenting distribution ψ for M, there is a pair $s_0, t_0 \in V(M)$, with $2^{d-1} - 1 \leq \text{dist}_M(s_0, t_0) \leq 2^d$ such that the expected number of steps of greedy routing from s_0 to t_0 is at least $\Omega(2^d)$ whenever $d < \sqrt{\log n}$. Let $d = \sqrt{\log n}/2$. To construct H, we connect the extremity u_n of P to the node t_0 of M (see Figure 1). In P, we use a slight modification $\hbar$ of the harmonic distribution h: $\hbar$ is exactly h except at node u_1 where $\hbar_{u_1}(s_0) = 1$ (i.e. for any trial of $\hbar$, the long-range contact of u_1 is s_0). Consider the augmented graph model $(H, \hbar \cup \psi)$, and set $e = \{u_1, s_0\}$.

In $(H, \hbar \cup \psi)$, greedy routing within P takes $O(\log^2 n)$ expected number of steps [23]. Let $H' = H \cup \{e\}$. We consider greedy routing using distances in H' between the two following sets:

$$S = \{u_2, \dots, u_{\sqrt{n}}\} \text{ and } T = \{u_{n-\sqrt{n}}, \dots, u_n\}.$$

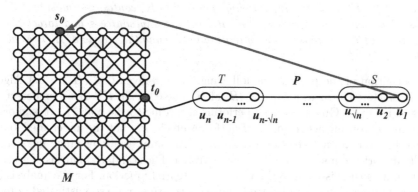

Fig. 1. Graph H in the proof of Property 1

Hence, for any $s \in S$ and $t \in T$, the shortest path from s to t in H' goes through e. Indeed, their shortest path in H is of length at least $n - 2\sqrt{n}$, while in H' it is of length at most $2\sqrt{n} + \text{dist}_H(s_0, t_0) + 2 \leq 2\sqrt{n} + 2^{\sqrt{\log n}/2} + 2$ using e, which is less than $n - 2\sqrt{n}$.

Let $\mathcal{B} = B_H(u_{n-\sqrt{n}}, 2\sqrt{n} + n^{1/d})$. For any node $x \in S$, the probability that the long-range contact of x is in $\mathcal{B}$ is at most $O\left(\frac{1}{\sqrt{n} \cdot \log n}\right)$. Therefore, the expected number of steps required to find such a link in S is at least $\Omega(\sqrt{n} \cdot \log n)$ which is larger than $|S|$. As a consequence, with constant probability, greedy routing from a node $s \in S$ to a node $t \in T$, using the distances in H', routes to u_1 and, from there to s_0. This implies that greedy routing from s to t will take at least as many steps as greedy routing from s_0 to t_0 within (M, ψ), that is $\Omega(2^{\sqrt{\log n}})$ expected number of steps, which is $\omega(\text{polylog}(n))$. $\square$

Property 1 illustrates that being aware of some of the long-range links may slow down greedy routing dramatically, at least for some source-destination pairs. Nevertheless, we show that algorithm EXTRACT is accurate enough for the undetected long-range links not to cause too much damage. Precisely, we show that for bounded growth graphs as well as for graphs of bounded doubling dimension, greedy routing using distances in H' can slow down greedy routing in (H, φ) only by a polylogarithmic factor.

Definition 4. *A graph G has (q_0, α)-expansion if and only if, for any node $u \in V(G)$, and for any $r > 0$, we have: $|B_G(u, r)| \geq q_0 \Rightarrow |B_G(u, 2r)| \leq 2^\alpha |B_G(u, r)|$. In the bulk of this paper, we will set $q_0 = O(1)$, and refer to α as the expanding dimension of G, and to 2^α as the growth rate of G.*

Definition 4 is inspired from Karger and Ruhl [21]. The only difference with Definition 1 in [21] is that we exponentiate the growth rate. Note that, according to Definition 4, a graph has bounded growth if and only if its expanding dimension is $O(1)$.

Theorem 1. *Let $(H, \varphi) \in \mathcal{M}(n, \beta)$ be such that H has (q_0, α)-expansion, with $q_0 = O(1)$ and $\alpha = O(1)$. Let $G \in (H, \varphi)$. Algorithm EXTRACT outputs (H', R')*

such that (a) $E(H) \subseteq H'$, (b) with high probability H' contains at most $\log^{2\beta+1} n$ links of stretch more than $\log^{\beta+1} n$, and (c) for any source s and target t, the expected number of steps of greedy routing in G using the metric of H' is at most $O(\log^{4+4\beta+(\beta+1)\alpha} n)$.

The intuition of the proof is the following. We are given $G \in (H, \varphi)$, but Algorithm EXTRACT returns a superset H' of H. The edges in $H' \setminus H$ are undetected long-range links. Greedy routing performs according to the map H'. It is known that greedy routing according to H performs efficiently, but the undetected long-range links create a distortion of the map. Actually, the long-range links that really distort the map are those of large stretch. The standard analysis of greedy routing uses the distance to the target as potential function. For the analysis of greedy routing using the distorted map H', we use a more sophisticated potential function that incorporates the number of undetected long-range links which belong to shortest paths between the current node and the target (cf. the notion of "concerned indices" in the proof).

Proof. The fact that $E(H) \subseteq H'$ and that with high probability H' contains at most $\log^{2\beta+1} n$ links of stretch more than $\log^{\beta+1} n$ is a direct consequence of Lemma 1. Recall that $S \subseteq R$ denotes the set of long-range links that are of stretch at least $\log^{\beta+1} n$. Let $H'' = H' \setminus S$. Note that the maximum stretch in H'' is $\log^{\beta+1} n$. For any $x \in V(H)$, let $L(x)$ denote the long-range contact of x. Let Z_u be the normalizing constant of the augmenting distribution at node u. We have the following claim.

Claim 3. *For any $u \in V(G)$, $Z_u \leq 2^\alpha \log n =_{\text{def}} Z_{\max}$.*

Let us analyze greedy routing in G from $s \in V(G)$ to $t \in V(G)$ using the distances in H'. Assume that $S = \{\{u_1, v_1\}, \dots, \{u_k, v_k\}\}$ is the set of the surviving long-range links (i.e. in $R \cap H'$) that have stretch more than $\log^{\beta+1} n$, v_i being the long-range contact of u_i for all $1 \leq i \leq k$. For the homogeneity of the notations, let $u_0 = v_0 = t$.

Let τ be the current step of greedy routing from s to t, and x the current node. We define the *concerned index* at step τ as the unique index j defined by:

$$j = \min_{i \in \{1,\dots,k\}} \{i \,|\, \text{dist}_{H'}(x, t) = \text{dist}_{H''}(x, u_i) + 1 + \text{dist}_{H'}(v_i, t)\}.$$

In other words, $\{u_j, v_j\}$ is the first surviving long-range link encountered along the shortest path from x to t in H'. If there is no such index, set $j = 0$.

Claim 4. *Let x be the current node of greedy routing, and j be the concerned index at the current step. If $x \in B_{H''}(u_j, r)$, but $\text{dist}_{H''}(x, u_j) > r/2$, and for some $r > 0$, then:*

$$\Pr\{L(x) \in B_{H''}(u_j, r/2)\} \geq \frac{1}{2^{4\alpha} \log^{1+\alpha(\beta+1)} n},$$

and if $L(x) \in B_{H''}(u_j, r/2)$ then greedy routing routes inside $B_{H''}(u_j, r/2)$ at the next step.

Claim 5. *Let x and x' be two nodes on the greedy route reached at respective steps τ and τ', $\tau < \tau'$. Assume that the concerned index at steps τ and τ' is the same, denoted by j, $j \leq k = |\mathcal{S}|$. If $x \in B_{H''}(u_j, r)$ for some $r > 0$, then $x' \in B_{H''}(u_j, r)$.*

For any $0 \leq i \leq \log n$, $0 \leq j \leq k$, and $\tau > 0$, let $\mathcal{E}_j^i(\tau)$ be the event: "greedy routing from s to t already entered $B_{H''}(u_j, 2^i)$ during the first τ steps". Note that, for any $0 \leq j \leq k$ and any $\tau > 0$, $\mathcal{E}_j^0(\tau) \subseteq \ldots \subseteq \mathcal{E}_j^{\log n}(\tau)$. We describe the current state of greedy routing at step τ by the event $\mathcal{E}_0^{i_0}(\tau) \cap \ldots \cap \mathcal{E}_k^{i_k}(\tau)$ where for every $0 \leq j \leq k$, $i_j = \min\{i \mid \mathcal{E}_j^i(\tau) \text{ occurs}\}$.

Note that greedy routing has reached t at step τ if and only if $\mathcal{E}_0^0(\tau)$ has occured. Clearly, at step 0 (in s), the event $\mathcal{E}_0^{\log n}(0) \cap \mathcal{E}_1^{\log n}(0) \ldots \cap \mathcal{E}_k^{\log n}(0)$ occurs.

Claim 6. *Assume that the state of greedy routing at step τ is $\mathcal{E}_0^{i_0}(\tau) \cap \ldots \cap \mathcal{E}_k^{i_k}(\tau)$, for some $i_0, \ldots i_k \in \{0, \ldots, \log n\}$. Then, after at most $(k+1) \cdot 2^{4\alpha} \log^{1+\alpha(\beta+1)} n$ steps on expectation, there exists an index $0 \leq \ell \leq k$ such that the state of greedy routing is $\mathcal{E}_0^{j_0}(\tau') \cap \ldots \cap \mathcal{E}_\ell^{j_\ell}(\tau') \ldots \cap \mathcal{E}_k^{j_k}(\tau')$, with $j_s \leq i_s$ for all s, and $j_\ell < i_\ell$, $\tau' > \tau$.*

Let X be the random variable counting the number of steps of greedy routing from s to t. As noticed before, $\mathbb{E}(X)$ is at most the expected number of steps τ to go from state $\mathcal{E}_0^{\log n}(0) \cap \mathcal{E}_1^{\log n}(0) \ldots \cap \mathcal{E}_k^{\log n}(0)$ to state $\mathcal{E}_0^0(\tau) \cap \mathcal{E}_1^{i_1}(\tau) \ldots \cap \mathcal{E}_k^{i_k}(\tau)$, for some $i_1, \ldots i_k \in \{0, \ldots, \log n\}$. From Claim 6, we get: $\mathbb{E}(X) \leq (k+1) \log n \cdot ((k+1) \cdot 2^{4\alpha} \log^{1+\alpha(\beta+1)} n)$. And, from Lemma 1, $\Pr\{k > \log^{2\beta+1} n\} \leq 1/n$. Therefore, we have:

$$\mathbb{E}(X) = \mathbb{E}(X \mid k \leq \log^{2\beta+1} n) \cdot \Pr\{k \leq \log^{2\beta+1} n\}$$
$$+ \mathbb{E}(X \mid k > \log^{2\beta+1} n) \cdot \Pr\{k > \log^{2\beta+1} n\}$$
$$\leq 2^{4\alpha} \log^{2+\alpha(\beta+1)\alpha+2(2\beta+1)} n + n \cdot (1/n) = O(\log^{4+4\beta+(\beta+1)\alpha} n).$$

$\square$

Remark. Graphs of bounded expanding dimension and graphs of bounded doubling dimension are very closely related. Indeed, it can be shown that, assigning a specific weight function to a graph of bounded doubling dimension (the doubling measure of its metric), it can be made bounded growth by considering the ball sizes with nodes multiplicity corresponding to their weight [18]. Moreover, this weight function can be computed in polynomial time [19]. This allows us to extend Theorem 1 to graphs of bounded doubling dimension, up to a constant factor change in the exponent of greedy routing performances.

4   Impossibility Results

Algorithm EXTRACT is an extreme case in the class of local maximum likelihood algorithms. Indeed, if $e = \{u, v\} \in E(H)$, one must have $\frac{1}{n}|N_G(u) \cap N_G(v)| \geq c$.

Hence, if $\frac{1}{n}|N_G(u) \cap N_G(v)| < c$, then $\Pr(G \mid e \in E(H)) = 0$, and therefore it would identify e as a long-range link. Algorithm EXTRACT is only failing in the detection of few long-range links with large stretch (Lemma 1) because, for a link $e = \{u, v\}$ with large stretch, $\Pr(\frac{1}{n} \cdot |N_G(u) \cap N_G(v)| \geq c)$ is small. We show that in absence of clustering, the number of long-range links with large stretch that are not detected can be much higher, for any local maximum likelihood algorithm.

This impossibility result even holds in the case of a $(2n+1)$-node cycle C_{2n+1} augmented using the harmonic distribution $h_u^{(n)}(v) = 1/(2H_n \cdot \mathrm{dist}_{H_n}(u, v))$, where $H_n = \sum_{i=1}^n \frac{1}{i}$ is the nth harmonic number, and even if the extraction algorithm is designed specifically for ring base graphs augmented with the harmonic distribution.

Note that $h^{(n)}$ is density-based, but C_{2n+1} has a clustering coefficient equal to zero. It was proved in [23] that greedy routing in $(C_{2n+1}, h^{(n)})$ performs in $O(\log^2 n)$ expected number of steps between any pair.

Theorem 2. *For any $0 < \varepsilon < 1/5$, any local maximum likelihood algorithm for recovering the base graph C_{2n+1} in $G \in (C_{2n+1}, h^{(n)})$ fails in the detection of an expected number $\Omega(n^{5\varepsilon}/\log n)$ of long-range links of stretch $\Omega(n^{1/5-\varepsilon})$.*

Due to lack of place, the proof of the theorem is omitted.

5   Conclusion

This paper is a first attempt to demonstrate the feasibility of recovering, at least partially, the base graph H and the long-range links R of an augmented graph $G = H + R$. Our methodology assumes some a priori knowledge about the structure of the base graph (of bounded doubling dimension, and with high clustering coefficient) and of the long-range links (resulting from a trial according to a density-based distribution). It would be interesting to check whether these hypotheses could be relaxed, and if yes up to which extend.

Acknowledgments. The authors are thankful to Dmitri Krioukov for having raised to them the question of how to extract the based graph of an augmented graph, and for having pointed to them several relevant references. They are also thankful to Augustin Chaintreau and Laurent Viennot for fruitful discussions.

References

1. Abraham, I., Gavoille, C.: Object location using path separators. In: 25th ACM Symp. on Principles of Distributed Computing (PODC), pp. 188–197 (2006)
2. Abraham, I., Malkhi, D., Dobzinski, O.: LAND: Stretch $(1 + \epsilon)$ locality aware networks for DHTs. In: ACM-SIAM Symposium on Discrete Algorithms (SODA) (2004)
3. Andersen, R., Chung, F., Lu, L.: Modeling the small-world phenomenon with local network flow. Internet Mathematics 2(3), 359–385 (2006)

4. Achlioptas, D., Clauset, A., Kempe, D., Moore, C.: On the bias of Traceroute sampling, or: power-law degree distributions in regular graphs. In: 37th ACM Symposium on Theory of Computing (STOC) (2005)
5. Aspnes, J., Diamadi, Z., Shah, G.: Fault-tolerant routing in peer-to-peer systems. In: 21st ACM Symp. on Principles of Distributed Computing (PODC), pp. 223–232 (2002)
6. Barabási, A., Albert, R.: Emergence of scaling in random networks. Science 286, 509–512 (1999)
7. Barrière, L., Fraigniaud, P., Kranakis, E., Krizanc, D.: Efficient routing in networks with long-range contacts. In: Welch, J.L. (ed.) DISC 2001. LNCS, vol. 2180, pp. 270–284. Springer, Heidelberg (2001)
8. Chung, F., Lu, L.: The small world phenomenon in hybrid power law graphs. Lect. Notes Phys. 650, 89–104 (2004)
9. Dodds, P., Muhamad, R., Watts, D.: An experimental study of search in global social networks. Science 301(5634), 827–829 (2003)
10. Duchon, P., Hanusse, N., Lebhar, E., Schabanel, N.: Could any graph be turned into a small-world? Theoretical Computer Science 355(1), 96–103 (2006)
11. Duchon, P., Hanusse, N., Lebhar, E., Schabanel, N.: Towards small world emergence. In: 18th Annual ACM Symposium on Parallel Algorithms and Architectures (SPAA), pp. 225–232 (2006)
12. Flammini, M., Moscardelli, L., Navarra, A., Perennes, S.: Asymptotically optimal solutions for small world graphs. In: Fraigniaud, P. (ed.) DISC 2005. LNCS, vol. 3724, pp. 414–428. Springer, Heidelberg (2005)
13. Fraigniaud, P.: Greedy routing in tree-decomposed graphs: a new perspective on the small-world phenomenon. In: Brodal, G.S., Leonardi, S. (eds.) ESA 2005. LNCS, vol. 3669, pp. 791–802. Springer, Heidelberg (2005)
14. Fraigniaud, P., Gavoille, C., Kosowski, A., Lebhar, E., Lotker, Z.: Universal augmentation schemes for network navigability: overcoming the $\sqrt{n}$-barrier. In: 19th Annual ACM Symposium on Parallelism in Algorithms and Architectures (SPAA) (2007)
15. Fraigniaud, P., Gavoille, C., Paul, C.: Eclecticism shrinks even small worlds. In: Proceedings of the 23rd ACM Symposium on Principles of Distributed Computing (PODC), pp. 169–178 (2004)
16. Fraigniaud, P., Lebhar, E., Lotker, Z.: A doubling dimension threshold $\Theta(\log \log n)$ for augmented graph navigability. In: Azar, Y., Erlebach, T. (eds.) ESA 2006. LNCS, vol. 4168, pp. 376–386. Springer, Heidelberg (2006)
17. Gupta, A., Krauthgamer, R., Lee, J.: Bounded geometries, fractals, and low- distortion embeddings. In: Proceedings of the 44th Annual IEEE Symposium on Foundations of Computer Science (FOCS), pp. 534–543 (2003)
18. Heinonen, J.: Lectures on analysis on metric spaces. Springer, Heidelberg (2001)
19. Har-Peled, S., Mendel, M.: Fast Construction of Nets in Low Dimensional Metrics, and Their Applications. SICOMP 35(5), 1148–1184 (2006)
20. Iamnitchi, A., Ripeanu, M., Foster, I.: Small-world file-sharing communities. In: 23rd Joint Conference of the IEEE Computer and Communications Societies (INFOCOM), pp. 952–963 (2004)
21. Karger, D., Ruhl, M.: Finding nearest neighbors in growth-restricted metrics. In: 34th ACM Symp. on the Theory of Computing (STOC), pp. 63–66 (2002)
22. Kay, S.M.: Fundamentals of Statistical Signal Processing: Estimation Theory, ch. 7. Prentice Hall, Englewood Cliffs (1993)
23. Kleinberg, J.: The small-world phenomenon: an algorithmic perspective. In: 32nd ACM Symp. on Theory of Computing (STOC), pp. 163–170 (2000)

24. Kleinberg, J.: Small-World Phenomena and the Dynamics of Information. Advances in Neural Information Processing Systems (NIPS) 14 (2001)
25. Kleinberg, J.: Complex networks and decentralized search algorithm. In: Nevanlinna prize presentation at the International Congress of Mathematicians (ICM), Madrid (2006)
26. Krioukov, D., Fall, K., Yang, X.: Compact routing on Internet-like graphs. In: 23rd Conference of the IEEE Communications Society (INFOCOM) (2004)
27. Kumar, R., Liben-Nowell, D., Tomkins, A.: Navigating Low-Dimensional and Hierarchical Population Networks. In: Azar, Y., Erlebach, T. (eds.) ESA 2006. LNCS, vol. 4168. Springer, Heidelberg (2006)
28. Liben-Nowell, D., Novak, J., Kumar, R., Raghavan, P., Tomkins, A.: Geographic routing in social networks. In: Proc. of the Natl. Academy of Sciences of the USA, vol. 102/3, pp. 11623–11628
29. Lebhar, E., Schabanel, N.: Searching for Optimal paths in long-range contact networks. In: Díaz, J., Karhumäki, J., Lepistö, A., Sannella, D. (eds.) ICALP 2004. LNCS, vol. 3142, pp. 894–905. Springer, Heidelberg (2004)
30. Manku, G., Naor, M., Wieder, U.: Know Thy Neighbor's Neighbor: The Power of Lookahead in Randomized P2P Networks. In: 36th ACM Symp. on Theory of Computing (STOC), pp. 54–63 (2004)
31. Martel, C., Nguyen, V.: Analyzing Kleinberg's (and other) Small-world Models. In: 23rd ACM Symp. on Principles of Distributed Computing (PODC), pp. 179–188 (2004)
32. Martel, C., Nguyen, V.: Analyzing and characterizing small-world graphs. In: 16th ACM-SIAM Symp. on Discrete Algorithms (SODA), pp. 311–320 (2005)
33. Martel, C., Nguyen, V.: Designing networks for low weight, small routing diameter and low congestion. In: 25th Conference of the IEEE Communications Society (INFOCOM) (2006)
34. Milgram, S.: The Small-World Problem. Psychology Today, pp. 60–67 (1967)
35. Newman, M.: The Structure and Function of Complex Networks. SIAM Review 45, 167–256 (2003)
36. Newman, M., Barabasi, A., Watts, D. (eds.): The Structure and Dynamics of Complex Networks. Princeton University Press, Princeton (2006)
37. Pastor-Satorras, R., Vespignani, A.: Evolution and Structure of the Internet: A Statistical Physics Approach. Cambridge University Press, Cambridge (2004)
38. Slivkins, A.: Distance estimation and object location via rings of neighbors. In: 24th Annual ACM Symposium on Principles of Distributed Computing (PODC), pp. 41–50 (2005)
39. Watts, D., Strogatz, S.: Collective dynamics of small-world networks. Nature 393, 440–443 (1998)

Computing Frequent Elements Using Gossip

Bibudh Lahiri* and Srikanta Tirthapura*

Iowa State University, Ames, IA, 50011, USA
{bibudh,snt}@iastate.edu

Abstract. We present algorithms for identifying frequently occurring elements in a large distributed data set using gossip. Our algorithms do not rely on any central control, or on an underlying network structure, such as a spanning tree. Instead, nodes repeatedly select a random partner and exchange data with the partner – if this process continues for a (short) period of time, the desired results are computed, with probabilistic guarantees on the accuracy. Our algorithm for frequent elements is built by layering a novel small space "sketch" of data over a gossip-based data dissemination mechanism. We prove that the algorithm converges to the approximate frequent elements with high probability, and provide bounds on the time till convergence. To our knowledge, this is the first work on computing frequent elements using gossip.

1 Introduction

We are increasingly faced with data-intensive decentralized systems, such as large scale peer-to-peer networks, server farms with tens of thousands of machines, and large wireless sensor networks. With such large networks comes increasing unpredictability; the networks are constantly changing, due to nodes joining and leaving, or due to node and link failures. *Gossip* is a type of communication mechanism that is ideally suited for distributed computation on such unstable, large networks. Gossip-based distributed protocols do not assume any underlying structure in the network, such as a spanning tree, so, there is no overhead of sub-network formation and maintenance. A gossip protocol proceeds in many "rounds" and in each round, a node contacts a few randomly chosen nodes in the system and exchanges information with them. The randomization inherently provides robustness, and surprisingly, often leads to fast convergence times. The use of gossip-based protocols for data dissemination and aggregation was first proposed by Demers *et al.* [1].

We focus on the problem of identifying *frequent data elements* in a network using gossip. Consider a large peer-to-peer network that is distributing content, such as news or software updates. Suppose that the nodes in the network (or the network administrators) wish to track the identities of the most frequently accessed items in the network. The relevant data for tracking this aggregate are the frequencies of accesses of different items. However, this data is distributed

---

* Supported in part by NSF grant CNS 0520102 and by ICUBE, Iowa State University.

A. Shvartsman and P. Felber (Eds.): SIROCCO 2008, LNCS 5058, pp. 119–130, 2008.
© Springer-Verlag Berlin Heidelberg 2008

throughout the network – in fact, even the number of accesses to a single item may not be available at any single point in the network. Our gossip-based algorithm for frequent elements can be used to track the most frequently accessed items in a low-overhead, decentralized manner. Another application of tracking frequent items is in the detection of a distributed denial of service (DDoS) attack, where many malicious nodes may team up to simultaneously send excessive traffic towards a single victim (typically a web server), so that legitimate clients are denied service. Detecting a DDoS attack is equivalent to finding that the total number of accesses to some server has exceeded a threshold. A distributed frequent elements algorithm can help by tracking the most frequently accessed web servers in a distributed manner, and noting if these frequencies are abnormally large. With a gossip-based algorithm this computation can proceed in a totally decentralized manner.

We consider two variants of the frequent elements problem, with absolute and relative thresholds. In the absolute threshold version, the task is to identify all elements whose frequency of occurrence is at least an absolute number (threshold), which is an user-defined parameter. In the relative threshold version, the task is to identify all elements whose frequency of occurrence is more than a certain fraction of the total size of the data, where the fraction (the relative threshold) is an user-defined parameter. In a distributed dynamic network, these two problems turn out to be rather different from each other.

Our algorithms work without explicitly tabulating the frequencies of different elements at any single place in the network. Instead, the distributed data is represented by a small space "sketch" that is propagated and updated via gossip. A sketch is a space-efficient representation of the input, which is specific to the aggregate being computed, and captures the essence of the data for our purposes. The space taken by the sketch can be tuned as a function of the desired accuracy. A complication with gossip is that since it is an unstructured form of communication, it is possible for the same data item to be inserted into the sketch multiple times as the sketch propagates. Due to this, a technical requirement is that the sketch should be able to handle duplicate insertions, i.e. it should be *duplicate-insensitive*. If the gossip proceeds long enough, the sketch can be used to identify all elements whose frequency exceeds the user defined threshold. At the same time, elements whose popularity is significantly below the threshold will be omitted (again, with high probability).

To summarize our contributions, we present the first work on computing frequent elements in a distributed data set using gossip. We present randomized algorithms for both the absolute threshold and the relative threshold versions of the problem. For each algorithm, we present a rigorous analysis of the correctness, and the time till convergence. Our analysis show that gossip-based algorithms converge quickly, and can maintain frequent elements in a network with a reasonable communication overhead. We also note that similar techniques can be used in estimating the number of distinct data items in the network. Due to space limitations, details of estimating the number of distinct elements are not presented here and can be found in the full version of the paper [2].

With a gossip protocol, communication is inherently randomized, and a node can never be certain that the results on hand are correct. However, the longer the protocol runs, the closer the results get to the correct answer, and we are able to quantify the time taken till the protocol converges to the correct answer, with high probability. Gossip algorithms are suitable for applications which can tolerate such relaxed consistency guarantees. Examples include a network monitoring application, which is running in the background, maintaining statistics about frequently requested data items, or the most frequently observed data in a distributed system. In such an application, a guaranteed accurate answer may not be required, and an approximate answer may suffice.

1.1 Related Work

Demers *et al.* [1] were the first to provide a formal treatment of gossip protocols (or "epidemic algorithms" as they called them) for data dissemination. Kempe, Dobra and Gehrke [3] proposed algorithms for computing the sum, average, approximately uniform random sampling and quantiles using uniform gossip. Their algorithm for quantiles are based on their algorithm for the sum – they choose a random element and count the number of elements that are greater and lesser than the chosen element, and recurse on smaller data sets until the median is found. Thus their algorithms need many instances of "sum" computations to converge before the median is found. A similar approach could potentially be used to find frequent elements using gossip. In contrast, our algorithms for frequent elements are not based on repeated computation of the sum, and converge faster.

Much recent work [4,5,6] has focused on computing "separable functions" using gossip. A separable function is one that can be expressed as the sum of individual functions of the node inputs. For example, the function "count" is separable, and so is the function "sum". However, the set of frequent elements is not a separable function. Hence, these techniques do not apply to our problem.

The problem of identifying frequent elements in data has been extensively studied [7,8,9] in the database, data streams and network monitoring communities (where frequent elements are often called "heavy-hitters"). The early work in this is due to Misra and Gries [7], who proposed a deterministic algorithm to identify frequent elements in a stream, followed by Manku and Motwani [8], who gave randomized and deterministic algorithms for tracking frequent elements in limited space. The above were algorithms for a centralized setting.

In a distributed setting, Cao and Wang [10] proposed an algorithm to find the top-k elements, where they first made a lower-bound estimate for the kth value, and then used the estimate as a threshold to prune away elements which should not qualify as top-k. Zhao *et al.* [11] proposed a sampling-based and a counting-sketch-based scheme to identify globally frequent elements. Manjhi *et al.* [12] present an algorithm for finding frequent items on distributed streams, through a tree-based aggregation. Keralapura, Cormode and Ramamirtham [13] proposed an algorithm for continuously maintaining the frequent elements over a network of nodes. The above algorithms sometimes assume the presence of a central node, or an underlying network structure such as a spanning tree [12,13], and hence are not applicable where the underlying network does not guarantee reliability or robustness.

1.2   Organization of the Paper

In Section 2, we state our model and the problem more precisely. The algorithm and analysis for the case of absolute threshold in the asynchronous time model is presented in Section 3 and the case of relative threshold is presented in Section 4. Due to space constraints, we only present sketches of the proofs. Detailed proofs can be found in the full version of the paper [2].

2   Model

We consider a distributed system with N nodes numbered from 1 to N. Each node i holds a single data item m_i. Without loss of generality, we assume that $m_i \in \{1, 2, \ldots, m\}$ is an integer representing an item identifier. For data item $v \in \{1, \ldots, m\}$, the frequency of v is denoted by f_v, and is defined as the number of nodes that have data item v, i.e. $f_v = |\{j \in [N] : m_j = v\}|$. Note that f_v may not be available locally at any node, in fact determining f_v itself requires a distributed computation. The task is to identify those elements that have large frequencies. We note that though we describe our algorithms for the case of one item per node, they can be easily extended to the case when each node has a (multi)set of items.

We consider the scenario of *uniform gossip*, which is the basic, and most commonly used model of gossip. Whenever a node i transmits, it chooses the destination of the message to be a node selected uniformly at random from among all the current nodes in the system. The selection of the transmitting node is done by the distributed scheduler, described later in this section. We consider two variants of the problem, depending on how the thresholds are defined.

Absolute Threshold. The user gives an absolute frequency threshold $k > 1$ and approximation error λ ($\lambda < k$). An item v is considered a frequent item if $f_v \geq k$, and v is an infrequent item if $f_v < k - \lambda$. Note that with a data set of N elements there may be up to N/k frequent elements according to this definition.

Relative Threshold. In some cases, the user may not be interested in an absolute frequency threshold, but may only be interested in identifying items whose relative frequency exceeds a given threshold. More precisely, given a relative threshold ϕ ($0 < \phi < 1$), approximation error ψ ($0 < \psi < \phi$), an item v is considered to be a frequent item if $f_v \geq \phi N$, and v is considered an infrequent item if $f_v < (\phi - \psi)N$. According to this definition, there may be no more than $1/\phi$ frequent items.

In a centralized setting, when all items are being observed at the same location, the above formulations of relative and absolute thresholds are equivalent, since the number of items N is known, and any absolute threshold can be converted into a relative threshold, or vice versa. However, in a distributed setting, a threshold for relative frequency cannot be locally converted by a node into a threshold on the absolute frequency, since the user in a large distributed system may not know the number of nodes or the number of data items in the system accurately enough. Thus, we treat these two problems separately. The lack of knowledge of the network size N does not, though, prevent the system from choosing gossip partners uniformly at random. For example, Gkantsidis *et al.* [14] show how random walks can provide a good approximation to uniform sampling for networks where the gap between the first and the second eigenvalues of the transition matrix is constant.

At the end of the gossip, the following probabilistic guarantees must hold, whether for absolute or relative thresholds. Let δ be a user-provided bound on the error probability $(0 < \delta < 1)$. With probability at least $(1 - \delta)$, every node reports every frequent item. With probability at least $(1 - \delta)$, no node reports an infrequent item. In other words, the latter statement implies that the probability that an infrequent item is incorrectly reported by a node in the system is less than δ. Note that we present randomized algorithms, where the probabilistic guarantees hold irrespective of the input.

Time Model. For gossip-based protocols, time is usually divided into non-overlapping rounds. We consider the *asynchronous* model, where in each round, a *single* source node, chosen uniformly at random out of all N nodes, transmits to another randomly chosen receiver. The time complexity is the number of rounds, or equivalently, the number of transmissions, since in each round there is only one transmission. We consider the synchronous model in the full version of the paper.

Performance Metrics. We evaluate the quality of our protocols via the following metrics: the *convergence time*, which is defined as the number of rounds of gossip till convergence, and the *communication complexity*, which is defined as the number of bytes exchanged till convergence.

3   Frequent Elements with an Absolute Threshold

We now present an algorithm in the asynchronous model for identifying elements whose frequency is greater than a user specified absolute threshold k. Let $S = \{m_i : i \in [N]\}$ denote the multi-set of all input values. The goal is to output all elements v such that $f_v \geq k$ without outputting any element v such that $f_v < k - \lambda$. We first describe the high level intuition.

Our algorithm is based on random sampling. The elements of S are sampled in a distributed manner, and the sampled elements are disseminated to all nodes using gossip – the cost of doing so is small, since the random sample is typically much smaller than the size of the population. The sampling also ensures that frequent elements are exchanged more often during the later gossip phase. Intuitively, suppose we sample each element from S into a set T with probability

$1/k$. For a frequent element v with $f_v \geq k$, we (roughly) expect one or more copies of v to be included within T. Similarly, for an infrequent element u with $f_u < k - \lambda$, we expect that no copy of u will be included in T. However, some infrequent elements may get "lucky" and may be included in T and similarly, some frequent elements may not make it to T. The probabilities of these events decrease as the sample size increases.

To refine this sampling scheme, we sample with a probability that is a little larger than $1/k$, say c/k for some parameter c. Finally, we select those elements that occur at least r times within T, for some parameter $r < c$ that will be decided by the analysis. It turns out that c and r will need to depend on the approximation error λ as well as the threshold k. The smaller λ is, the greater should be the sampling probability, since we need to make a more precise distinction between the frequencies of frequent and infrequent elements. In the actual algorithm, we use a sampling probability of $\frac{12k}{\lambda^2} \ln \frac{2}{\delta}$ – note that this is $\Omega(\frac{1}{k})$ since $\lambda < k$ and hence $\frac{k}{\lambda^2} > \frac{1}{k}$.

The precise algorithm for sampling and gossip is shown in Figure 1. There are three parts to this algorithm (and all others that we describe). The first part is the *Initialization*, where each node initialized its own sketch, which is usually through drawing a random sample. The next part is the *Gossip* portion, where the nodes in the system exchange sketches with each other. The algorithm only describes what happens during each round of gossip – it is implicit that such computations repeat forever. The third part is the *Query*, where we describe how a query for frequent elements is answered using the sketch. The accuracy of the result improves as further rounds of gossip occur. Through our analysis, we give a bound on the number of rounds after which frequent elements are likely to be found at all nodes.

Input: Data item m_i, error probability δ, frequency threshold k, approximation error λ

1. **Initialization**
   (a) Choose ρ as a uniformly distributed random number in $(0, 1)$.
   (b) If $\rho < \frac{12k}{\lambda^2} \ln \frac{2}{\delta}$ then $S_i \leftarrow \{(i, m_i)\}$, else $S_i \leftarrow \Phi$ /* null set */
2. **Gossip**
   In each round of gossip:
   (a) If sketch S_j received from node j then $S_i \leftarrow S_i \cup S_j$
   (b) If node i is selected to transmit, then select node j uniformly at random from $\{1, \ldots, N\}$ and send S_i to j
3. **Query**
   When asked for the frequent elements, report all data items which occur more than $r = \frac{12k^2}{\lambda^2}(1 - \frac{\lambda}{2k}) \ln \frac{2}{\delta}$ times in S_i as frequent elements.

Fig. 1. Gossip algorithm at node i for finding the frequently occurring elements with an absolute threshold k

3.1 Analysis

We now analyze the correctness and the time complexity (proof sketches only) of the algorithm in Figure 1.

Lemma 1. False Negative. *If v is an element with $f_v \geq k$, then with probability at least $1 - \delta$, v is returned as a frequent element by every node after $20N \ln N$ rounds.*

Sketch of proof: A false negative can occur in one of two ways. (1)Either too few copies of v were sampled during initialization or (2)The sampled copies of v were not disseminated to all nodes during the gossip. We show that the first event is unlikely by an analysis of the sampling process using Chernoff bounds. We show that the second event is also very unlikely through an analysis of the asynchronous gossip in Lemma 4. □

Lemma 2. False Positive. *If u is an element with $f_u \leq k - \lambda$, where $k^{\frac{3}{4}} \leq \lambda < k$, then the probability that u is returned by some node as a frequent element is no more than δ.*

Sketch of proof: A false positive can occur if both the following events occur: (1)r or more copies of u were sampled initially and (2)all r copies of u reach some node in the network through gossip. We show that the first event is very unlikely, if $f_v \leq k - \lambda$, and hence the intersection of the events is also unlikely. □

Now that we have proved Lemmas 1 and 2, it is natural to ask what happens to an element with frequency in the range $[k - \lambda, k)$ of length λ. With our algorithm, such elements could be reported as frequent items, or not. Clearly, a smaller value of λ implies less uncertainty, but this comes at the cost of increased sampling probability, and hence greater communication complexity of gossip. For example, with $k = 10^8$ and $\lambda = 5 \times 10^6$, the approximation error with respect to k is 5%. All elements with frequency greater than 10^8 will be reported (w.h.p) and all elements with frequency below 0.95×10^8 will not be reported (w.h.p., once again), and the sampling probability is approximately $4.8 \times 10^{-5} \times \ln \frac{2}{\delta}$. This is the fraction of input items that are gossiped through the network in finding the frequent elements in the distributed data set.

Analysis of the Gossip. We now shift our attention to the gossip mechanism itself. Let $\mathcal{T}$ denote the multi-set of all items sampled during initialization $\mathcal{T} \subseteq S$ and $|\mathcal{T}| \leq N$. Consider a single sampled item $\theta \in \mathcal{T}$. Let T_θ be defined as the number of rounds till θ has been disseminated to all nodes in the network.

Lemma 3. $E[T_\theta] = 2N \ln N + O(N)$.

Sketch of proof: Let ξ_t be the set of nodes that have θ after t rounds. Thus ξ_0 has only one node (the one that sampled θ during the initialization step). For $t = 1 \ldots N - 1$, let random variable X_t be the number of rounds required to increase $|\xi|$ from t to $t + 1$. We can write $T_\theta = X_1 + X_2 + \cdots + X_{N-1}$. By noting that each X_t is a geometric random variable and using linearity of expectation we can arrive at the desired result. Further details are in the full version. □

Our proof for high-probability bounds on T_θ use the following result about a sharp concentration for the *coupon collector* problem. Suppose there are coupons of M distinct types, and one has to draw coupons (with replacement) at random until at least one coupon of each type has been collected. Initially, it is very easy to select a type not yet chosen, but as more and more types get chosen, it becomes increasingly difficult to get a coupon of a type not yet chosen. The following result can be found in standard textbooks (for example, Motwani and Raghavan [15]).

Theorem 1 (Folklore). *Let the random variable $\mathcal{C}$ denote the number of trials to collect at least one coupon of each of M types. Then, for any constant $c \in \mathcal{R}$, $\lim_{M \to \infty} \Pr[\mathcal{C} > M \ln M + cM] = 1 - e^{-e^{-c}}$.*

Lemma 4. $\lim_{N \to \infty} \Pr(T_\theta > 20N \ln N) = O(\frac{1}{N^2})$

Sketch of proof: The dissemination of θ by gossip can be divided into two phases. The first phase starts when θ is at a single node and continues until it has reached $\frac{N}{2}$ distinct nodes. The second phase starts after θ has reached $\frac{N}{2}$ nodes and continues until it reaches N nodes. In the first phase, in each round of gossip, it is less likely to find a source node that has θ and at the same time, it is more likely to find a destination that does not have θ. Once θ has reached $\frac{N}{2}$ nodes, the situation reverses. We analyze the number of rounds required for these two phases separately. For each phase, we bound the random variable that defines the number of rounds in the phase by a simpler random variable that can be analyzed with the help of a coupon-collector type of argument. Combining the results from the two phases yields the desired result.   □

For an item v, let T_v denote the number of rounds required to disseminate all copies of v to all nodes.

Lemma 5. $\lim_{N \to \infty} \Pr[T_v > 20N \ln N] = O(\frac{1}{N})$

Proof. The proof follows from Lemma 4 using a union bound, and the fact that there are no more than N copies of v.

Lemmas 1, 2 and 5 together lead to the following theorem about the correctness of the algorithm.

Theorem 2. *Suppose the distributed algorithm in Figure 1 is run for $20N \ln N$ rounds. Then, with probability at least $1 - \delta$, any data item with k or more occurrences will be identified as a frequent element at every node. With probability at least $1 - \delta$, any data item with less than $k - \lambda$ occurrences will not be identified as a frequent element at any node.*

Communication Complexity. We next analyze the communication complexity, i.e. the number of bytes transmitted during the gossip. During the algorithm, the sizes of the messages exchanged start from one item and grow as the algorithm progresses. To avoid the complexity of dealing with different message sizes, we separately analyze the total number of bytes contributed to gossip by each sampled item, and add these contributions together. Consider any sampled item θ. We assume that transmitting any item (i, m_i) takes a constant number of bytes. Let random variable $\mathcal{B}$ denote the number of bytes it takes to disseminate θ among all nodes.

Lemma 6. $E[\mathcal{B}] = O(N \ln N)$

Sketch of proof: Let ξ_t be the set of nodes that have θ after t rounds. In each round of gossip, θ may or may not be transmitted. Further each time θ is transmitted, $|\xi|$ increases only if the destination of the message is a node which is not already in ξ_t. We analyze $\mathcal{B}$ as the number of transmissions of θ till ξ includes all nodes. The details of the proof, using conditional probabilities, are presented in the full version. $\qquad\qquad\qquad\qquad\qquad\qquad\qquad\qquad\qquad\qquad$ $\square$

We can similarly get a high probability bound on $\mathcal{B}$ (proofs in full version).

Lemma 7. $\Pr[\mathcal{B} > 3N \ln N] = O(\frac{1}{N^2})$.

Let $\mathcal{Y}$ denote the total number of bytes that need to be exchanged for the whole protocol until the frequent elements have been identified. By combining Lemma 7 with an estimate on the number of sampled items, we get the following result about the communication complexity of the algorithm in Figure 1.

Theorem 3 (Communication Complexity for Absolute Threshold)
With high probability, $\mathcal{Y} = O(\frac{N^2 k}{\lambda^2} \ln \frac{1}{\delta} \ln N)$

4 Frequent Elements with Relative Threshold

Given thresholds ϕ and ψ, where $\psi < \phi$, the goal is to identify all elements v such that $f_v \geq \phi N$ and no element u such that $f_u < (\phi - \psi)N$. Unlike the case of absolute threshold, there is no fixed probability that a node can use to sample data elements locally. For the same relative frequency threshold, the absolute frequency threshold (ϕN), as well as the approximation error (ψN) increases with N. Thus if ϕ and ψ are kept constant and N increases, then a smaller sampling probability will suffice, because of the analysis in 3. Since we do not have prior knowledge of N, we need a more "adaptive" method of sampling where the sampling probability decreases as more elements are encountered during gossip.

To design our sketch, we use an idea similar to *min-wise independent permutations* [16]. Each data item $m_i, i = 1 \ldots N$ is assigned a weight w_i, which is a random number in the unit interval $(0, 1)$. The algorithm maintains a sketch T of (m_i, w_i) tuples that have the t smallest weights w_i. The value of t can be decided independent of the population size N. The intuition is that if an element v has a large relative frequency, then v must occur among the tuples with the smallest weight. Maintaining these minimum-weight elements through gossip is easy, and if we choose a large enough sketch, the likelihood of a frequent element appearing in the sketch a sufficient number of times is very high. We identify a value m as a frequent item if there are at least $(\phi - \frac{\psi}{2})t$ tuples in T with $m_i = m$; otherwise, m is not identified as a frequent element. The algorithm for the asynchronous model is described in Figure 2. The threshold t is chosen to be $O(\frac{1}{\psi^2} \ln(\frac{1}{\delta}))$.

Input: Data item m_i; error probability δ, relative frequency threshold ϕ, approximation error $\psi < \phi$

1. **Initialization:**
   (a) $t \leftarrow \frac{128}{\psi^2} \ln(\frac{3}{\delta})$
   (b) Choose w_i as a uniformly distributed random number in the real interval $(0, 1)$; set $S_i \leftarrow \{(m_i, w_i)\}$
2. **Gossip**
   In each round of gossip:
   (a) If sketch S_j is received from node j then
      i. $S_i \leftarrow S_i \cup S_j$
      ii. If $|S_i| > t$ then retain t elements of S_i with smallest weights
   (b) If node i is selected to transmit, then select node j uniformly at random and send S_i to j
3. **Query**
   When queried for the frequent elements, report every value v such that at least $(\phi - \frac{\psi}{2})t$ (value, weight) tuples exist in S_i with value equal to v

---

Fig. 2. Gossip algorithm at node i for finding the frequently occurring elements with a relative threshold

4.1   Analysis

The proofs of most of the following lemmas appear in the full version. Let τ denote the tth minimum element among the N random values $\{w_i, i = 1 \ldots N\}$. The next lemma shows that τ is sharply concentrated around $\frac{t}{N}$.

Lemma 8. *For $t = \frac{128}{\psi^2} \ln(\frac{3}{\delta})$, τ satisfies the following properties: (1) $\Pr[\tau < \frac{t}{N}(1 - \frac{\psi}{4})] < \frac{\delta}{3}$ and
(2) $\Pr[\tau > \frac{t}{N}(1 + \frac{\psi}{4})] < \frac{\delta}{3}$*

We now present a bound on the dissemination time of the smallest weights. Let T_t denote the time taken for the t smallest weights to be disseminated to all nodes.

Lemma 9. $\Pr[T_t > 20N \ln N] \leq O(\frac{1}{N})$.

Proof. The proof follows by using the union bound along with Lemma 4.

The following lemmas provide upper bounds on the probabilities of finding a false negative and a false positive respectively, by the algorithm described in Figure 2.

Lemma 10. *Suppose the distributed algorithm in Figure 1 is run for $20N \ln N$ rounds. Then, if v is a frequent element, i.e. $f_v \geq \phi N$, then with probability at least $1 - \delta$, v is identified by every node as a frequent element.*

Sketch of proof: Two events need to happen for v to be recognized as a frequent element. (1) Enough copies of v must occur among the t smallest weights, and (2) The t smallest weight elements must be disseminated to all nodes via gossip. In the full proof, we show that both these events are very likely. □

Lemma 11. *Suppose the distributed algorithm in Figure 1 is run for $20N \ln N$ rounds. If u is an infrequent element, i.e. $f_u < (\phi - \psi)N$, then, with probability at least $1 - \delta$, u is not identified by any node as a frequent element.*

Sketch of proof: A false positive can happen if both the following events occur. (1) There are an unusually high number of copies of u among the elements with the τ smallest weights, and (2) all these copies are disseminated to all nodes. We show that the first event is highly unlikely, and so is the probability of a false positive. □

Combining Lemmas 11, 10 and 9 we get the following theorem.

Theorem 4. *Suppose the distributed algorithm in Figure 2 is run for $20N \ln N$ rounds. Then, with probability at least $1 - \delta$, any data item with ϕN or more occurrences will be identified as a frequent item at every node. Similarly, with probability at least $1 - \delta$, any data item with less than $(\phi - \psi)N$ occurrences will not be identified as a frequent item at any node.*

Since the size of the sketch at any time during gossip is at most t, we get the following result on the communication complexity, using Lemma 9.

Theorem 5. *The number of bytes exchanged by the algorithm in Figure 2 till the frequent elements are identified is at most $O(\frac{1}{\psi^2} \ln(\frac{1}{\delta})N \ln N)$, with probability at least $1 - O(\frac{1}{N})$.*

References

1. Demers, A.J., Greene, D.H., Hauser, C., Irish, W., Larson, J., Shenker, S., Sturgis, H.E., Swinehart, D.C., Terry, D.B.: Epidemic algorithms for replicated database maintenance. In: Proceedings of the Principles of Distibuted Computing (PODC), pp. 1–12 (1987)
2. Lahiri, B., Tirthapura, S.: Computing frequent elements using gossip. Technical report, Dept. of Electrical and Computer Engineering, Iowa State University (April 2008), http://archives.ece.iastate.edu/archive/00000415/01/gossip.pdf
3. Kempe, D., Dobra, A., Gehrke, J.: Gossip-based computation of aggregate information. In: Proceedings of the 44th Symposium on Foundations of Computer Science (FOCS), pp. 482–491 (2003)
4. Boyd, S.P., Ghosh, A., Prabhakar, B., Shah, D.: Gossip algorithms: design, analysis and applications. In: Proceedings of the IEEE Conference on Computer Communications (INFOCOM), pp. 1653–1664 (2005)

5. Boyd, S.P., Ghosh, A., Prabhakar, B., Shah, D.: Randomized gossip algorithms. IEEE Transactions on Information Theory 52(6), 2508–2530 (2006)
6. Mosk-Aoyama, D., Shah, D.: Computing separable functions via gossip. In: Proceedings of the Twenty-Fifth Annual ACM Symposium on Principles of Distributed Computing (PODC), pp. 113–122 (2006)
7. Misra, J., Gries, D.: Finding repeated elements. Science of Computer Programming 2(2), 143–152 (1982)
8. Manku, G.S., Motwani, R.: Approximate frequency counts over data streams. In: Proceedings of 28th International Conference on Very Large Data Bases (VLDB), pp. 346–357 (2002)
9. Karp, R.M., Shenker, S., Papadimitriou, C.H.: A simple algorithm for finding frequent elements in streams and bags. ACM Trans. Database Syst. 28, 51–55 (2003)
10. Cao, P., Wang, Z.: Efficient top-k query calculation in distributed networks. In: Proceedings of the Twenty-Third Annual ACM Symposium on Principles of Distributed Computing (PODC), pp. 206–215 (2004)
11. Zhao, Q., Ogihara, M., Wang, H., Xu, J.: Finding global icebergs over distributed data sets. In: Proceedings of the Twenty-Fifth ACM SIGACT-SIGMOD-SIGART Symposium on Principles of Database Systems (PODS), pp. 298–307 (2006)
12. Manjhi, A., Shkapenyuk, V., Dhamdhere, K., Olston, C.: Finding (recently) frequent items in distributed data streams. In: Proceedings of the 21st International Conference on Data Engineering (ICDE), pp. 767–778 (2005)
13. Keralapura, R., Cormode, G., Ramamirtham, J.: Communication-efficient distributed monitoring of thresholded counts. In: Proceedings of the ACM SIGMOD International Conference on Management of Data (SIGMOD), pp. 289–300 (2006)
14. Gkantsidis, C., Mihail, M., Saberi, A.: Random walks in peer-to-peer networks. In: Proceedings of the 23rd Conference of the IEEE Communications Society (INFOCOM) (2004)
15. Motwani, R., Raghavan, P.: Randomized Algorithms. Cambridge University Press, Cambridge (1995)
16. Broder, A.Z., Charikar, M., Frieze, A.M., Mitzenmacher, M.: Min-wise independent permutations (extended abstract). In: Proceedings of the ACM Symposium on Theory of Computing (STOC), pp. 327–336 (1998)

Maintaining Consistent Transactional States without a Global Clock

Hillel Avni[1,3] and Nir Shavit[1,2]

[1] Tel-Aviv University, Tel-Aviv 69978, Israel
[2] Sun Microsystems Laboratories, 1 Network Drive, Burlington MA 01803-0903
[3] Freescale Semiconductor Israel Ltd., 1 Shenkar Street, Herzliya 46725, Israel
hillel.avni@gmail.com

Abstract. A crucial property required from software transactional memory systems (STMs) is that transactions, even ones that will eventually abort, will operate on consistent states. The only known technique for providing this property is through the introduction of a globally shared version clock whose values are used to tag memory locations. Unfortunately, the need for a shared clock moves STM designs from being completely decentralized back to using centralized global information.

This paper presents *TLC*, the first thread-local clock mechanism for allowing transactions to operate on consistent states. TLC is the proof that one can devise coherent-state STM systems without a global clock.

A set of early benchmarks presented here within the context of the TL2 STM algorithm, shows that TLC's thread-local clocks perform as well as a global clock on small scale machines. Of course, the big promise of the TLC approach is in providing a decentralized solution for future large scale machines, ones with hundreds of cores. On such machines, a globally coherent clock based solution is most likely infeasible, and TLC promises a way for transactions to operate consistently in a distributed fashion.

1 Introduction

The question of the inherent need for global versus local information has been central to distributed computing, and will become central to parallel computing as multicore machines, now in the less than 50 core range, move beyond bus based architectures and into the 1000 core range. This question has recently arisen in the context of designing state-of-the-art software transactional memories (STMs).

Until recently, STM algorithms [1,2,3,4,5] allowed the execution of "zombie" transactions: transactions that have observed an inconsistent read-set but have yet to abort. The reliance on an accumulated read-set that is not a valid snapshot [6] of the shared memory locations accessed can cause unexpected behavior such as infinite loops, illegal memory accesses, and other run-time misbehavior. Overcoming zombie behavior requires specialized compiler and runtime support, and even then cannot fully guarantee transactional termination [7].

A. Shvartsman and P. Felber (Eds.): SIROCCO 2008, LNCS 5058, pp. 131–140, 2008.
© Springer-Verlag Berlin Heidelberg 2008

In response, Reigel, Felber, and Fetzer [8] and Dice, Shalev, and Shavit [7] introduced a *global clock* mechanism as a means of guaranteeing that transactions operate on consistent states. Transactions in past STM systems [1,2,3,4,5] typically updated a tag in the lock-word or object-record associated with a memory location as a means of providing transactional validation. In the new global clock based STMs [7,9,10] instead of having transactions locally increment the tags of memory locations, they update them with a time stamp from the globally coherent clock. Transactions provide consistency (recently given the name *opacity* [11]) by comparing the tags of memory locations being read to a value read from the global clock at the transaction's start, guaranteeing that the collected read-set remains coherent.

Unfortunately, this globally shared clock requires frequent remote accesses and introduces invalidation traffic, which causes a loss of performance even on small scale machines [7]. This problem will most likely make global clocks infeasible on large scale machines with hundreds of cores, machines that no longer seem fictional [12,13].

To overcome this problem, there have been suggestions of distributing the global clock (breaking the clock up into a collection of shared clocks) [14,15], or of providing globally coherent clock support in hardware [16]. The problem with schemes that aim to distribute the global clock is that the cost of reading a distributed clock grows with the extent to which it is distributed. The problem with globally coherent hardware clocks, even of such hardware modifications were to be introduced, is that they seem to be limited to small scale machines.

This paper presents *TLC*, the first *thread-local* clock mechanism that allows transactions to operate on consistent states. The breakthrough TLC offers is in showing that one can support coherent states without the need for a global notion of time. Rather, one can operate on coherent states by validating memory locations on a *per thread* basis. TLC is a painfully simple mechanism that has the same access patterns as prior STMs that operate on inconsistent states [1,3,2,4,5]: the only shared locations to be read and written are the tags associated with memory locations accessed by a transaction. This makes TLC a highly distributed scheme by its very nature.

1.1 TLC in a Nutshell

Here is how TLC works in a nutshell. As usual, a *tag* containing a time-stamp (and other information such as a lock bit or HyTM coordination bit) is associated with each transactional memory location. In TLC, the time-stamp is appended with the ID of the thread that wrote it. In addition, each thread has a *thread local clock* which is initially 0, and is incremented by 1 at the start of every new transaction. There is also a *thread local array* of entries, each recording a time-stamp for each other thread in the system. We stress that this array is local to each thread and will never be read by others.

Without getting into the details of a particular STM algorithm, we remind the reader that transactions in coherent-state STMs [7,10,9] typically *read* a location by first checking its associated tag. If the tag passes a *check*, the location is consistent with locations read earlier and the transaction continues. If it is not, the transaction is aborted. Transactions *write* a memory location by *updating* its associated tag upon commit.

Here is how TLC's *check* and *update* operations would be used in a transaction by a given thread i:

1. *Update*(location) Write to the locations tag my current transaction's new local clock value together with my ID i.
2. *Check*(location) Read the location's tag, and extract the ID of the thread j that wrote it. If the location's time-stamp is higher than the current time-stamp stored in my local array for the thread j, update entry j and abort my current transaction. If it is less than or equal to the stored value for j, the state is consistent and the current transaction can continue.

This set of operations fits easily into the global-clock-based schemes in many of today's STM frameworks, among them McRT [10], TinySTM [17], or TL2 [7], as well as hardware supported schemes such as HyTM [1] and SigTM [9].

How does the TLC algorithm guarantee that a transaction operates on a consistent read-set? We argue that a TLC transaction will always fail if it attempts to read a location that was written by some other transaction after it started. For any transaction by thread i, if a location is modified by some thread j after the start of i's transaction, the first time the transaction reads the location written by j, it must find the associated time-stamp larger than its last recorded time-stamp for j, causing it to abort.

An interesting property of the TLC scheme is that it provides natural locality. On a large machine, especially NUMA machines, transactions that access a particular region of the machine's memory will only ever affect time-stamps of transactions accessing the same region. In other words, the interconnect traffic generated by any transaction is limited to the region it accessed and goes no further. This compares favorably to global clock schemes where each clock update must be machine-wide.

The advantages of TLC come at a price: it introduces more false-aborts than a global clock scheme. This is because a transaction by a thread j may complete a write of some location completely before a given transaction by i reads it, yet i's transaction may fail because its array recorded only a much older time-stamp for j.

As our initial benchmarks show, on small scale state of the art multicore machines, the benefits of TLC are overshadowed by its higher abort rate. We did not have a 1000 node NUMA machine to test TLC on, and so we show that on an older generation 144 node NUMA machine, in benchmarks with a high level of locality, TLC can significantly outperform a global clock. Though this is by no way an indication that one should use TLC today, it is an indicator of its potential on future architectures, where a global clock will most likely be costly or even impossible to implement.

2   An STM Using TLC

We now describe an STM implementation that operates on consistent states without the need for a global clock by using the TLC algorithm. Our choice STM is the TL2 algorithm of Dice, Shalev, and Shavit [7], though TLC could fit in other STM frameworks such as McRT [10], or TinySTM [17] as well as hardware supported schemes such as HyTM [1] and SigTM [9]. We will call this algorithm *TL2C*.

Recall that with every transacted memory location TL2 associates a special versioned write-lock. The time-stamp used in the TL2C algorithm will reside in this lock. The structure of the TL2C algorithm is surprisingly simple. Each time-stamp written will be tagged with the ID of the thread that wrote it. Each thread has:

- a thread local *TLClock*, initially 0, which is incremented by 1 at the start of every new transaction, and
- a thread local *CArray* of entries, each entry of which records a time-stamp for each other thread in the system.

The clock has no shared components.

In the TL2C algorithm, as in the original TL2, the write-lock is updated by every successful lock-release, and it is at this point that the associated time-stamp is updated. The algorithm maintains thread local read-and write-sets as linked lists. The read-set entries contain the address of the lock. The write-set entries contain the address of the memory location accessed, the value to be written, and the address of the associated lock. The write-set is kept in chronological order to avoid write-after-write hazards. During transactional writing operations the read-set is checked for coherency, then write set is locked, and then the read-set is rechecked. Obviously aborts that happen before locking are preferable.

We now describe how TL2C, executes in commit mode a sequential code fragment that was placed within a transaction. The following sequence of operations is performed by a writing transaction, one that performs writes to the shared memory. We will assume that a transaction is a writing transaction. If it is a read-only transaction this can be annotated by the programmer, determined at compile time, or heuristically inferred at runtime.

1. Run through a speculative execution in a TL2 manner collecting read and write sets. A load instruction sampling the associated lock is inserted before each original load, which is then followed by post-validation code which is different than in the original TL2 algorithm. If the lock is free, a TL2C *check* operation is performed. It reads the location's time-stamp from the lock, and extracts the ID of the thread j that wrote it. If the location's time-stamp is higher than the current time-stamp stored in its *CArray* for the thread j, it updates entry j and aborts the transaction. If it is less than or equal to the stored value for j, the state is consistent and the speculative execution continues.

2. Lock the write set: Acquire the locks in any convenient order using bounded spinning to avoid indefinite deadlock. In case not all of these locks are successfully acquired, the transaction fails.

3. Re-validate the read-set. For each location in the read-set, first check it was not locked by another other thread. It might have been locked by the local thread if it is a part of both the read and write sets. Then complete the TL2C *check* for the location, making sure that its time stamp is less than the associated thread j's entry in the *CArray*. In case the *check* fails, the jth entry of the *CArray* transaction is aborted. By re-validating the read-set, we guarantee that its memory locations have not been modified while steps 1,2 and 3 were being executed.

4. Increment the local *TLClock*.

5. Commit and release the locks. For each location in the write-set, store to the location the new value from the write-set and *update* the time-stamp in the location's lock to the value of the *TLClock* before releasing it. This is done using a simple store.

Note that the updating of the time-stamps in the write-locks requires only a simple store operation. Also, notice that the local *TLClock* is only updated once it has been determined that the transaction will successfully commit.

The key idea of the above algorithm is the maintaining of a consistent read-set by maintaining a local view of each thread's latest time-stamp, and aborting the transaction every time a new time-stamp value is detected for a given thread. This prevents any concurrent modifications of the locations in the read set since a thread's past time-stamp was determined in an older transaction, so if the change occurs within the new transaction the new time stamp will be detected as new. This allows detection to proceed on a completely local basis. It does however introduce false aborts, aborts by threads that completed their transaction long before the current one, but will cause it to fail since the time-stamp recorded for them in the *CArray* was not current enough.

We view the above TL2C as a proof of concept, and are currently testing various schemes to improve its performance even on today's machines by reducing its abort rate.

3   Proof of the TL2C Algorithm

We outline the correctness argument for the TLC algorithm in the context of the TL2C algorithm. Since the TLC scheme is by construction wait-free, we only to argue safety. The proof of safety amounts to a simple argument that transactions always operate on a consistent state.

We will assume correctness of the basic underlaying TL2 algorithm as proven in [7]. In our proof argument, we refer to the TL2C algorithm's steps as they were defined in Section 2. Given the assumption that TL2 operates correctly, we need only prove that in both steps (1) and (3) in which the algorithm collects a read-set using TLC, this set forms a coherent snapshot of the memory, one that can be linearized at the start (first read) of the given collection phase.

We recall that every TL2 transaction that writes to at least one variable, can be serialized at the point in which it acquired all the locks on the locations it is about to write. Consider any collection phase (of read and write sets), including reads and writes by memory by a transaction of thread i in either step (1) or (3). For every location read by i, let the transaction that wrote to it last before it was read by i be one performed by a thread j. If j's transaction was not serialized before the start of the current collection, then we claim the collection will fail and the transaction by i will be aborted. The reason for this is simple. The last value stored in the *CArray* of i for j was read in a prior transaction of i, one that must have completed before i started the current transaction. Thus, since j increments its *TLClock* before starting its new transaction, if j's transaction was not linearized before i, then the value it wrote was at least one greater than the one recorded for j in the *CArray* of i. Thus i will detect an inconsistent view and abort its current transaction.

4   Empirical Performance Evaluation

The type of large scale NUMA multicore machine on which we believe one will benefit from the TLC approach is still several years ahead. We will therefore present two sets of benchmarks to allow the reader to gauge the limitations of the TLC approach on today's architectures and its potential benefits on future ones.

- The first benchmark is a performance comparison of the TL2C algorithm to the original TL2 algorithm with a version GV5 global clock (see [7] for details, the key idea of GV5 is to avoid frequent increments of the shared clock by limiting these accesses to aborted transactions.) on a 32-way Sun UltraSPARC T1™ machine. This is a present day single chip multi-core machine based on the Niagara architecture that has 8 cores, each supporting 4 multiplexed hardware threads.

  Our benchmark is the standard concurrent red-black tree algorithm, written by Dave Dice, taken from the official TL2 release. It was in turn derived from the `java.util.TreeMap` implementation found in the Java 6.0 JDK. That implementation was written by Doug Lea and Josh Bloch. In turn, parts of the Java TreeMap were derived from the Cormen et al [18].

- The second benchmark is a performance comparison of the TL2C algorithm to the TL2 algorithm on a Sun E25K™ system, an older generation NUMA multiprocessor machine with 144 nodes arranged in clusters of 2, each of which sits within a cluster of 4 (so there are 8 cores per cluster, and 18 clusters in the machine), all of which are connected via a large and relatively slow switch.

  Our benchmark is a synthetic work-distribution benchmark in the style of Shavit and Touitou [19]. The benchmark picks random locations to modify, in our case 4 per transaction, and has overwhelming fraction of operations within the cluster and a minute fraction outside it. This is intended to mimic

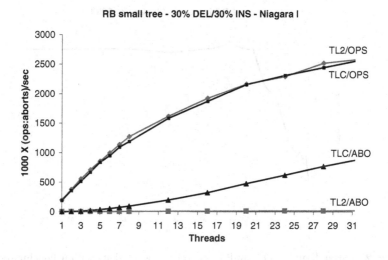

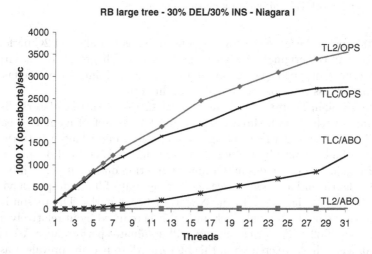

Fig. 1. Throughput of TL2 and TL2C on a Red-Black Tree with 30% puts, 30% deletes. The figure shows the throughput and the abort rate of each algorithm.

the behavior of future NUMA multicore algorithms that will make use of locality but will nevertheless have some small fraction of global coordination.

The graph of an execution of small (1000 nodes) and large red-black trees (1 million nodes) appears in Figure 1. The operation distribution was 30% puts, 30% deletes, and 40% gets. To show that the dominant performance factor in terms of TLC is the abort rate, we plot it on the same graph.

As can be seen, in both cases the smaller overhead of the TLC mechanism in the TL2C algorithm is shadowed by the increased abort rate. On the smaller tree the algorithms perform about the same, yet on the larger one the price of the

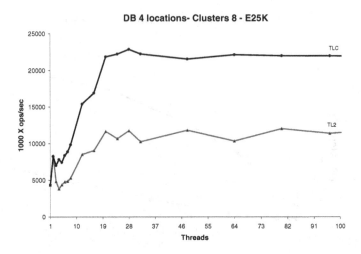

Fig. 2. Throughput of TL2 and TL2C on the work distribution benchmark in which most of the work is local within a cluster of 8 nodes

aborts is larger because the transactions are longer, and so TL2C performs more poorly than the original TL2 with a global clock. This result is not surprising as the overhead of the GV5 clock mechanism is very minimal given the fast uniform memory access rates of the Niagara I architecture.

The graph in Figure 2 shows the performance of the artificial work-distribution benchmark where each thread picks a random subset of memory locations out of 2000 to read and write during a transaction, mimicking a pattern of access that has high locality by having an overwhelming fraction of operations happen within a cluster of 8 nodes and a minute fraction outside it. As can be seen, the TL2C algorithm has about twice the throughput of TL2, despite having a high abort rate (not shown) as in the Niagara I benchmarks. The reason is that the cost of accessing the global clock, even if it is reduced by in relatively infrequent accesses in TL2's GV5 clock scheme, still dominates performance. We expect the phenomena which we created in this benchmark to become prevalent as machine size increases. Algorithms, even if they are distributed across a machine, will have higher locality, and the price of accessing the global clock will become a dominant performance bottleneck.

5   Conclusion

We presented a novel decentralized local clock based implementation of the coherence scheme used in the TL2 STM. The scheme is simple, and can greatly reduce the overheads of accessing a shared location. It did however significantly increase the abort rate in the microbenhmarks we tested. Variations of the algorithm that we tried, for example, having threads give other threads hints, proved too expensive given the simplicity of the basic TLC mechanism: they reduced the abort rate but increased the overhead. The hope is that in the future, on larger distributed machines, the cost of the higher abort rate will be offset by the reduction in the cost that would have been incurred by using a shared global clock.

References

1. Damron, P., Fedorova, A., Lev, Y., Luchangco, V., Moir, M., Nussbaum, D.: Hybrid transactional memory. In: ASPLOS-XII: Proceedings of the 12th international conference on Architectural support for programming languages and operating systems, pp. 336–346. ACM, New York (2006)
2. Dice, D., Shavit, N.: What really makes transactions fast? In: TRANSACT ACM Workshop (to appear, 2006)
3. Ennals, R.: Software transactional memory should not be obstruction-free (2005), http://www.cambridge.intel-research.net/~rennals/notlockfree.pdf
4. Harris, T., Fraser, K.: Language support for lightweight transactions. SIGPLAN Not. 38(11), 388–402 (2003)
5. Saha, B., Adl-Tabatabai, A.R., Hudson, R.L., Minh, C.C., Hertzberg, B.: Mcrt-stm: a high performance software transactional memory system for a multi-core runtime. In: PPoPP 2006. Proceedings of the eleventh ACM SIGPLAN symposium on Principles and practice of parallel programming, pp. 187–197. ACM, New York (2006)
6. Afek, Y., Attiya, H., Dolev, D., Gafni, E., Merritt, M., Shavit, N.: Atomic snapshots of shared memory. J. ACM 40(4), 873–890 (1993)
7. Dice, D., Shalev, O., Shavit, N.: Transactional locking ii. In: Dolev, S. (ed.) DISC 2006. LNCS, vol. 4167, pp. 194–208. Springer, Heidelberg (2006)
8. Riegel, T., Fetzer, C., Felber, P.: Snapshot Isolation for Software Transactional Memory. In: Proceedings of the First ACM SIGPLAN Workshop on Languages, Compilers, and Hardware Support for Transactional Computing (2006)
9. Minh, C.C., Trautmann, M., Chung, J., McDonald, A., Bronson, N., Casper, J., Kozyrakis, C., Olukotun, K.: An effective hybrid transactional memory system with strong isolation guarantees. In: ISCA 2007. Proceedings of the 34th annual international symposium on Computer architecture, pp. 69–80. ACM, New York (2007)
10. Wang, C., Chen, W.-Y., Wu, Y., Saha, B., Adl-Tabatabai, A.-R.: Code generation and optimization for transactional memory constructs in an unmanaged language. In: CGO 2007. Proceedings of the International Symposium on Code Generation and Optimization, Washington, DC, USA, pp. 34–48. IEEE Computer Society, Los Alamitos (2007)
11. Guerraoui, R., Kapalka, M.: On the correctness of transactional memory (2007)
12. Sun Microsystems, Advanced Micro Devices: Tokyo institute of technology (tokyo tech) suprecomputer (2005)
13. Systems, A.: Azul 7240 and 7280 systems (2007)
14. Dice, D., Moir, M., Lev, Y.: Personal communication (2007)
15. Felber, P.: Personal communication (2007)

16. Riegel, T., Fetzer, C., Felber, P.: Time-based transactional memory with scalable time bases. In: 19th ACM Symposium on Parallelism in Algorithms and Architectures (SPAA) (2007)
17. Felber, P., Fetzer, C., Riegel, T.: Dynamic Performance Tuning of Word-Based Software Transactional Memory. In: Proceedings of the 13th ACM SIGPLAN Symposium on Principles and Practice of Parallel Programming (PPoPP) (2008)
18. Cormen, T.H., Leiserson, C.E., Rivest, R.L.: Introduction to Algorithms. MIT Press, Cambridge (1990); COR th 01:1 1.Ex
19. Shavit, N., Touitou, D.: Software transactional memory. Distributed Computing 10(2), 99–116 (1997)

Equal-Area Locus-Based Convex Polygon Decomposition

David Adjiashvili and David Peleg*

Department of Computer Science and Applied Mathematics,
The Weizmann Institute of Science, Rehovot, Israel 76100
david.peleg@weizmann.ac.il

Abstract. This paper presents an algorithm for convex polygon decomposition around a given set of locations. Given an n-vertex convex polygon P and a set X of k points positioned arbitrarily inside P, the task is to divide P into k equal area convex parts, each containing exactly one point of X. The problem is motivated by a terrain covering task for a swarm of autonomous mobile robots. The algorithm runs in time $O(kn + k^2 \log k)$.

1 Introduction

Motivation. Consider a swarm X of k mobile autonomous robots (cf. [3,5,13,19]) assigned the task of exploring an unknown region (cf. [15]), represented as a convex polygon P. Designing an algorithm for performing this task calls for developing ways of dividing the task among the robots, by partitioning the region into subregions and assigning each robot to a subregion. The problem of subdividing a given polygon in the plane has been studied extensively, and several variants of it have been examined. In our context, it may be convenient for load-balancing purposes that the resulting subregions be of equal size. Moreover, for efficiency purposes, it may be convenient that the resulting subregions be convex. How can such a subdivision be achieved?

This problem is straightforward if the parts are allowed to differ in size. Even the problem of partitioning P into k *equal area* parts around the points of X is easy if the parts are allowed to be non-convex. The task becomes harder, however, if it is required also that the parts be convex.

This leads to the following geometric problem. The input is a k-*configuration* $\langle P, X \rangle$ consisting of a *convex polygon* P with n vertices and a set of k points $X = \{x_1, ..., x_k\}$ in P. Let S be the area of the polygon P, and let $\sigma = S/k$. The task is to find a subdivision of $\langle P, X \rangle$ into k equal-area *1-configurations* $\langle P_i, \{x_i\} \rangle$, referred to also as *atomic configurations*, namely, k equal-area convex subregions $\{P_1, ..., P_k\}$ around the k points. Formally, for each $i = 1, ..., k$, the area of P_i is *exactly* σ, $x_i \in P_i$, and P_i is *convex*.

---

* Supported in part by a grant from the Israel Science Foundation.

A. Shvartsman and P. Felber (Eds.): SIROCCO 2008, LNCS 5058, pp. 141–155, 2008.
© Springer-Verlag Berlin Heidelberg 2008

We call this kind of subdivision, depicted in the figure, a *locus-based convex subdivision* of $\langle P, X \rangle$.

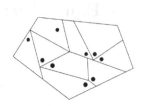

In general, the requirement that the parts be convex cannot be achieved in case P is a concave polygon. This is clearly evident for $k = 1$, though examples can be constructed for any $k \geq 1$, of a polygon that cannot be divided into k or fewer convex subregions.

In fact, a-priori it is not clear that a locus-based convex subdivision exists for every k-configuration $\langle P, X \rangle$ even when P is convex. In particular, its existence can be derived from the *ham sandwich* theorem [11,18] for k values that are powers of 2 (i.e., $k = 2^s$ for integral s) by recursive ham sandwich cuts, but not for other k values. Moreover, a closely related area concerns *equitable subdivisions* of the plane. The main existential result proved here (albeit not its algorithmic aspects) can also be derived from the *Equitable Subdivision Theorem* [9,17], which states that for integers $r \geq 1$, $b \geq 1$ and $k \geq 2$, if the set R contains rk red points and the set B contains bk blue points, then there exists a subdivision $X_1 \cup X_2 \ldots \cup X_k$ of the plane into k disjoint convex polygons such that every X_i contains exactly r red points and b blue points. Thus our main contribution is in presenting a polynomial time algorithm for the problem.

Our results. In this paper we prove that any configuration $\langle P, X \rangle$ enjoys a locus-based convex subdivision; more importantly, we present a $O(kn + k^2 \log k)$ time algorithm for computing such a subdivision.

We do not concentrate our discussion on specific families of configurations, although it is evident that the subdivision can be found more efficiently in certain simple point configurations. In particular, one can obtain a convex subdivision more efficiently if all points lie on a single line.

The solution we provide is recursive, namely, in each stage a large polygon is divided into a number of smaller *convex* parts, which are *balanced*, in the sense that the ratio between the number of points and the area in each part is the same. Each part is further divided recursively. More formally, a μ-*split* of the k-configuration $\langle P, X \rangle$ for $X = \{x_1, ..., x_k\}$ and integer μ is a subdivision of $\langle P, X \rangle$ into $\mu \geq 2$ smaller configurations $\langle P_j, X_j \rangle$ for $j = 1, .., \mu$, namely, a subdivision of P into μ convex subregions $P_1, .., P_\mu$ and a corresponding partition of X into $X_1, .., X_\mu$ of cardinalities $k_1, ..., k_\mu$ respectively, such that for every $1 \leq j \leq \mu$,

1. the area of P_j is $k_j \cdot \sigma$, and
2. the points of X_j are in P_j.

Our algorithm is based on recursively constructing a μ-split, for $\mu \geq 2$, for each obtained configuration, until reaching 1-configurations. The method consists of two main parts. The first component tries to achieve a 2-split, i.e., a division of P into two convex balanced regions. If this fails, then the second component applies a technique for achieving a 3-split or a 4-split, i.e., dividing P into three or four convex balanced regions simultaneously.

We remark that our algorithm can be extended to handle the 3-dimensional version of the problem as well. The resulting algorithm and its analysis are deferred to the full paper. Another extention of our algorithm which is not described here is a decomposition for a general continuous distribution over P.

The following section is devoted to basic definitions, facts and techniques utilized by the algorithm. This section also describes the first component of the algorithm, which attempts to compute a 2-split. Section 3 deals with configurations that do not admit a 2-split by the first component of our algorithm. In this case, the solution proposed computes a μ-split for $\mu \leq 4$. Section 4 summarizes the algorithm and provides a complexity analysis.

Related work. Very recently, we learned that a result similar to ours was obtained independently (and at about the same time) in [12], using similar methods and with a bound of $O(k(k + n) \log(k + n))$ on the time complexity. A number of related problems were studied in the field of computational geometry. In particular, algorithms for polygon decomposition under different constraints were developed (cf. [14]). The problem of decomposing a simple polygon into k equal-area subregions with the constraint that each subregion has to contain one of k distinguished points on its boundary was studied in [16]. Here, we solve a similar problem in a convex polygon, with the additional constraint that every subregion has to be convex as well. Another algorithm for a different polygon decomposition problem, where a simple polygon with k sites on its boundary has to be decomposed into k equal-area subregions, each containing a single site on its boundary, is presented in [7]. An approximation algorithm for supplying a polygonal demand region by a set of stationary facilities in a load-balanced way is presented in [4]. The cost function discussed is based on the distances between the facilities and their designated regions. Problems related to equitable subdivisions have also been studied in [1,2,6,8,10].

2 Preliminaries

2.1 Basic Definitions

Let us start with common notation and some definitions to be used later on. For a point x_0 inside P and another point x_1, denote the directed ray originating at x_0 and passing through x_1 by $\overrightarrow{x_0, x_1}$. We use directed rays instead of lines in order to clearly distinguish one side as the *left* side of the line (namely, the side to our left when standing at x_0 and facing x_1) and the other as the *right* side.

Let x be a point inside P and y_1, y_2 be two points on the boundary of P. Let $\rho_1 = \overrightarrow{x, y_1}$ and $\rho_2 = \overrightarrow{x, y_2}$. The *slice* defined by ρ_1 and ρ_2, denoted by $\mathsf{Slice}(\rho_1, \rho_2)$, is the portion of P enclosed between ρ_1, ρ_2 and the boundary of P, *counterclockwise* to ρ_1.

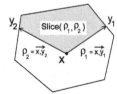

Let P' be a subregion of P and $\rho = \overrightarrow{x, y}$ be a ray inside P. We use the following notation.

- Area(P') denotes the area inside P'.
- Points(P') denotes the number of points inside P' *including on the boundary*.
- $\overline{\text{Points}}(P')$ denotes the number of points *strictly* inside P'.
- Points(ρ) denotes the number of points on ρ (between x and y).
- Excess(P') = Area(P') $- \sigma \cdot$ Points(P').
- $\overline{\text{Excess}}(P')$ = Area(P') $- \sigma \cdot \overline{\text{Points}}(P')$.

We say the region P' is *balanced* if Excess(P') = 0, *dense* if Excess(P') < 0 and *sparse* if Excess(P') > 0.

2.2   Dividing Lines

This subsection explains the notion of a *dividing* line and establishes a couple of its properties.

The following lemma and corollary establish the fact that a 2-configuration can always be 2-split using a single separating straight line. This provides the base of our inductive proof for the existence of a locus-based convex subdivision for any k-configuration.

Lemma 1. *For a given polygon P and $X = \{x_1, x_2\}$, there exists a straight line ℓ that divides P into two equal-area subregions, with one point in each subregion.*

Proof. Let $s = \overline{x_1, x_2}$ be the line segment that connects the two points, and let p be a point on it. Let $\bar{\ell}$ be the horizontal line that goes through p. Define the function $f : [0, \pi] \to \mathbb{R}$ as

$$f(\theta) = \text{area of } P \text{ under } \bar{\ell} \text{ after rotating } P \text{ by an angle of } \theta \text{ around } p.$$

Clearly, f is a continuous function. Furthermore, $f(\pi) = S - f(0)$. If $f(0) = S/2$ then we are done with $\ell = \bar{\ell}$. Now assume without loss of generality that $f(0) < S/2$. This means that $f(\pi) > S/2$. Since f is continuous, by the Mean Value theorem, there exists an angle $\hat{\theta} \in [0, \pi]$ such that $f(\hat{\theta}) = S/2$. Due to the choice of p, we are guaranteed that the points x_1 and x_2 are not on the same side of the appropriate line. (See Figure 1). ∎

Corollary 1. *Every 2-configuration has a locus-based convex subdivision.*

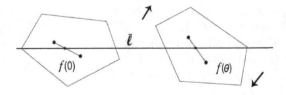

Fig. 1. Rotating P by θ degrees, in proof of Lemma 1

We proceed with the following basic definition.

Definition 1. *A line l that intersects P is called a* dividing line *if it does not go through any point of X, and the two resulting parts of P are both balanced.*

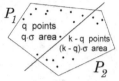

A dividing line clearly yields a 2-split of the configuration. Therefore, finding a dividing line is a natural approach for establishing a recursive solution to the problem. Our algorithm always attempts first to find a dividing line that will satisfy the inductive step. Unfortunately, the existence of a dividing line is not always guaranteed, as illustrated by the following example.

Example: Consider a 3-configuration $\langle P, X \rangle$ where P is a square of area S and the three points $X = \{x_1, x_2, x_3\}$ are positioned densely around the center of the square. A dividing line should thus separate one point from the other two, so the areas of the respective parts should be σ and 2σ for $\sigma = S/3$.

The location of the points dictates, however, that any dividing line will necessarily pass close to the square center. The area on both sides of the dividing line can be made arbitrarily close to $1.5\sigma = S/2$ by positioning the points arbitrarily close to the center.

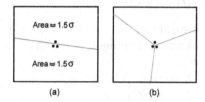

It follows that a dividing line does not exist in this case (Figure (a)). Hence such configurations require a different approach, such as dividing the polygon into convex subregions as in Figure (b).

2.3   Vertical Scans and Hull Scans

As established in this subsection, certain configurations $\langle P, X \rangle$ are actually guaranteed to have a dividing line. We consider a vertical scan procedure. Intuitively, a scan of a planar region is the process of sweeping a straight line over the region in a continuous fashion, scanning the region as it progresses. Formally, one can define a *scan* of a planar region as a pair $\langle L, S \rangle$ such that L is a set of straight lines in the plane and $S : [a, b] \rightarrow L$ is a *continuous progress function*, which associates with each $t \in [a, b]$ a unique line $S(t) \in L$. We also associate with each t some measured quantity $\varphi(t)$, given by the scan function $\varphi : [a, b] \rightarrow \mathbb{R}$. In practice, scans are implemented as discrete processes, sampling only a specific predefined finite set of *event points* corresponding to line positions along the sweep process.

In this section we define two types of scans of the k-configuration $\langle P, X \rangle$, namely, the *vertical scan* and the *hull scan*. See Figure 2. A *polar scan* around a fixed point x is defined in the next section.

Vertical Scan: Let L be the set of vertical lines (lines of the form $x = \hat{x}$), and let $Left(\hat{x})$ (respectively, $Right(\hat{x})$) denote the region inside P to the left (respectively, right) of the vertical line $x = \hat{x}$. Let $\hat{x}_l$ and $\hat{x}_r$ be the x-coordinates

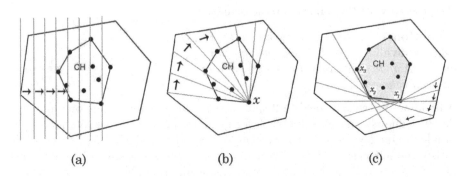

Fig. 2. (a) Vertical scan. (b) Polar scan around x. (c) Hull scan around x_1 and x_2.

of the leftmost and the rightmost points inside P, respectively. During the scan, for any coordinate t in the range $[\hat{x}_l, \hat{x}_r]$, the scan function φ returns the value $\varphi(t) = \mathsf{Excess}(Left(t))$.

Lemma 2. *Let* $X = \{x_1, ..., x_k\}$ *be ordered by increasing x-coordinates, and assume a dividing line does not exist for* $\langle P, X \rangle$. *Then one of the following must be true:*

1. $\mathsf{Area}(Left(x_1)) > \sigma$.
2. $\mathsf{Area}(Right(x_k)) > \sigma$.

Proof. Let $\bar{x}_1, ..., \bar{x}_k$ be the x-coordinates of $x_1, ..., x_k$, respectively. Assume, toward contradiction, that both statements are false, i.e., $\mathsf{Area}(Left(x_1)) < \sigma$ and $\mathsf{Area}(Right(x_k)) < \sigma$ (as clearly, equality yields a dividing line). We claim that in this case there exists some $\bar{x} \in [\bar{x}_1, \bar{x}_k]$ such that the vertical line $x = \bar{x}$ is a dividing line. To see this, note that φ is continuous and monotonically increasing in every open interval $(\bar{x}_i, \bar{x}_{i+1}^-)$ for $i = 1, .., k-1$. Furthermore, if statement 1 is false, then $\varphi(\bar{x}_1 + \epsilon) < 0$ for sufficiently small ϵ, and if statement 2 is false then $\varphi(\bar{x}_k) > 0$. Notice that the jumps in the function φ (which occur only at coordinates $\bar{x}_i$) are always to a lower value. Therefore, there exists an index i such that $\varphi(\bar{x}_i) < 0$ and $\varphi(x_{i+1}^-) > 0$. Therefore, by the Mean Value theorem, there exists some $\bar{x} \in (\bar{x}_i, x_{i+1}^-)$ such that $\varphi(\bar{x}) = 0$, implying that $x = \bar{x}$ is a dividing line; a contradiction. ∎

Corollary 2. *If* $\mathsf{Area}(Left(x_1)) \leq \sigma$ *and* $\mathsf{Area}(Right(x_k)) \leq \sigma$, *then a dividing line (and hence a 2-split) exists for the configuration* $\langle P, X \rangle$.

Denote by $x(l)$ the x-coordinate of the vertical line l. By the definition of a dividing line, if the line l partitions P into two regions, $Left(x(l))$ and $P \setminus Left(x(l))$, then $\varphi(l) = \mathsf{Excess}(Left(x(l))) = 0$ means that l is a dividing line. Plotting the function φ as the line l moves from left to right, we observe that it is composed of continuous monotonically increasing segments and jumps to a lower value. The jumps occur whenever the scan line hits a point of X. A dividing line l satisfies $\varphi(l) = \mathsf{Excess}(Left(x(l))) = 0$ if l does not contain a point. A dividing line can also occur when l reaches a point. In this case either Excess($Left(x(l))$) = 0 or $\overline{\text{Excess}}(Left(x(l)))$ = 0 holds. This means that a crossing of the *zero* value can either happen in the continuous part, creating a dividing line, or on a point where φ jumps from a positive value to a negative one, which means the scan did not cross a dividing line. If on jump the φ jumps to $\varphi(l) = -1$, then l is a dividing line as well. We elaborate on handling the latter case in Subsection 2.4.

This result can actually be generalized to any continuous scanning procedure of the (not necessarily convex) region. (The next section describes a *polar scan* of the polygon P.)

We next describe a different kind of scanning procedure, that involves the *convex hull* of the set of points inside P.

A *tangent* to a convex region C is a straight line that intersects C in exactly one point. An *edge line* of a convex C is a line that contains an edge of C.

Let $CH(X)$ denote the convex hull of the set $X = \{x_1, ...x_k\}$. Assume without loss of generality that $x_1, .., x_s$ are the points on $CH(X)$ in clockwise order, and let l be a tangent to $CH(X)$ that intersects it at x_1. The polygon P is divided by l into two regions, one of which contains the entire $CH(X)$. Let $Out(l)$ be the other region.

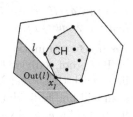

We consider the following scanning procedure.

Hull Scan: Start rotating l around x_1 in the counterclockwise direction until it reaches x_2. Now continue rotating the line around x_2 until it reaches x_3, and so on. The scan ends when it completes a wrap around the convex hull (i.e., once it reaches the original tangent l). We refer to the (infinitely many) lines obtained along this process as the *scan lines* of X, and denote the set of scan lines by $SL(X)$. Note that $SL(X)$ consists of precisely the tangents and edge lines of $CH(X)$.

For convenience, we assume that initially the configuration is rotated so that the original tangent l is aligned with the x-axis, and every other line reached during the process is represented by the angle it forms with l. Due to the cyclic nature of the process and the fact that we end up with l, the entire process can be represented by the progress function $S : [0, 2\pi] \to SL(X)$, with

$$S(\theta) = \text{the scan line that forms an angle of } \theta \text{ with } l$$

and taking the scan function to be

$$\varphi(\theta) = \text{Area}(\text{Out}(S(\theta))).$$

Note that $S(\theta)$ is a tangent to some point for every $\theta \in [0, \pi]$, except for the passages from one point to the other, where $S(\theta)$ is an edge line. Furthermore, $\overline{\text{Points}}(\text{Out}(S(\theta))) = 0$ for every $\theta \in [0, 2\pi]$.

Most importantly, the scanning procedure is again continuous, in the sense that Area($\text{Out}(S(\theta))$) is a continuous function. This implies the following straightforward lemma.

Lemma 3. *If* $\mathsf{Area}(\mathsf{Out}(S(\alpha_1))) \leq \sigma$ *for some* $\alpha_1 \in [0, 2\pi]$ *and* $\mathsf{Area}(\mathsf{Out}(S(\alpha_2)))$ $\geq \sigma$ *for some* $\alpha_2 \in [0, 2\pi]$, *then the configuration has a dividing line.*

Proof. Assume without loss of generality that $\alpha_1 \leq \alpha_2$. Since $\mathsf{Area}(\mathsf{Out}(S(\theta)))$ is a continuous function, according to the Mean Value Theorem there exists some $\beta \in [\alpha_1, \alpha_2]$ such that $\mathsf{Area}(\mathsf{Out}(S(\beta))) = \sigma$. Since $\overline{\mathsf{Points}}(\mathsf{Out}(S(\theta))) = 0$ and the line $S(\beta)$ contains at least one point, we can assign $\mathsf{Out}(S(\beta))$ to this point. Both new regions are convex, since they were obtained by dividing a convex region with a straight line. We conclude that $S(\beta)$ is a dividing line. ∎

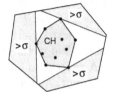

Combining Corollary 2 and Lemma 3 leads to the following characterization.

A configuration $\langle P, X \rangle$ is called *compact* iff $\mathsf{Area}(\mathsf{Out}(l)) > \sigma$ for every tangent or edge line l of $CH(X)$.

Lemma 4. *If* $\langle P, X \rangle$ *is a non-compact configuration then a dividing line (and hence a 2-split) exists.*

Proof. By assumption, there is a scan line l of $CH(X)$ such that $\mathsf{Area}(\mathsf{Out}(l)) \leq \sigma$. The proof is divided into two cases. First assume that there also exists a scan line $\hat{l}$ such that $\mathsf{Area}(\mathsf{Out}(\hat{l})) \geq \sigma$. In this case, Lemma 3 guarantees that a hull scan will find a dividing line. Now suppose that $\mathsf{Area}(\mathsf{Out}(\hat{l})) \leq \sigma$ for every scan line $\hat{l}$. In this case, a vertical scan starting from l is guaranteed to find a dividing line, by Corollary 2. ∎

The more complex case is, therefore, that of a compact configuration, where the convex hull of the set of points is *'too far'* from the boundary of the polygon P. This is exactly the case of the example from Subsection 2.2, in which no dividing line exists. In this case, a recursive solution should try to achieve a μ-split for $\mu > 2$, i.e., divide the problem into *three or more* smaller subproblems.

2.4 The Function 'Excess' and Semi-dividing Lines

In this subsection we focus on the function Excess and its properties on compact configurations. As shown in the previous subsection, one cannot always expect to find a dividing line. Nevertheless, we would like to follow a vertical scan procedure on P from left to right, as defined in Subsection 2.3. Since the configuration is compact, φ will reach a value higher than σ before reaching x_1, the first point of X. In addition, just after passing the last point, x_n, φ will reach a value smaller than $-\sigma$.

This means that during the scan, φ must cross *zero* somewhere in between. If the crossing occurs in a continuous part of φ, then it yields a dividing line. The problematic case is when the crossing occurs on a "jump", namely, the crossing line $\hat{l}$ goes through some point $x_i \in X$ and satisfies $0 < \varphi(\hat{x}(l)) = \mathsf{Excess}(Left(\hat{x}(l))) < \sigma$. Such lines, called *semi-dividing lines*, will be useful in a later stage of the algorithm.

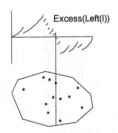

Example: Let us return to our example from Subsection 2.2, consisting of a square and three points positioned closely around the center of the square. In this case, any line that passes between the points is a semi-dividing line. Consider one such line. Let P_1 be the side with one point, and P_2 the side with the other two points. Since all points are so close to the middle, we have that $\mathsf{Area}(P_1) \approx \mathsf{Area}(P_2) \approx 1.5\sigma$. This means that $\mathsf{Excess}(P_1) \approx -0.5\sigma$ and $\mathsf{Excess}(P_2) \approx 0.5\sigma$. Therefore, this is a semi-dividing line.

Note that we may want to consider polar scan procedures (where the scan line rotates around some point, instead of moving in some fixed direction) as well as vertical scans. The same definitions apply to both kinds.

The following sections deal solely with compact configurations, and describe a way to achieve a μ-split of the configuration for $\mu \geq 2$.

3   A Solution for Compact Configurations

3.1   Solution Strategy

In this section we describe an algorithmic way to handle the inductive step in the case of a compact configuration. The first step is to choose a vertex x of the convex hull $CH(X)$. For concreteness, choose the one with the lowest y-coordinate. Let E_l and E_r be the two edges of $CH(X)$ incident to x. Assume without loss of generality that E_l is to the left of x and E_r is to its right. The straight lines containing E_l and E_r intersect the boundary of P at four points, p_1, p_2, p_3, p_4, in polar order on P.

Define three sectors around x:
LeftSlice = Slice$(\overrightarrow{x,p_1}, \overrightarrow{x,p_2})$,
MidSlice = Slice$(\overrightarrow{x,p_2}, \overrightarrow{x,p_3})$,
RightSlice = Slice$(\overrightarrow{x,p_3}, \overrightarrow{x,p_4})$.

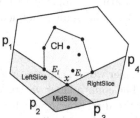

The following fact is quite straightforward.

Fact 1. $\overline{\mathsf{Points}}(\mathsf{LeftSlice}) = \overline{\mathsf{Points}}(\mathsf{MidSlice}) = \overline{\mathsf{Points}}(\mathsf{RightSlice}) = 0$.

Furthermore, since the configuration is compact, the following must hold.

Fact 2

(1) $\mathsf{Area}(\mathsf{LeftSlice}) + \mathsf{Area}(\mathsf{MidSlice}) > \sigma$,
(2) $\mathsf{Area}(\mathsf{MidSlice}) + \mathsf{Area}(\mathsf{RightSlice}) > \sigma$.

However, we cannot assume anything about the individual areas of LeftSlice, MidSlice and RightSlice. The actual amount of area in each individual part affects the way the polygon is divided and the inductive step is completed.

We now classify the space of possible compact configurations into eight cases, (s_1, s_2, s_3), where $s_1 = $ '+' if Area(LeftSlice) $\geq \sigma$ and $s_1 = $ '-' otherwise, and similarly for s_2 and s_3 with respect to MidSlice and RightSlice respectively. For example, $(+, +, -)$ represents the case where Area(LeftSlice) $\geq \sigma$, Area(MidSlice) $\geq \sigma$ and Area(RightSlice) $< \sigma$. Noticing the symmetry between LeftSlice and Right-Slice, it follows that the case $(+, +, -)$ is equivalent to $(-, +, +)$, and $(+, -, -)$ is equivalent to $(-, -, +)$. Therefore, we actually have only six different scenarios to analyze. The major part of our analysis involves showing how a μ-split can be obtained for these six cases. This is achieved in the following subsections, each of which describes the construction of a μ-split for some subset of the six cases. As a result we get the following.

Lemma 5. *Every compact configuration has a μ-split for $\mu \geq 2$.*

Combining Lemmas 4 and 5 yields our main theorem.

Theorem 3. *Every configuration has a locus-based convex subdivision, constructible in polynomial time.*

3.2   Polar Scans

Before starting with the analysis of the different cases, we describe another necessary tool, namely, the polar scan.

Clockwise Polar Scan around the point x: Let L be the collection of all straight lines that intersect the point x. Fix $l_0 \in L$ and let l_θ be the line obtained by rotating l_0 clockwise around x by θ degrees. The progress function $S : [0, \theta_e] \to \mathbb{R}$ of a polar scan is defined for some $\theta_e > 0$ as

$$S(\theta) = l_\theta.$$

In other words, the scan rotates the line l_0 clockwise until it reaches the line l_{θ_e}.

The counterclockwise polar scan around the point x is defined similarly, except for the direction in which the line l_0 is rotated.

Next we would like to consistently name one intersection of every scan line with P as the 'Top' intersection. Let $Top(l_0)$ be the intersection of l_0 with the boundary of P that is of larger y-coordinate. If l_0 is horizontal, then $Top(l_0)$ is the intersection with the larger x-coordinate. For every other line l_θ in the scan, the definition of $Top(l_\theta)$ preserves the continuity of the path $p(\theta) = Top(l_\theta)$. In particular, when the scan line passes the horizontal line that contains x for the first time, $Top(l_\theta)$ becomes the intersection of l_θ with P's boundary, that is of *lower* y-coordinate.

We can now define $Left(l)$ (respectively, $Right(l)$) for a scan line l in a polar scan to be the part of P that is to the left (respectively, right) of l when standing at x and facing $Top(l)$. Let the scan function $\varphi : [0, \theta_e] \to \mathbb{R}$ be defined as

$$\varphi(\theta) = \text{Excess}(Left(l_\theta)).$$

The following lemma guarantees that under conditions similar to those in Corollary 2, a dividing line will be found in a polar scan around p in case p is a vertex of the polygon P. In fact, the lemma holds even for non-convex polygons, so long as they are completely visible from p (i.e., for every point $q \in P$, there is a straight line connecting p and q, that is completely contained in P). This fact will be useful in the analysis to follow.

Lemma 6. *Let $\langle S, L \rangle$ be a clockwise polar scan around a vertex p of P, which scans every point in P (every point is touched by some scan line). Let $\{x_1, ..., x_k\}$ be the order in which the points in X intersect the scan lines. Denote by l_{θ_1} and l_{θ_2} the first scan lines that intersect x_1 and x_k, respectively. Assume that*

1. $\mathsf{Area}(Left(l_{\theta_1})) < \sigma$,
2. $\mathsf{Area}(Right(l_{\theta_2})) < \sigma$.

Then a dividing line will be found in the scan.

We skip the proof of Lemma 6, since it is based on a repetition of the ideas in the proof of Lemma 2.

For convenience we enumerate the points in $X \setminus \{x\}$ in clockwise polar order around x, starting with the point that sits on the edge E_l (exactly to the left of x). Let $x_1, x_2, .., x_{k-1}$ be the enumeration. If two points appear on the same ray originating at x, then we order them according to their distance from x, the closer one first. Hereafter, we denote prefixes of the set X in this polar order by

$$X_m = \{x_1, x_2, ..., x_m\}.$$

According to this notation, S_l contains exactly the set of points X_t for some $t > 0$.

3.3   Case Analysis

Solving cases $(+, +, +)$ *and* $(+, -, +)$: In both the $(+, +, +)$ and $(+, -, +)$ cases, the left and right slices are larger than σ. This fact can be used to achieve a 3-split. Start by performing a polar scan procedure around x. The first scan line is aligned with E_l and the procedure ends before reaching E_r. Throughout the procedure, x *is not assigned to either part of P.* The effect of this is that we have an overall excess of σ in area. If the scan finds a dividing line, hence a 2-split, then we are done by Subsection 2.4.

Otherwise, the scan finds *semi-dividing line,* $\hat{l}$. Let $Left(\hat{l})$ and $Right(\hat{l})$ be the regions to the left and right of $\hat{l}$, respectively. Due to the excess area in the scan procedure, we have

Fig. (*)

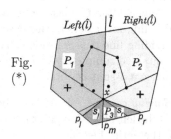

$$0 < \mathsf{Excess}(Left(\hat{l})) < \sigma,$$
$$0 < \mathsf{Excess}(Right(\hat{l})) < \sigma,$$
$$\mathsf{Excess}(Left(\hat{l})) + \mathsf{Excess}(Right(\hat{l})) = \sigma.$$

Clearly, $\hat{l}$ has to pass inside MidSlice (since it intersects $CH(X)$). Since both Area(LeftSlice) $\geq \sigma$ and Area(RightSlice) $\geq \sigma$, we can *peel off* the excess area from $Left(\hat{l})$ and $Right(\hat{l})$ by removing a slice of the appropriate size from each. The two slices, denoted S_l and S_r respectively, will form a convex region of area σ, that will be assigned to x. More formally, Let p_m be the point of intersection between $\hat{l}$ and the boundary of P inside MidSlice (see Figure (*) above). Let p_l and p_r be the points on the boundary of P defining S_l and S_r, namely, such that $S_l = \mathsf{Slice}(\overrightarrow{x,p_l}, \overrightarrow{x,p_m})$ and $S_r = \mathsf{Slice}(\overrightarrow{x,p_m}, \overrightarrow{x,p_r})$, and the two slices satisfy

$$\mathsf{Area}(S_l) = \mathsf{Excess}(Left(\hat{l})),$$
$$\mathsf{Area}(S_r) = \mathsf{Excess}(Right(\hat{l})).$$

Clearly, removing S_l and S_r from $Left(\hat{l})$ and $Right(\hat{l})$, respectively, will create three balanced regions, $P_1 = Left(\hat{l}) \setminus S_l$, $P_2 = Right(\hat{l}) \setminus S_r$ and $P_3 = S_l \cup S_r$ with $\mathsf{Area}(P_1) = q\sigma$, $\mathsf{Area}(P_2) = (k - q - 1)\sigma$, $\mathsf{Area}(P_3) = \sigma$, for some $0 < q < k - 2$. Since Area(LeftSlice) $\geq \sigma$ and Area(RightSlice) $\geq \sigma$ we are guaranteed that S_l and S_r *do not touch* $CH(X)$, except for the point x, and therefore P_3 does not contain other points. Finally, P_3 is convex since the configuration is compact. (so if $S_l \cup S_r$ was concave, then the tangent line containing $\overrightarrow{x,p_r}$ would have area less than σ on the smaller side, thus contradicting the compactness assumption.) Figure (*) above depicts the decomposition.

In conclusion, in this case we obtain a 3-split.

Solving case $(-,-,-)$: Unlike the solution in the previous section, in this case we assign a region for x as the first step. This region consists of the entire MidSlice, but since Area(MidSlice) $< \sigma$, we have to add a segment from LeftSlice.

The part to be added to MidSlice will be separated by a ray $\overrightarrow{x,\hat{p}}$, where $\hat{p}$ is a point on the boundary of P inside LeftSlice . We will not assign LeftSlice entirely, though, since Area(MidSlice) + Area(LeftSlice) $> \sigma$ (as the configuration is compact). The region assigned to x, denoted P_1, is clearly convex.

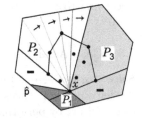

The remaining region, $P \setminus P_1$, is clearly concave, thus has to be divided as well. To do this, we perform a polar scan procedure around x. The first and last scan lines will form the borders between P_1 and $P \setminus P_1$. Since Area(RightSlice) $< \sigma$ and Area(LeftSlice) $< \sigma$, the area scanned before the first point and the area scanned after the last point are both smaller than σ. Lemma 6 yields that a dividing line exists in this case, thus it will be found by the scan. Note that any dividing line passes inside $CH(X)$. Therefore, it passes inside MidSlice, which is contained completely in P_1. In conclusion, all three resulting regions are convex, hence a 3-split is obtained.

Solving case $(-,-,+)$: Let q be the point on the vertex of $CH(X)$, immediately to the left of x. The point q occurs on the edge of LeftSlice. We start by assigning q the entire LeftSlice.

Since Area(LeftSlice) $< \sigma$, this area is too small, and
we add another slice from MidSlice, much in the same
way we added a slice of LeftSlice in the previous sec-
tion, using a ray $\overrightarrow{x,p}$. Again, we do not assign MidSlice
entirely since Area(MidSlice) + Area(LeftSlice) $> \sigma$.

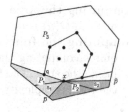

Next we assign x the remaining part of MidSlice. It is clearly too small, as
MidSlice itself is too small. Therefore, we add a slice from RightSlice, again using
a ray $\overrightarrow{x,p}$. Since Area(RightSlice) $\geq \sigma$, we do not need all of RightSlice, and
certainly do not cut a piece of $CH(X)$.

Clearly, the two regions assigned to q and x are convex. Their union, however,
is *concave*, thus, due to the shape of the union of the two regions, the remaining
part is convex. Again, we obtained a 3-split of P.

Solving cases $(-, +, +)$ *and* $(-, +, -)$: These cases are the most involved ones,
and their analysis is deferred to the full paper.

4   The Overall Algorithm

This section presents the algorithm summarizing the case analysis described in
the previous section. The algorithm has two main stages. Stage I tries to divide
the given polygon P into two convex balanced parts by means of a dividing line.
Lemma 4 gives a sufficient condition for the existence of a dividing line. Our
algorithm checks this condition in two separate procedures. It first attempts to
find a dividing line by a vertical scan of P. If this fails, then a hull scan is
performed. Failure in the second attempt implies that $\langle P, X \rangle$ is necessarily a
compact configuration.

Stage II, which is reached in case of failure in Stage I, divides the polygon
into μ balanced convex parts, $2 \leq \mu \leq 4$, following the case analysis described
in Section 3. Lemma 5 guarantees that such a decomposition can be obtained in
this case. The resulting smaller subregions are then further divided recursively.

Finally, we analyze the complexity of the algorithms. We first discuss the
complexity of some of the basic procedures utilized by the algorithm. We assume
that P is given as an ordered sequence of its vertices.

Convex Hull. The complexity of finding the convex hull of a set of k points is
known to be $O(k \log k)$.

Vertical scan. The set of event points in our scan procedures is the set of
vertices and points. These need to be sorted. We required the vertices of
P to be sorted to begin with, so sorting all event points should take only
$O(n+k \log k)$ (by sorting the k points and merging the two sorted sequences).
The scan itself takes $O(n+k)$ time, thus the total complexity is $O(n+k \log k)$.

Hull scan. Once we computed the convex hull of X, this scan can be performed
in $O(n)$ time (assuming that the convex hull computation returns the order
of the points on the convex hull).

Computing the area of a slice. This can be done using a scan of the vertices of P, in $O(n)$ time.

Computing the function Excess in a slice. The computation of Excess requires computing the area inside the slice, as well as counting the points inside the slice. The former can be done in $O(n)$ and the latter in $O(k)$, thus the total complexity is $O(n + k)$.

Denote by $T(n, k)$ the time complexity of the algorithm for a configuration $\langle P, X \rangle$ when P has n vertices and X contains k points. Assume first that stage I of the algorithm was executed (i.e., P was divided into two parts by a dividing line). In this case there are two possibilities. If the dividing line was found by the vertical scan, then

$$T(n, k) = T(\hat{n}, \hat{k}) + T(n - \hat{n} + 2, k - \hat{k}) + O(n + k \log k)$$

for some $0 < \hat{k} < k$ and $0 < \hat{n} < n$. Otherwise, the dividing line was found during the hull scan, in which case

$$T(n, k) = T(\hat{n}, k - 1) + T(n - \hat{n} + 2, 1) + O(n + k \log k)$$

for some $0 < \hat{n} < n$.

Now assume that stage II of the algorithm was executed. In this case the algorithm executes one of several procedures and divides P into μ convex areas, $2 \leq \mu \leq 4$. The most time consuming of the different procedures is the last one, with $\mu = 2$, and it satisfies

$$T(n, k) = T(n_1 + 3, k_1) + T(n_2 + 3, k_2) + O(n + k \log k)$$

for some positive n_1, n_2 and k_1, k_2 such that $n_1 + n_2 = n$ and $k_1 + k_2 = k$.

Notice that the total number of divisions performed by the algorithm is at most $k - 1$. Therefore, the total time complexity can be bounded by $O(k(\bar{N} + k \log k))$, where $\bar{N}$ is the maximal number of vertices in any polygon throughout the execution of the algorithm. Clearly $\bar{N} < 3k + n$, yielding that

$$T(n, k) = O(kn + k^2 \log k).$$

Theorem 4. *For any configuration $\langle P, X \rangle$, our algorithm constructs a locus-based convex subdivision in time $O(kn + k^2 \log k)$.*

References

1. Akiyama, J.M., Kaneko, A., Kano, M., Nakamura, G., Rivera-Campo, E., Tokunaga, S., Urrutia, J.: Radial perfect partitions of convex sets. In: Akiyama, J., Kano, M., Urabe, M. (eds.) JCDCG 1998. LNCS, vol. 1763, pp. 1–13. Springer, Heidelberg (2000)
2. Akiyama, J.M., Nakamura, G., Rivera-Campo, E., Urrutia, J.: Perfect division of a cake. In: Proc. 10th Canadian Conf. on Computational Geometry, pp. 114–115 (1998)

3. Arkin, R.C.: Behavior-Based Robotics. MIT Press, Cambridge (1998)
4. Aronov, B., Carmi, P., Katz, M.J.: Minimum-cost load-balancing partitions. In: Proc. 22nd ACM Symp. on Computational geometry (SoCG), pp. 301–308 (2006)
5. Balch, T., Parker, L.E. (eds.): Robot Teams: From Diversity to Polymorphism. A.K. Peters (2001)
6. Barany, I., Matousek, J.: Simultaneous partitions of measures by k-fans. Discrete and Computational Geometry, pp. 317–334 (2001)
7. Bast, H., Hert, S.: The area partitioning problem. In: Proc. 12th Canadian Conf. on Computational Geometry (CCCG), pp. 163–171 (2000)
8. Bereg, S.: Equipartitions of measures by 2-fans. Discrete and Computational Geometry, pp. 87–96 (2005)
9. Bespamyatnikh, S., Kirkpatrick, D., Snoeyink, J.: Generalizing ham sandwich cuts to equitable subdivisions. Discrete and Computational Geometry, pp. 605–622 (2000)
10. Bespamyatnikh, S.: On partitioning a cake. In: Akiyama, J., Kano, M. (eds.) JCDCG 2002. LNCS, vol. 2866, pp. 60–71. Springer, Heidelberg (2003)
11. Beyer, W.A., Zardecki, A.: The early history of the ham sandwich theorem. American Mathematical Monthly, pp. 58–61 (2004)
12. Carlsson, J.G., Armbruster, B., Ye, Y.: Finding equitable convex partitions of points in a polygon (Unpublished manuscript, 2007)
13. Choset, H.: Coverage for robotics - a survey of recent results. Annals of Mathematics and Artificial Intelligence 31, 113–126 (2001)
14. de Berg, M., van Kreveld, M., Overmars, M., Schwarzkopf, O.: Computational Geometry: Algorithms and Applications. Springer, Heidelberg (1997)
15. Fox, D., Ko, J., Konolige, K., Limketkai, B., Schulz, D., Stewart, B.: Distributed multi-robot exploration and mapping. In: Proc. IEEE 2006 (2006)
16. Hert, S., Lumelsky, V.J.: Polygon area decomposition for multiple-robot workspace division. International Journal of Computational Geometry and Applications 8, 437–466 (1998)
17. Ito, H., Uehara, H., Yokoyama, M.: 2-dimension ham sandwich theorem for partitioning into three convex pieces. In: Akiyama, J., Kano, M., Urabe, M. (eds.) JCDCG 1998. LNCS, vol. 1763, pp. 129–157. Springer, Heidelberg (2000)
18. Steinhaus, H., et al.: A note on the ham sandwich theorem. Mathesis Polska, pp. 26–28 (1938)
19. Peleg, D.: Distributed coordination algorithms for mobile robot swarms: New directions and challenges. In: Pal, A., Kshemkalyani, A.D., Kumar, R., Gupta, A. (eds.) IWDC 2005. LNCS, vol. 3741, pp. 1–12. Springer, Heidelberg (2005)

On the Power of Local Orientations[*]

Monika Steinová

Department of Computer Science, ETH Zurich, Switzerland
monika.steinova@inf.ethz.ch

Abstract. We consider a network represented by a simple connected undirected graph with N anonymous nodes that have local orientations, i.e. incident edges of each vertex have locally-unique labels – port names.

We define a pre-processing phase that enables a right-hand rule using agent (RH-agent) to traverse the entire graph. For this phase we design an algorithm for an agent that performs the precomputation. The agent will alter the network by modifying the local orientations using a simple operation of exchanging two local labels in each step. We show a polynomial-time algorithm for this precomputation that needs only one pebble and $O(\log N)$ memory in the agent.

Furthermore we design a similar algorithm where the memory that the agent uses for the precomputation is decreased to $O(1)$. In this case, the agent is not able to perform some operations by itself due to the lack of memory and needs support from the environment.

Keywords: Mobile computing, local orientation, right-hand rule.

1   Introduction

The problem of visiting all nodes of a graph arises when searching for data in a network. A special case of this problem, when nodes need to be visited in a periodic manner, is often required in network maintenance. In this paper, we consider the task of periodic exploration by a mobile entity (called *robot* or *agent*) that is equipped with a small memory.

We consider an undirected graph that is anonymous, i.e. nodes in the graph are neither labeled nor marked. The way how we allow the agent to perceive the environment are the *local orientations*: while visiting a node v, the agent is able to distinguish between incident edges that are labeled by numbers $1 \ldots d_v$, where d_v is the degree of the vertex v. A local orientation uniquely determines the ordering of the incident edges of a vertex.

Dobrev et al. [7] considered the problem of perpetual traversal (an agent visits every node infinitely many times in a periodic manner) in an anonymous undirected graph $G = (V, E)$. The following traversal algorithm called *right-hand rule* is fixed: *"Start by taking the edge with the smallest label. Afterwards, whenever you come to a node, continue by taking the successor edge (in the local*

---

[*] This research was done as a part of author's Master Thesis in Comenius University, Bratislava, Slovakia.

orientation) to the edge via which you have arrived". They showed that in every undirected graph the local orientation can be assigned in such a way that the agent obeying the right-hand rule can traverse all vertices in a periodic manner. Moreover, they designed an algorithm that, using the knowledge of the entire graph G, is able to precompute this local orientation so that the agent visits every node in at most $10|V|$ moves.

Further investigation of this problem was done by Ilcinkas and by Gąsieniec et al. [12,11]. In [12] the author continues in the study of the problem by changing the traversal algorithm and thus decreasing the number of moves in perpetual traversal. By making the agent more complex (a fixed finite automaton with three states) the author designed an algorithm that is able to preprocess the graph so that the period of the traversal is at most $4|V| - 2$. In the most recent publication [11] the authors use a larger fixed deterministic finite automaton, and are thereby able to decrease the period length to at most $3.75|V| - 2$.

1.1   Our Results

In this paper we have returned to the original problem investigated by Dobrev et al. in [7] – we want to preprocess the graph so that an agent obeying the right-hand rule will visit all vertices. We are answering the questions: Is it possible to do the changes of the local orientation locally? How much memory do we need to make these local changes?

We consider a simple undirected anonymous graph $G = (V, E)$ with a local orientation. We design an algorithm for a single agent A, that is inserted into the graph and performs the precomputation so that later an agent B obeying the right-hand rule is able to visit all nodes in a periodic manner. The agent A recomputes the local orientations in polynomial time using $O(\log |V|)$ bits of memory and one pebble. Furthermore, we modify this algorithm using rotations of local orientations to decrease the memory in the agent to a constant, at the expense of losing the termination detection.

1.2   Related Work

The task of searching and exploring the unknown environment is studied under many different conditions. The environment is modeled either as geometric plane with obstacles or as graph-based, in which the edges give constraints on possible moves between the nodes. The first approach is often used to model landscape with possible obstacles where a single or multiple mobile entities are located. They navigate in the environment cooperatively or independently using sensors, vision, etc. to fulfill a particular mission. For more details see for example the survey [15].

The graph-based approach was investigated under various assumptions. The directed version where the mobile entities (called *robots* or *agents*) have to explore a strongly connected directed graph was extensively studied in the literature ([1,8,5]). In this scenario, agents are able to move only along the directed edges of the graph. The undirected version where the edges of a graph can be traversed in both directions was studied in [6, 13, 14, 2, 9]. The labeling of the graph is studied under two different assumptions: either it is assumed that nodes of the graph have unique labels and the agent is able to recognize them (see [5, 13]) or it is assumed that nodes are anonymous (see [16]) with no identifiers. In the second case, the common requirement is that agents can locally distinguish between incident edges.

As the graph can be large and there can be more agents independently or cooperatively operating in it, various complexity measures were considered. Therefore not only the time efficiency but also both the local memory of the agent and its ability to mark the graph is investigated. The agent traversing the graph is often modeled as a finite automaton. This basic model allows the agent to use a constant amount of memory that is represented by its states. In 1978 Budach [4] proved that it is not possible to explore an arbitrary graph by a finite automaton without marking any node. Later Rollik [16] (improved by Fraigniaud et al. in [10]) proved that no finite group of finite automata can cooperatively explore all cubic planar graphs. These negative results lead to increasing the constant memory the finite automaton may use. The agent is given a *pebble* that can be dropped in a node to mark it and later taken and possibly moved to different nodes. Bender et al. [3] show that strongly connected directed graphs with n vertices can be traversed by an agent with a single pebble that knows the upper bound on the number of vertices, or with $\Theta(\log \log n)$ pebbles if no upper bound is known.

1.3 Outline of the Paper

In Section 2 we introduce the notation used and give basic definitions and properties. Section 3 contains the algorithm for precomputing the graph in a local manner with $O(\log |V|)$ memory in the agent and one pebble in the graph. A modification of the algorithm from Section 3 is presented in Section 4. Here, rotations of the local orientation in vertices are used to decrease the memory that agent needs for the precomputation. On one hand, the memory complexity is minimized but on the other hand, the termination detection using so little memory remains an open problem. However, for example marking the initial edge in the graph can be used to terminate this algorithm. Section 5 contains the conclusion and the discussion about open problems.

2 Notation and Preliminaries

Let G be a simple, connected, undirected graph. The degree of vertex v will be denoted d_v. In our model we assume that each vertex is able to distinguish its incident edges by assigning unique labels to them. Note that by such an assignment each edge has two labels – one in each endpoint. For simplicity we assume that for a vertex v these labels are $1, 2, \ldots d_v$. This labeling is called *local orientation* and denoted π_v.

Note that the existence of a local orientation in every vertex is a very natural requirement. Indeed, in order to traverse the graph, agents have to distinguish between incident edges. The labels of the incident edges in a vertex v define a natural cyclic ordering, where $succ_v(e) = (\pi_v(e) \bmod d_v) + 1$ is the successor function and the corresponding predecessor function $pred_v(e)$ is defined similarly.

We want to construct a cycle in G that contains all vertices of G and satisfies the *right-hand rule* (RH-rule for short): "*If the vertex v is entered by the edge e, leave it using an edge with label $succ_v(e)$.*" We will call the traversal according to the RH-rule *RH-traversal*. More precisely we want to construct a cycle that contains all vertices such that RH-traversal started in any vertex of G by edge with label 1 leads to a visit of all nodes of G. We try to construct the cycle in a local manner with minimization of memory requirements for computations. The local computations are done by inserting an agent into the graph G and altering the local orientation by using a small amount of memory.

Rotation of a local orientation π_v by k steps is a local orientation π'_v such that $(\pi_v(e) + k) \bmod d_v + 1 = \pi'_v(e)$. An obvious Lemma follows.

Lemma 1. *Rotations of local orientation in a vertex of a simple, undirected, connected graph G do not have any influence on traversals according to the RH-rule in G. Formally, π_v and π'_v define the same $succ_v(\cdot)$ function.*

In certain situations we will need to speak about traversing edges in a certain direction. In such cases we will call the directed edges *arcs* and undirected edges *edges*. In the further text, we will denote the number of vertices of the graph by $N = |V|$ and the number of its edges by $M = |E|$.

Note that RH-traversal is reversible and therefore the trajectory of RH-traversal is always cycle. These trajectories are called *RH-cycles*. The initial arc of an RH-traversal unambiguously determines a RH-cycle. The graph G is always a union of disjoint RH-cycles.

3   Algorithm *MergeCycles*

Our algorithm will connect multiple RH-cycles into a single RH-cycle that passes through all vertices of graph G. This is done by applying rules *Merge3* and *EatSmall* from [7] in a local manner. In this section we will discuss the rules, their implementation and we show some interesting properties of the algorithm. In the beginning we choose one RH-cycle and call it the *witness cycle*. By applying rules *Merge3* and *EatSmall* the witness cycle is prolonged and new vertices are added. We show that if no rule can be applied in any vertex of the witness cycle then it contains all vertices of graph G. To know when to terminate the algorithm, we will count the number of steps while no rule could be applied.

The desired output of this algorithm is a new labeling of the edges in graph G such that the RH-traversal started in any vertex using the edge with label 1 will visit all vertices in G. Note that after the final witness cycle is constructed, our preprocessing agent can traverse the witness cycle once and rotate labeling in each vertex so that the outgoing arc with label 1 will be an arc of the witness cycle.

3.1   Rules Merge3 and EatSmall

Rule Merge3: [7] To apply this rule, we have to find three different RH-cycles and then connected them by exchanging labels so that the remaining RH-cycles stay unchanged. More precisely, let x_1, x_2 and x_3 be three incoming arcs to a vertex v that define three different RH-cycles C_1, C_2 and C_3 respectively. Then we change the ordering of the edges in v so that the successor of x_2 becomes the successor of x_1, successor of x_3 becomes successor of x_2, and successor of x_1 becomes successor of x_3, while keeping the rest of the relative ordering edges unchanged. For illustration see Figure 1.

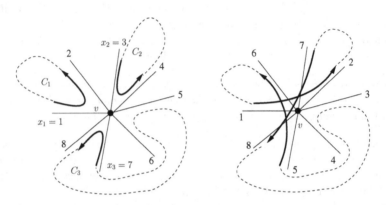

Fig. 1. Applying rule *Merge3*

To apply Merge3, we need to identify three arcs that enter v and determine three different RH-cycles (C_1, C_2 and C_3). Note that we entered v while RH-traversing the witness cycle. Therefore we pick the arc used to enter v as one of those three arcs. The remaining two arcs are found by sequentially testing all incoming arcs in the order given by the current local orientation in v. The pebble is used to mark the processed vertex. To check whether two incoming arcs e_1 and e_2 define different RH-cycles, the agent RH-traverses the cycle defined by e_2 and checks whether it encounters e_1 before returning to e_2. The agent will either find three different RH-cycles and merge them together, or it will ensure that no three different RH-cycles pass through vertex v.

Note that rule *Merge3* can only be applied finitely many times, as after each application of the rule *Merge3* the number of cycles in the graph decreases by two. If there are three different RH-cycles in vertex v, our approach always detects these cycles and rule *Merge3* is applied. Notice that for these operations we need one pebble and $O(1)$ local variables of size $O(\log N)$ bits in the agent's memory. The time complexity of one application of the rule *Merge3* in a vertex v is $O(Md_v)$.

Rule EatSmall [7]: To apply this rule in a vertex v, we have to find two different RH-cycles where the vertex v appears in one of them at least twice. More precisely, let x and y be two incoming arcs to a vertex v that define two

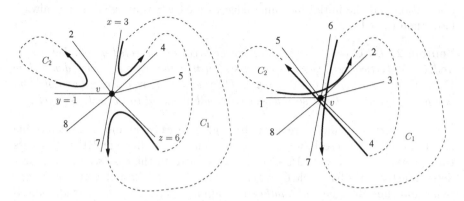

Fig. 2. Applying rule *EatSmall*. The edge y is the incoming edge via which the agent entered the vertex v and thus it is a part of the witness cycle.

different RH-cycles C_1 and C_2 respectively; let z be the incoming arc to the vertex v by which C_1 returns to v after leaving via the successor of arc x. Then we modify the ordering of the edges in v such that the successor of x becomes the successor of y, the old successor of y becomes the successor of z, and the old successor of z becomes the successor of x, while preserving the order of the remaining edges. The application of the rule is shown in Figure 2. Note that the rule is applied on an ordered triplet of edges (y, x, z).

The local version of rule *EatSmall* in a vertex v is similar to the *Merge3* rule. Note that blindly applying the rule *EatSmall* may lead to deadlocks when a part of a cycle will be transferred back and forth between two cycles. To prevent the deadlocks, we will always use our witness cycle as the cycle C_2 in our algorithm.

Note that the rule *EatSmall* can only be applied finitely many times as after an application of the rule *EatSmall* the witness cycle is prolonged. By similar argumentation as for rule *Merge3*, if two cycles C_1 and C_2 with required properties pass through vertex v, their are found and the rule *EatSmall* is applied. Again, one pebble and $O(1)$ local variables of size $O(\log N)$ are needed, and the time complexity of one application of the rule *EatSmall* in a vertex v is $O(Md_v)$.

The rule *EatSmall* has to be applied with care. More precisely, according to Figure 2, if the edge z (the edge via which the agent enters for the second time the vertex v in the RH-traversal of the cycle C_1 that started via the edge x) is the successor of edge y, the rule *EatSmall* will fail. The problem that occurs here is that setting the successor of z to be the successor of y means setting the successor of z to z and that is not allowed. However, our algorithm processes edges incident with a vertex in order given by the local orientation, starting by the arc of the witness cycle (in Figure 2 the arc y). Thus we will never encounter such a situation. In sequential testing of edges, the agent first picks edge x and then verifies the existence of edge z. If z is found, it will be the edge that was not processed at that point by the agent. Thus in the ordering given by the local orientation and the initial incoming edge to v edge z is an edge that is always later than the edge x.

Lemma 2. *Let $G = (V, E)$ be a simple undirected connected graph. Let $v \in V$ and $x \in V$ be two neighbouring vertices such that the witness cycle (denoted by W) passes through v but not through x. Then, either the rule Merge3 or EatSmall can be applied in vertex v so that vertex x will be added to the witness cycle.*

Proof. As vertices v and x are neighbouring, the arcs $\overrightarrow{(v, x)}$ and $\overleftarrow{(v, x)}$ determine two RH-cycles. If these RH-cycles are different, three different RH-cycles pass through vertex v and rule *Merge3* can be applied. In the case where both arcs determine the same RH-cycle C, we claim that C passes through vertex v at least twice and thus the rule *EatSmall* can be applied here. By the contradiction: if the RH-cycle C passes through vertex v only once, the successive arc of $\overrightarrow{(x, v)}$ is arc $\overleftarrow{(x, v)}$. Then by the definition of the cyclical ordering $succ(\cdot)$, vertex v has degree 1 and that is the contradiction.

Lemma 3. *Execute the algorithm MergeCycles on a simple connected undirected graph G and denote the resulting witness cycle by W. For any RH-cycle C in the resulting graph there is a vertex w such that $w \in C$ and $w \in W$.*

Proof. As the graph is connected, there exists a path between a vertex in witness cycle W and a vertex in C. Therefore by Lemma 2 rules *Merge3* and *EatSmall* were applied and all vertices on this path belong to W.

Lemma 4. *During our algorithm MergeCycles in each vertex v of a simple connected undirected graph $G = (V, E)$ we apply all the rules Merge3 and EatSmall at the first time when the vertex is processed. In other words, once we finish processing a particular vertex v for the first time, this vertex is done – no rule applications in v will be possible in the future.*

Proof. Follows from previous discussion.

Theorem 1. *Let G be a simple connected undirected graph. Suppose that the algorithm MergeCycles already terminated on G. Let W be the witness cycle constructed by the algorithm. Then W contains all the vertices in G.*

Proof. By contradiction. Let $v \notin W$. Then there is a path between a vertex in W and v. Take the first vertex x on this path that is not in W (its predecessor y is in W). By Lemma 2 either *Merge3* or *EatSmall* can be applied in y.

The time complexity of our algorithm *MergeCycles* is $O(M^2 \Delta)$, where Δ is the maximal degree of a vertex. The RH-cycle consists of at most all edges of graph in each direction and thus its length is $O(M)$. The termination detection is solved by comparing the number of the consecutive vertices of the cycle where no rule is applied with the length of the witness cycle. This can be done in $O(\log N)$ memory. To sum up, we used a few variables of size $O(\log N)$ in the agent and a single pebble.

4   Algorithm *MergeCycles+* Using Constant Memory

In this section we present the modification of algorithm *MergeCycles* where rotating local orientations will enable us to decrease the memory needed in the agent to apply rules *Merge3* and *EatSmall*. For now, we will assume that the vertices of graph G have degrees greater than one.

Definition 1. *Let $G = (V, E)$ be a simple undirected connected graph. Denote incident edges of vertex v by $e_1, \ldots, e_{d_v}$, so that $e_i = (v, u_i)$. Let π_v, π_{u_i} be the local orientations in vertices v, u_i respectively. We will call label $\pi_v(e_i)$ the inner label of the edge e_i in vertex v and the label $\pi_{u_i}(e_i)$ the outer label of the edge e_i in vertex v.*

As by Lemma 1 the rotations of local orientation do not have any influence on RH-rule, we will use these rotations to store information in local orientations. The general idea of the algorithm *MergeCycles+* is to set all outer labels of the incident edges of a vertex v to 1 and then find and mark representative arcs that are used in rules *Merge3* and *EatSmall* by outer label 2.

4.1   Memory Needed by Our Algorithm

When trying to identify and minimize the memory requirements, we need to be more precise on how the local orientation changes are realized. Note that we can not get rid of the variable that our agent uses for storing the label of the incoming edge. By losing of this information the agent lost the sense of direction and it will not be able to distinguish between edges in G. Note that we will only use this as a read-only variable. Moreover, depending on the hardware realization, this variable does not even have to be stored in the agent's memory – the agent may be able to determine its value on demand from its environment. Thus this variable will not be counted in the agent's memory requirements.

In our algorithm the following local operations are necessary: rotation of the entire local orientation, application of rule *Merge3* and application of rule *EatSmall*. To realize these operations, the agent is able to use two primitives that relabel edges in the current vertex. The first primitive rotates the local orientation in vertex v by one. By multiple uses of this primitive the agent is able to rotate the local orientation so that the label of the incoming edge is 1 or 2. The second primitive operates in a vertex where three incident edges are marked by outer label 2 and the rest have outer label 1. It picks the three marked edges and performs changes in the local orientation as it is done in rules *Merge3* and *EatSmall*.[1] (As we know that the edges for either rule are tested sequentially, their correct order can be determined from the local orientation.)

The last part of the algorithm *MergeCycles* which forces the agent to use $\Omega(\log N)$ memory is the knowledge of the length of the witness cycle and the counter for the number of vertices visited in a row in which neither the rule

---

[1] The application of the rules *Merge3* and *EatSmall* are in fact the same, the difference is only in the conditions the edges have to satisfy.

Merge3 nor rule *EatSmall* could be applied. The price for getting rid of these computations is that we lose the termination detection – the precomputing agent will RH-traverse the witness cycle forever.

As the agent will execute the algorithm *MergeCycles+* forever, by attempting to apply the modified rules of *Merge3* and *EatSmall*, the local orientations will be changing. This does not match the desired output of the algorithm *MergeCycles*: There is no guarantee that if the RH-agent starts via the edge with label 1, then it will traverse the witness cycle. (Note that e.g. by marking the starting edge of the algorithm, termination detection can be solved easily and then the agent obeying the RH-rule can start the traversal via the marked edge and be sure to traverse the witness cycle.)

4.2   Basic Instructions

The entire algorithm for our agent can be specified using the following basic instructions (where e is the edge used to enter vertex v):

- place/pick up pebble, test for its presence in v
- leave vertex v via e
- leave vertex v via $succ_v(e)$
- two tests: whether the edge e has inner label 1 or 2
- apply one of the primitives discussed in Subsection 4.1

Using these instructions we can build the following procedures (again, where $e = (u_k, v)$ is the incoming edge via which the agent entered v):

- rotate local orientation so that $\pi_v(e) = 1$, or so that $\pi_v(e) = 2$
- two tests: $\pi_{u_k}(e) = 1$? and $\pi_{u_k}(e) = 2$?
- remember the incoming edge e by setting $\pi_v(e) = 1$
- rotate local orientation in u_k so that $\pi_{u_k}(e) = 1$ or $\pi_{u_k}(e) = 2$
- traverse all u_i in the order given by local orientation π_v
- for all $i \in \{1, \ldots, d_v\}$ set $\pi_{u_i}(v, u_i) = 1$ and afterwards find e
  (This is realized as follows: rotate π_v so that $\pi_v(e) = 1$, set $\pi_{u_k}(e) = 1$ and then repeatedly traverse via $succ_v(e)$, set the outer label of the edge to 1, and return back to v until the incoming edge has the inner label 1.)

We will show how (according to the order given by the local orientation) we can process all edges incident with a vertex v. During this processing we will use suitable rotations of local orientations in such a way that whenever we enter v we will be able to identify the initial and the currently processed edge. We start by placing a pebble into v, setting outer label 2 to the initial edge and outer label 1 to the remaining edges. We set the inner label of the initial edge to 1 and we start to process it. Whenever we are going to process the next edge, we rotate the local orientation by one so that the currently processed edge always has the inner label 1. If after the rotation of local orientation we find out that the edge to process already has outer label 2, we know that we processed all incident edges and we are done.

The test for two disjoint RH-cycles passing through vertex v is done as follows: The agent sets the outer labels of incident edges so that two tested edges e_1 and e_2 (in the incoming direction they represent two tested RH-cycles) have outer label 2 and the rest of incident edges have outer label 1. Then it rotates the local orientation in v so that e_1 has inner label 1, inserts pebble to v and RH-traverses RH-cycle via the successor of edge e_1. Sooner or later the agent will enter v via edge e_1 (and recognize this thanks to the pebble and the local orientation). The cycles are disjoint iff the agent did not enter v via e_2 during this RH-traversal.

The discussed operations will now be used to describe the modified rules *Merge3* and *EatSmall*.

4.3   Changing the Rule *Merge3* to *Merge3+*

The general idea is to set all outer labels of the incident edges of vertex v to 1, sequentially test edges to find three representatives of different RH-cycles and finally to apply the instruction to change the local orientation.

The simplified description of the rule *Merge3+* applied in a vertex v follows (for illustration see Figure 3):

– Mark the incoming edge e with inner label 1.
– Sequentially traverse all incident edges of v and set their outer label to 1.
– Set outer label of e to 2 (this will remain unchanged during the rule *Merge3+*), denote the RH-cycle represented by incoming arc of e as C_1.

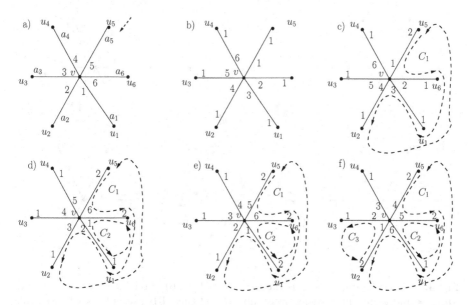

Fig. 3. a) initial state, edge e_5 used to enter v b) outer labels set to 1   c) witness cycle C_1 denoted by outer label 2   d) edge e_6 tested, a disjoint RH-cycle C_2 found   e) edge e_1 tested, the cycle is found to be C_1, thus e_1 will be unmarked now   f) edge e_2 tested, RH-cycle C_3 found, rule *Merge3* can be applied to merge the three cycles

- Sequentially process incident edges of v starting with the edge $succ_v(e)$. While doing this, rotate local orientation so that the processed edge is always marked by inner label 1. Find the first edge e' such that its incoming arc represents a RH-cycle C_2 different from C_1 and mark it by outer label 2.
- Continue in the process of marking the processed edge by inner label 1 with the edge $succ_v(e')$ and find the first edge e'' such that its incoming arc represents a RH-cycle C_3 different from C_1 and C_2. Mark it by outer label 2.
- If these three edges are found, the call of instruction for changing the local orientation is made. Otherwise we reach an edge with outer label 2 (edge e). In this case, the agent found out that three disjoint cycles are not present.

4.4   Changing the Rule *EatSmall* to *EatSmall+*

The change of rule *EatSmall* to *EatSmall+* is similar to the change in Subsection 4.3. The idea is same as the one in previous subsection: we split edges into two partitions – those that will be used in the rule *EatSmall* (with outer label 2) and the rest (with outer label 1) and apply the corresponding primitive. Figure 4 illustrates the application of the rule *EatSmall+*.

4.5   Summary

In this section we will summarize the modifications of the algorithm *MergeCycles* to the algorithm with constant memory for the agent – *MergeCycles+*.

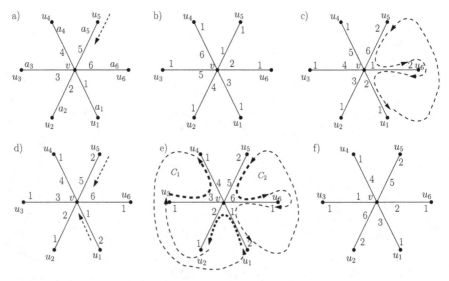

Fig. 4. a) initial state, edge e_5 is the incoming edge   b) outer labels are set to 1   c) edges e_5 and e_6 are marked by outer label 2 and their RH-cycles tested for the rule *EatSmall*, answer is negative   d) edge e_6 is unmarked (outer label set to 1), edge e_1 is processed now   e) the answer is positive – two RH-cycles that are needed in the rule *EatSmall* were found and the rule *EatSmall* can be applied   f) after the application of the rule *EatSmall*

Until now, we assumed that the vertices of graph G have degrees greater than one. We will now explain how to handle vertices with degree one. When we start dealing with vertex v, we first need to process neighbouring vertices with degree 1 and then we can ignore them in the applications of the rules *Merge3* and *EatSmall*. In the case when a neighbouring vertex u of vertex v has degree 1 and the edge (u, v) is not the edge we just used to enter v, we check whether u is present on the witness cycle and if not, we add it by using the rule *EatSmall*. In the case when edge $e = (u, v)$ is the edge via which agent entered v, either the agent already visited more than one vertex (and thus already visited v and processed it at that time, see Lemma 4), or we can find the first incident edge of v with degree greater than one and use it as the incoming one. The only case when such an edge does not exist is if the topology of the graph is a star. This can be easily checked in the beginning of the entire algorithm.

By Lemma 4 and the fact that in *MergeCycles+* the RH-cycles are added to the witness cycle so that they are traversed in the first RH-traversal of the formed witness cycle, it is clear that to build the whole witness cycle the preprocessing agent only has to RH-traverse the witness cycle once. After the agent returns to

---

Algorithm 1. Algorithm *MergeCycles+*

1: check whether the graph is a star, if so, finish the algorithm, any local orientation forms the witness cycle
2: start the RH-traversal via edge with label 1

3: **procedure** RH-TRAVERSAL(v,e)   // *agent entered vertex v via $e = (u, v)$*
4:     **if** vertex u has degree 1 **then**
5:         **if** the agent already visited two or more vertices **then**
6:             continue with RH-traversal via edge $succ_v(e)$
7:         **else**
8:             execute this function with $e = pred_v(e)$
9:         **end if**
10:     **else**
11:         rotate local orientation so that $\pi_v(e) = 1$
12:         **for all** neighbouring vertices w **do**
13:             set the outer label of edge (v, w) to 1
14:         **end for**
15:         **for all** neighbouring vertices w **do**
16:             **if** $d_w = 1$ and w is not on the witness cycle **then**
17:                 apply *EatSmall* to add w to witness cycle
18:             **end if**
19:         **end for**
20:         **while** it is possible **do**
21:             ignoring neighbours with degree 1, apply *Merge3+* and *EatSmall+*
22:         **end while**
23:         continue the RH-traversal via $succ_v(e)$
24:     **end if**
25: **end procedure**

---

the initial arc where the algorithm *MergeCycles+* begun, no rule can be applied in its future RH-traversal.

The time complexity of this algorithm is polynomial. The agent doing the precomputations needs one pebble and $O(1)$ memory. The local orientation is used to store a few bits of information, at most $O(\log N)$ at any time.

5   Conclusions, Open Problems, and Further Research

We designed an algorithm that can create a right-hand rule cyclic walk of length $O(|E|)$ that contains all vertices of the given connected simple undirected graph. This goal is achieved by changes in the local orientations. The algorithm is performed by an agent that only uses $O(\log |V|)$ memory and a single pebble. We have discussed properties of this algorithm and we have shown how it is possible to further decrease the memory requirements of the agent. We believe that the main contribution of this paper is showing that local orientations can be used for storing information and decreasing the agent's memory requirements.

The termination detection in algorithm *MergeCycles+* remains an open problem. Further research can be focused on the amount of the memory that is needed to create a witness cycle of length $O(|V|)$. Another interesting question that needs further research is the amount of useful information that can be stored at once in the local orientations during the execution of an algorithm.

Acknowledgements

I would like to thank Michal Forišek for many valuable notes, helpful comments and detailed suggestions during the preparation of this manuscript, to Ján Oravec for fruitful discussions and to Rastislav Královič for introducing the topic and guiding me in my Master Thesis.

References

1. Albers, S., Henzinger, M.R.: Exploring unknown environments. In: STOC, pp. 416–425 (1997)
2. Awerbuch, B., Betke, M., Rivest, R.L., Singh, M.: Piecemeal graph exploration by a mobile robot. Information and Computation 152(2), 155–172 (1999)
3. Bender, M.A., Fernández, A., Ron, D., Sahai, A., Vadhan, S.: The power of a pebble: exploring and mapping directed graphs. In: STOC 1998. Proceedings of the thirtieth annual ACM symposium on Theory of computing, pp. 269–278. ACM, New York (1998)
4. Budach, L.: Automata and labyrinths. Math. Nachr. 86, 195–282 (1978)
5. Deng, X., Papadimitriou, C.H.: Exploring an unknown graph. J. Graph Theory 32(3), 265–297 (1999)
6. Dessmark, A., Pelc, A.: Optimal graph exploration without good maps. Theor. Comput. Sci. 326(1-3), 343–362 (2004)

7. Dobrev, S., Jansson, J., Sadakane, K., Sung, W.-K.: Finding short right-hand-on-the-wall walks in graphs. In: Pelc, A., Raynal, M. (eds.) SIROCCO 2005. LNCS, vol. 3499, pp. 127–139. Springer, Heidelberg (2005)
8. Fraigniaud, P., Ilcinkas, D.: Digraphs exploration with little memory. In: Diekert, V., Habib, M. (eds.) STACS 2004. LNCS, vol. 2996, pp. 246–257. Springer, Heidelberg (2004)
9. Fraigniaud, P., Ilcinkas, D., Peer, G., Pelc, A., Peleg, D.: Graph exploration by a finite automaton. Theor. Comput. Sci. 345(2-3), 331–344 (2005)
10. Fraigniaud, P., Ilcinkas, D., Rajsbaum, S., Tixeuil, S.: Space lower bounds for graph exploration via reduced automata. In: Pelc, A., Raynal, M. (eds.) SIROCCO 2005. LNCS, vol. 3499, pp. 140–154. Springer, Heidelberg (2005)
11. Gąsieniec, L., Klasing, R., Martin, R., Navarra, A., Zhang, X.: Fast periodic graph exploration with constant memory. In: Prencipe, G., Zaks, S. (eds.) SIROCCO 2007. LNCS, vol. 4474, pp. 26–40. Springer, Heidelberg (2007)
12. Ilcinkas, D.: Setting port numbers for fast graph exploration. In: Flocchini, P., Gąsieniec, L. (eds.) SIROCCO 2006. LNCS, vol. 4056, pp. 59–69. Springer, Heidelberg (2006)
13. Panaite, P., Pelc, A.: Exploring unknown undirected graphs. In: SODA 1998. Proceedings of the ninth annual ACM-SIAM symposium on Discrete algorithms, Philadelphia, PA, USA, Society for Industrial and Applied Mathematics, pp. 316–322 (1998)
14. Panaite, P., Pelc, A.: Impact of topographic information on graph exploration efficiency. Networks 36(2), 96–103 (2000)
15. Rao, N., Kareti, S., Shi, W., Iyenagar, S.: Robot navigation in unknown terrains: Introductory survey of non-heuristic algorithms (1993)
16. Rollik, H.A.: Automaten in planaren graphen. In: Proceedings of the 4th GI-Conference on Theoretical Computer Science, London, UK, pp. 266–275. Springer, Heidelberg (1979)

Best Effort and Priority Queuing Policies for Buffered Crossbar Switches

Alex Kesselman[1], Kirill Kogan[2], and Michael Segal[3]

[1] Google, Inc.
alx@google.com
[2] Cisco Systems, South Netanya, Israel
and
Communication Systems Engineering Dept., Ben Gurion University, Beer-Sheva, Israel
kkogan@cisco.com
[3] Communication Systems Engineering Dept., Ben Gurion University, Beer-Sheva, Israel
segal@cse.bgu.ac.il

Abstract. The buffered crossbar switch architecture has recently gained considerable research attention. In such a switch, besides normal input and output queues, a small buffer is associated with each crosspoint. Due to the introduction of crossbar buffers, output and input contention is eliminated, and the scheduling process is greatly simplified. We analyze the performance of switch policies by means of competitive analysis, where a uniform guarantee is provided for all traffic patterns. The goal of the switch policy is to maximize the weighted throughput of the switch, that is the total value of packets sent out of the switch. For the case of unit value packets (Best Effort), we present a simple greedy switch policy that is 4-competitive. For the case of variable value packets, we consider the Priority Queueing (PQ) mechanism, which provides better Quality of Service (QoS) guarantees by decreasing the delay of real-time traffic. We propose a preemptive greedy switch policy that achieves a competitve ratio of 18. Our results hold for any value of the switch fabric *speedup*. Moreover, the presented policies incur low overhead and are amenable to efficient hardware implementation at wire speed. To the best of our knowledge, this is the first work on competitive analysis for the buffered crossbar switch architecture.

1 Introduction

The main task of a *router* is to receive a packet from the input port, to find its destination port using a routing table, to transfer the packet to that output port, and finally to transmit it on the output link. The switching fabric in a router is responsible for transferring packets from the input ports to the output ports. If a burst of packets destined to the same output port arrives, it is impossible to transmit all the packets immediately, and some of them must be buffered inside the switch (or dropped).

A. Shvartsman and P. Felber (Eds.): SIROCCO 2008, LNCS 5058, pp. 170–184, 2008.

A critical aspect of the switch architecture is the placement of buffers. In the output queueing (OQ) architecture, packets arriving from the input lines immediately cross the switching fabric, and join a queue at the switch output port. Thus, the OQ architecture allows one to maximize the throughput, and permits an accurate control of packet latency. However, in order to avoid contention, the internal speed of an OQ switch must be equal to the sum of all the input line rates. The recent developments in networking technology produced a dramatic growth in line rates, and have made the internal speedup requirements of OQ switches difficult to meet. This has in turn generated great interest in the input queueing (IQ) switch architecture, where packets arriving from the input lines are queued at the input ports. The packets are then extracted from the input queues to cross the switching fabric and to be forwarded to the output ports.

It is well-known that the IQ architecture can lead to low throughput, and it does not allow the control of latency through the switch. For example, for random traffic, uniformly distributed over all outputs, the throughput (i.e. the average number of packets sent in a time unit) of an IQ switch has been shown to be limited to approximately 58% of the throughput achieved by an OQ switch [16]. The main problem of the IQ architecture is head-of-line (HOL) blocking, which occurs when packets at the head of various input queues contend on a specific output port of the switch. To alleviate the problem of HOL blocking, one can maintain at each input a separate queue for each output. This technique is known as virtual output queueing (VOQ).

Another method to get the delay guarantees of an IQ switch closer to that of an OQ switch is to increase the *speedup S* of the switching fabric. A switch is said to have a speedup S, if the the switching fabric runs S times faster than each of the input or the output lines. Hence, an OQ switch has a speedup of N (where N is the number of input/output lines), while an IQ switch has a speedup of 1. For values of S between 1 and N packets need to be buffered at the inputs before switching as well as at the outputs after switching. In order to combine the advantages of both OQ and IQ switches, combined input-output queued (CIOQ) switches make a tradeoff between the crossbar speedup and the complexity of scheduling algorithms they usually have fixed small speedup of two, and thus need buffer space at both input and output side. This architecture has been extensively studied in the literature, see e.g. [8,11,12].

Most of CIOQ switches use a crossbar switching fabric with a centralized scheduler. While it is theoretically possible to build crossbar schedulers that give 100% throughput [23] or rate and delay guarantees [8,14] they are considered too complex to be practical. No commercial backbone router today can make hard guarantees on throughput, rate or delay. In practice, commercial systems use heuristics such as iSLIP [22] with insufficient speedup to give guarantees. Perhaps the most promising way of obtaining guaranteed performance has been to use maximal matching with a speedup of two in the switch fabric [11]. The parallel matching process can be characterized by three phases: request, grant, and accept. Therefore, the resolution time would be the time spent in each of the phases plus the transmission delays for the exchange of request, grant, and accept information. Unfortunately, schedulers based on matching do not perform well with the increase of the switch speed due to the communication and arbitration complexity.

A solution to minimize the scheduling overhead is to use buffers in the crosspoints of the crossbar fabric, or buffered crossbar. The adoption of internal buffers drastically improves the overall performance of the switch. The main benefit of the buffered crossbar switch architecture is that each input and output port can make efficient scheduling decisions independently and in parallel, eliminating the need for a centralized scheduler. As a result, the scheduler for a buffered crossbar is much simpler than that for a traditional unbuffered crossbar [9]. Note that the number of buffers is proportional to the number of crosspoints, that is $O(N^2)$. However, the crosspoint buffers are typically very small.

Buffered crossbar switches recently received significant research attention. Javidi et al. [15] demonstrated that a buffered crossbar switch with no speedup can achieve 100% throughput under a uniform traffic. Nabeshima [24] introduced buffered crossbar switches with VOQs and proposed a scheme based on the oldest cell first (OCF) arbitration at the input as well as the crosspoint buffers. Chuang et al. [9] described a set of scheduling algorithms to provide throughput, rate and delay guarantees with a moderate speedup of 2 and 3.

In the previous research, the scheduling policies for the buffered crossbar switch architecture were analyzed by means of simulations that assumed particular traffic distributions. However, Internet traffic is difficult to model and it does not seem to follow the traditional Poisson arrival model [26,27]. In this work we do not assume any specific traffic model and rather analyze our policies against arbitrary traffic using competitive analysis [25,6], which provides a uniform throughput guarantee for all traffic patterns. In competitive analysis, the online policy A is compared to the optimal clairvoyant offline policy OPT that knows the entire input sequence in advance. The *competitive ratio* of a policy A is the maximum, over all sequences of packet arrivals σ, of the ratio between the the total value of packets sent by OPT out of σ, and that of A.

1.1   Our Results

We consider a buffered crossbar switch with crosspoint buffers of arbitrary capacity. The switch policy controlling the switch consists of two components: a buffer management policy that controls admission to buffers, and a scheduling policy that is responsible for the transfer of packets from input to crosspoint buffers and from crosspoint buffers to output buffers. The goal of the switch policy is to maximize the weighted throughput of the switch. When all packets have a unit value, this corresponds to the number of packets sent out of switch. When packets have variable values, this corresponds to the total value of the sent packets.

First we study the case of unit value packets, which abstracts the *Best Effort* model [10]. We present a simple greedy policy that is 4-competitive. Then we study *Priority Queueing* (PQ) buffers, where packets of the highest priority must be forwarded first. We assume that each packet has an intrinsic value designating its priority, which abstracts the *Diffirentiated Services* (DiffServ) model [7]. We propose a preemptive greedy switch policy that achieves a competitive ratio of 18. Our results hold for any value of the *speedup*. Moreover, the proposed policies have low implementation overhead and can operate at high speeds. We are not aware of any previous work on the competitive analysis of buffered crossbar switches.

1.2 Related Work

Kesselman et al. [17] studied preemptive policies for FIFO buffers in OQ switches and introduce a new bounded-delay model. Competitive analysis of preemptive and non-preemptive scheduling policies for shared memory OQ switches was given by Hahne et al. [13] and Kesselman and Mansour[19], respectively. Kesselman et al. [18] studied the throughput of local buffer management policies in a system of merge buffers.

Azar and Richter [4] presented a 4-competitive algorithm for a weighted multi-queue switch problem with FIFO buffers. An improved 3-competitive algorithm was given by Azar and Richter [5]. Albers and Schmidt [2] proposed a deterministic 1.89-competitive algorithm for the case of unit-value packets. Azar and Litichevskey [3] derived a 1.58-competitive algorithm for switches with large buffers. Recently, Albers and Jacobs [1] gave an experimental study of new and known online packet buffering algorithms.

Kesselman and Rosén [20] studied CIOQ switches with FIFO buffers. For the case of packets with unit values, they presented a switch policy that is 3-competitive for any speedup. For the case of packets with variable values, they proposed two switch policies achieving a competitive ratio of $4S$ and $8 \min(k, 2 \log \beta)$, where S is the speedup of the switch, k is the number of distinct packet values and β is the ratio between the largest and the smallest value. Azar and Richter [5] obtained a 8-competitive policy for CIOQ switches with FIFO buffers for the latter case. Kesselman and Rosén [21] considered the case of CIOQ switches with PQ buffers and proposed a policy that is 6-competitive for any speedup.

1.3 Paper Organization

The rest of the paper is organized as follows. The model description appears in Section 2. Unit and variable value packets are analyzed in Section 3 and Section 4, respectively. We conclude with Section 5.

2 Model Description

We consider an $N \times N$ buffered crossbar switch (see Fig 1). Packets, of equal size, arrive at input ports, and each packet is labeled with the output port on which it has to leave the switch. For a packet p, we denote by $V(p)$ its value. The switch has three levels of buffering: each input i maintains for each output

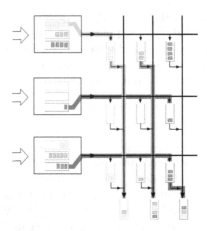

Fig. 1. An example of a buffered crossbar switch

j a separate queue $VOQ_{i,j}$ of capacity $BI_{i,j}$; each crosspoint corresponding to input i and output j maintains a queue $CQ_{i,j}$ of capacity $BC_{i,j}$; each output j maintains a queue OQ_j of capacity BO_j. We denote the length of queue q by $|q|$. Sometimes we use "*" to refer to all queue indices in range $[1, N]$.

The *buffering model* defines in which order packets should be fetched out of the buffer. We consider the First-In-First-Out (FIFO) model under which packets must leave the buffer in the order of their arrivals and the Priority Queuing (PQ) model under which packets of the highest value (priority) must be forwarded first.

We divide time into discrete steps, where a step is the arrival time between two packets at an input port. That is, during each time step one packet can arrive at each input port and one packet can be sent out of each output port.

We divide each time step into three phases. The first phase is the *transmission phase* during which the first packet from each non-empty output queue is sent on the output link. The second phase is the *arrival phase*. During "arrival" phase at most one packet arrives at each input port. The third phase is the *scheduling phase*, which consists of so called *input* and *output* subphases. During the input scheduling subphase each input port may transfer one packet from a virtual output queue to the corresponding crosspoint queue. During the output scheduling subphase each output port can fetch one packet from a crosspoint queue. Notice that a packet arriving at the input port i and destined to the output port j passes through three buffers before it leaves the switch, namely $VOQ_{i,j}$, $CQ_{i,j}$ and OQ_j.

In a switch with a speedup of S, up to S packets can be removed from any input port and up to S packets can be added to each output port during the scheduling phase. This is done in (up to) S *scheduling cycles*, where each cycle comprises input and output scheduling subphases.

Suppose that the switch is managed by a policy A. We estimate the effectiveness of a switch policy by means of *competitive analysis*. In competitive analysis, the online policy is compared to the optimal offline policy OPT, which knows the entire input sequence in advance. The aim of a switch policy is to maximize the total value of the packets sent out of the switch. Let σ be a sequence of packets arriving at the input ports of the switch. We denote by $V^A(\sigma)$ the total value of packets transmitted by A under the input sequence σ. The competitive ratio is defined as follows.

Definition 1. *An online switch policy A is said to be c-competitive if for every input sequence of packets σ , $V^{OPT}(\sigma) = cV^A(\sigma)$, where c is a constant independent of σ.*

3 Unit Value Packets

In this section we consider the case of unit value packets. We define a simple *Gready Unit Switch Policy* (see Figure 2). Note that *GU* never drops accepted packets implementing back pressure at all buffering levels inside the switch.

Greedy Unit Switch Policy (GU)

Transmission Phase : Transmit the first packet from each non-empty output queue.

Arrival Phase : Accept the arriving packet p if there is free space in the buffer. Drop p in case the buffer is full.

Scheduling Phase :

   Input Subphase: for each input i choose an arbitrary head-of-line packet p if any in $VOQ_{i,j}^{GU}$ such that $CQ_{i,j}^{GU}$ is not full and transfer it to $CQ_{i,j}^{GU}$.

   Output Subphase: for each output j if OQ_j^{GU} is not full choose an arbitrary head-of-line packet p if any in $CQ_{i,j}^{GU}$ and transfer it to OQ_j^{GU}.

Fig. 2. GU Switch Policy for Unit Size and Value Packets

We show that the *GU* policy is 4-competitive for any speedup. To analyze the throughput of the *GU* policy we introduce some helpful definitions. The next definition concerns packets that *OPT* may deliver during a time step while *GU* does not.

Definition 2. *For a given switch policy A, a packet sent by OPT from output port j at time t is said to be* extra *if A does not transmit a packet from output port j at this time.*

Next we define a wider class of so called *potential extra* packets that encompass extra packets.

Definition 3. *For a given policy A, a packet p located at queue Q of OPT is called* potential extra *if the number of packets in Q preceding p with respect to the FIFO order is greater than the length of the corresponding queue of A.*

Clearly, each extra packet should eventually become potential extra prior to transmission. We will map every potential extra packet to a packet sent by GU, in such a way that each GU packet is mapped to at most three potential extra packets. This mapping technique was first introduced in [13]. We need some auxiliary claims. First we will show that no new potential extra packets appear during the transmission phase.

Claim. No new potential extra packets appear during a transmission phase.

Proof. Consider an OPT output queue OQ_j^{OPT}. If OQ_j^{OPT} is empty at the beginning of the transmission phase, then we are done. Otherwise, OPT transmits a packet out of OQ_j^{OPT} and thus the difference between $|OQ_j^{OPT}|$ and $|OQ_j^{GU}|$ cannot increase.

Now we will show that the number of potential extra packets does not increase during the arrival phase as well.

Claim. The number of potential extra packets does not increase during an arrival phase.

Proof. Consider a virtual output queue $VOQ_{i,j}^{OPT}$. We argue that the difference between $|VOQ_{i,j}^{OPT}|$ and $|VOQ_{i,j}^{GU}|$ cannot increase unless $VOQ_{i,j}^{GU}$ is full. It follows from the fact that GU greedily accepts all arriving packets if the input buffer is not full. Obviously, $VOQ_{i,j}^{OPT}$ may contain no potential extra packets if $VOQ_{i,j}^{GU}$ is full.

In the following claim we bound the number of new potential extra packets that may appear during an input scheduling subphase.

Claim. Consider an input scheduling subphase. For any input port i, the number of new potential extra packets in the virtual output queues $VOQ_{i,*}^{OPT}$ and crosspoint queues $CQ_{i,*}^{OPT}$ that appear at the end of this subphase is at most two.

Proof. New potential extra packets may appear only in $VOQ_{i,j}^{OPT}$ if GU transfers a packet from $VOQ_{i,j}^{GU}$ and in $CQ_{i,k}^{OPT}$ if OPT transfers a packet to this queue provided that $j \neq k$. Thus, at most two new potential extra packets may occur.

The next claim limits the number of new potential extra packets that may occur during an output scheduling subphase.

Claim. Consider an output scheduling subphase. For any output port j, the number of new potential extra packets in the crosspoint queues $CQ_{*,j}^{OPT}$ and output queue OQ_j^{OPT} that appear at the end of this subphase is at most one.

Proof. Consider an output scheduling subphase t_{os}. We have that one new potential extra packet may appear in $CQ_{i,j}^{OPT}$ if GU transfers a packet from this queue. Notice that if a new potential extra packet appears in OQ_j^{OPT}, then all crossbar queues $CQ_{*,j}^{GU}$ must have been empty at the beginning of t_{os}. In this case, the potential extra packet appearing in OQ_j^{OPT} must have already been a potential extra packet in a queue $CQ_{i,j}^{OPT}$ at the beginning of t_{os}. That establishes the claim.

The mapping routine presented in Figure 3 maps all potential extra packets to the packets sent by GU (we will show in the sequel that the routine is feasible). The routine runs at each (sub)phase, and adds some mappings according to the actions of GU and OPT.

Mapping Routine:

- *Step 1: Arrival Phase.* For each $VOQ_{i,j}^{OPT}$, if OPT accepts a packet that becomes potential extra, this packet replaces in the mapping the preceding packet in $VOQ_{i,j}^{OPT}$ that ceases to be potential extra.
- *Step 2: Scheduling Phase.* (The next sub-steps are repeated S times for each scheduling cycle.)
  - *Sub-Step 2.1: Input Scheduling Subphase.* For each input port i, map new potential extra packet(s) in the virtual output queues $VOQ_{i,*}^{OPT}$ and crosspoint queues $CQ_{i,*}^{OPT}$ to the packet transferred by GU from input port i.
  - *Sub-Step 2.2: Output Scheduling Subphase.* For each output port j, map new potential extra packet in the crosspoint queues $CQ_{*,j}^{OPT}$ and output queue OQ_j^{OPT} to the packet transferred by GU to output port j.

Fig. 3. Mapping Routine for the GU policy

We make the following observation concerning potential extra packets.

Observation 1. *All potential extra packets are mapped by the mapping routine.*

The observation is due to the fact that the routine runs during all but the transmission phase. Notice that by Claim 3, no new potential extra packets appear during a transmission phase. The next lemma shows that the mapping routine is feasible and at most three potential extra packets are mapped to a packet transmitted out of the switch by GU.

Lemma 1. *The mapping routine is feasible and no GU packet is mapped more than three times by the mapping routine prior to transmission out of the switch for any value of the speedup S.*

Proof. According to Claim 3, the number of potential extra packets in the virtual output queues does not increase during the arrival phase. Therefore, if a new potential extra packet appears in $VOQ_{i,j}^{OPT}$, it must be the case that another packet in $VOQ_{i,j}^{OPT}$ ceases to be potential extra at the end of the arrival phase. Hence, Step 1 of the mapping routine is feasible and no new mappings are added to GU packets.

Claim 3 implies that the number of new potential extra packets in the virtual output queues and crosspoint queues corresponding to an input port that appear at the end of the input scheduling subphase is at most two. We argue that no new potential extra packet may appear if GU does not transmit a packet from this input port. In this case, the potential extra packet appearing in a crosspoint queue of OPT must have already been potential extra at the beginning of the input scheduling subphase under consideration. Thus, Step 2.1 of the mapping routine is feasible and no GU packet is mapped more than twice.

By Claim 3, the number of new potential extra packets in the crosspoint queues and output queue corresponding to an output port that appear at the end of the output scheduling subphase is at most one. We claim that no new potential extra packet may appear if GU does not transfer a packet to this output port. In this case, the potential extra packet appearing in the output queue of OPT must have already been potential extra at the beginning of the output scheduling subphase under consideration. Therefore, Step 2.2 of the mapping routine is feasible and no GU packet is mapped more than once.

Note the GU does not drop packets that have been admitted to the switch. Therefore, the mapping is persistent and all mapped GU packets are eventually sent out of the switch. Furthermore, no GU packet is mapped more than three times in total.

Now we will show that GU achieves a competitive ratio of 4.

Theorem 2. *The competitive ratio of GU is at most* 4 *for any speedup value.*

Proof. Fix an input sequence σ. Evidently, the number of packets sent by OPT is bounded by the number of packets sent by GU plus the number of extra packets. Observe that every extra packet at first becomes a potential extra packet prior to transmission. By Lemma 1, the number of extra packets is bounded by three times the number of packets transmitted by GU. In this way we obtain, $V^{OPT}(\sigma) \leq 4V^{GU}(\sigma)$.

4 Variable Value Packets

In this section we study the case of variable value packets under the Priority Queueing buffering model. Remember that packets of the highest value have a strict priority over packets with lower values and are always forwarded first. The goal of the switch policy is to maximize the total value of packets that cross the switch.

Next we define the *Preemptive Greedy Variable Switch Policy* (see Figure 4). The rationale behind PGV is that it preempts packets inside the switch only to serve significantly more valuable packets (twice the value of the evicted packet). Although the Priority Queueing mechanism may violate the global FIFO order, it still maintains the FIFO order within each individual flow consisting of packets with the same value.

We will show that PGV achieves a competitive ratio of 18 for any speedup. In order to show the competitive ratio of PGV we will assign value to the packets

---

Preemptive Greedy Variable Switch Policy (PGV)

Transmission Phase : For each non-empty output queue, transmit the first packet in the FIFO order with the *largest value*.

Arrival Phase : Accept an arriving packet p if there is a free space in the corresponding virtual output queue $VOQ_{i,j}^{PGV}$. Drop p if $VOQ_{i,j}^{PGV}$ is full and $V(p)$ is less than the minimal value among the packets currently in $VOQ_{i,j}^{PGV}$. Otherwise, drop from $VOQ_{i,j}^{PGV}$ a packet p' with the minimal value and accept p. We say that p preempts p'.

Scheduling Phase :

   Input Subphase: For each input port i do the following. For each virtual output queue $VOQ_{i,j}^{PGV}$, choose the packet p that is the first packet in the FIFO order among the packets with the largest value if any. If $CQ_{i,j}^{PGV}$ is not full, mark p as eligible. Otherwise, consider a packet p' with the smallest value in $CQ_{i,j}^{PGV}$. If $V(p) \geq 2V(p')$, then mark p as eligible (p will preempt p' if selected for transmission). Among all the eligible packets in $VOQ_{i,*}^{PGV}$, select an arbitrary packet p'' with the largest value and transfer p'' to the corresponding crosspoint queue preempting a packet with the smallest value from that queue if necessary.

   Output Subphase: For each output port j do the following. For each crosspoint queue $CQ_{i,j}^{PGV}$, choose the packet that is the first packet in the FIFO order among the packets with the largest value if any. Among all chosen packets in $CQ_{*,j}^{PGV}$, select an arbitrary packet p with the largest value. if OQ_j^{PGV} is not full, then transfer p to OQ_j^{PGV}. Otherwise, consider a packet p' with the smallest value in OQ_j^{PGV}. If $V(p) \geq 2V(p')$, then preempt p' and transfer p to OQ_j^{PGV}.

Fig. 4. PGV Switch Policy for Priority Queuing Model

sent by PGV so that no packet is assigned more than 18 times its value and then show that the value assigned is indeed at least $V^{OPT}(\sigma)$. Our analysis is done along the lines of the work in [21], which studies Priority Queuing (PQ) buffers for CIOQ switches.

For the analysis, we assume that OPT maintains $FIFO$ order and never preempts packets. Notice that any schedule of OPT can be transformed into a non-preemptive $FIFO$ schedule of the same value.

Lemma 2. *For any finite sequence σ, the value of OPT in the non-FIFO model equals the value of OPT in the FIFO model.*

The proof of Lemma 2 is similar to that for CIOQ switches [20].

The assignment routine presented on Figure 5 specifies how to assign value to the packets sent by PGV. Observe that the routine assigns some value only to packets that are scheduled out of the virtual output queues and crosspoint queues. Furthermore, if a packet is preempted, then the total value assigned to it is re-assigned to the packet that preempts it.

Now we demonstrate that the routine is feasible and establish an upper bound on the value assigned to a single PGV packet.

> – **Step 1** Assign to each packet scheduled by PGV during the input scheduling subphases of t_s *once* its own value; assign to each packet scheduled by PGV during the output scheduling subphases of t_s *twice* its own value.
>
> For each input port i, let p' be the packet scheduled by OPT from $VOQ_{i,j}^{OPT}$ if any during the input scheduling subphase of t_s. Let p be the first packet with the largest value in $VOQ_{i,j}^{PGV}$ if any or a dummy packet with zero value otherwise.
>
> – **Step 2** If $V(p') \leq V(p)$ and p is not eligible for transmission, then proceed as follows. Consider the beginning of the output scheduling subphase that takes place during a scheduling cycle t'_s when OPT schedules p' from $CQ_{i,j}^{OPT}$ and let p'' be the first packet with the largest value in $CQ_{i,j}^{PGV}$ if any or a dummy packet with zero value otherwise.
>   - **Sub-Step 2.1** If $V(p'') \geq V(p)/2$ and p'' is not eligible for transmission at the beginning of the output scheduling subphase of t'_s, let $\hat{p}$ be the packet that will be sent out of OQ_j^{PGV} at the same time at which OPT will send p' from OQ_j^{OPT} (we will later show that $\hat{p}$ exists and its value is at least $V(p')/4$)). Assign the value of p' to $\hat{p}$.
>   - **Sub-Step 2.2** If $V(p'') < V(p)/2$, consider the set of packets with value at least $V(p')/2$ that are scheduled by PGV from $CQ_{i,j}^{PGV}$ prior to t'_s. Assign the value of $V(p')$ to a packet in this set that has not previously been assigned any value by Sub-Step 2.2 (we will later show that such a packet exists).
> – **Step 3** If $V(p') > V(p)$ then proceed as follows:
>   - **Sub-Step 3.1** If p' was already scheduled by PGV, then assign the value of $V(p')$ to p'.
>   - **Sub-Step 3.2** Otherwise, consider the set of packets with value at least $V(p')$ that are scheduled by PGV from $VOQ_{i,j}^{PGV}$ prior to the scheduling cycle t_s. Assign the value of $V(p')$ to a packet in this set that is not in $VOQ_{i,j}^{OPT}$ at the beginning of this subphase, and has not previously been assigned any value by either Sub-Step 3.1 or Sub-Step 3.2 (we will later show that such a packet exists).
> – **Step 4** If a packet q preempts a packet q' at a crosspoint or output queue of PGV, re-assign to q the value that was or will be assigned to q'.

Fig. 5. Assignment Routine for PGV policy - executed for every scheduling cycle t_s

Lemma 3. *The assignment routine is feasible and the value of each packet scheduled by OPT is assigned to a PGV packet so that no PGV packet is assigned more than 18 times its value. The result holds for any value of the speedup.*

Proof. First we show that the assignment routine as defined is feasible. Step 1, Sub-Step 3.1 and Step 4 are trivially feasible. Consider Sub-Steps 2.1, 2.2 and 3.2.

Sub-Step 2.1. Let p'' be the first packet with the largest value in $CQ_{i,j}^{PGV}$ at the beginning of the output scheduling subphase of t'_s and suppose that p'' is not eligible for transmission. If $V(p'') \geq V(p)/2$ then, by the definition of PGV, the minimal value among the packets in OQ_j^{PGV} is at least $V(p'')/2 \leq V(p)/4$

and OQ_j^{PGV} is full. Thus, during the following BO_j time steps PGV will send packets with value of at least $V(p'/4)$ out of OQ_j^{PGV}. The packet p' scheduled by OPT from $VOQ_{i,j}^{OPT}$ will be sent from OQ_j^{OPT} in one of these time steps (recall that by our assumption OPT maintains FIFO order). Since $V(p') \leq V(p)$, we have that the packet $\hat{p}$ of PGV as specified in Step 2.1 indeed exists, and its value is at least $V(p')/4$.

Sub-Step 2.2. If $V(p'') < V(p)/2$, then evidently PGV scheduled at least $BC_{i,j}$ packets with value at least $V(p')/2$ out of $CQ_{i,j}^{PGV}$ during $[t_s, t'_s)$. By the construction, at most $BC_{i,j} - 1$ of these packets have been assigned some value by Sub-Step 2.2. That is due to the fact that p' is still present in $CQ_{i,j}^{OPT}$ at the beginning of the output scheduling subphase of t'_s and by our assumption OPT maintains FIFO order. Henceforth, one of these packets must be *available* for assignment, i.e., it has not been assigned any value by Sub-Step 2.2 prior to t'_s.

Sub-Step 3.2. First note that if this case applies, then the packet p' (scheduled by OPT from $VOQ_{i,j}^{OPT}$ during the input scheduling subphase of t_s) is dropped by PGV from $VOQ_{i,j}^{PGV}$ during the arrival phase $t_a < t_s$. Let $t'_a \geq t_a$ be the last arrival phase before t_s at which a packet of value at least $V(p')$ is dropped from $VOQ_{i,j}^{PGV}$. Since the greedy buffer management policy is applied to $VOQ_{i,j}^{PGV}$, it contains $BI_{i,j}$ packets with value of at least $V(p)$ at the end of t'_a. Let P be the set of these packets. Note that $p' \notin P$ because it has been already dropped by PGV by this time. We have that in $[t'_a, t_s)$, PGV has actually scheduled all packets from P, since in $[t'_a, t_s)$ no packet of value at least $V(p')$ has been dropped, and at time t_s all packets in $VOQ_{i,j}^{PGV}$ have value less than $V(p')$. We show that at least one packet from P is *available* for assignment, i.e., it has not been assigned any value by Step 3 prior to t_s and is not currently present in $VOQ_{i,j}^{OPT}$. Let x be the number of packets from P that are present in $VOQ_{i,j}^{OPT}$ at the end of the scheduling cycle t_s. By the construction, these x packets are unavailable for assignment. From the rest of the packets in P, a packet is considered available for assignment unless it has been already assigned a value by Step 3. Observe that a packet from P can be assigned a value by Step 3 only during $[t'_a, t_s)$ (when it is scheduled). We now argue that OPT has scheduled at most $BI_{i,j} - 1 - x$ packets out of $VOQ_{i,j}$ in $[t'_a, t_s)$, and thus P contains at least one available packet. To see this observe that the x packets from P that are present in $VOQ_{i,j}^{OPT}$ at the beginning of the scheduling cycle t_s, were already present in $VOQ_{i,j}^{OPT}$ at the end of the arrival phase t'_a. The same applies to the packet p' (recall that $p' \notin P$). Since OPT maintains FIFO order, all the packets that OPT scheduled out of $VOQ_{i,j}^{OPT}$ in $[t'_a, t_s)$ were also present in $VOQ_{i,j}^{OPT}$ at the end of the arrival phase t'_a. Therefore, the number of such packets is at most $BI_{i,j} - 1 - x$ (recall that the capacity of $VOQ_{i,j}$ is $BI_{i,j}$). We obtain that at least one packet from P is available for assignment at Sub-Step 3.2 since $|P| = BI_{i,j}$, x packets are unavailable for assignment because they are present in $VOQ_{i,j}^{OPT}$ and at most $BI_{i,j} - 1 - x$ packets are unavailable because they have been already assigned some value by Step 3.

We show that the value of each packet scheduled by OPT is assigned to a PGV packet. Note that the assignment routine handles all packets scheduled by OPT out of the virtual output queues. The only two cases left uncovered by Step 2 and Step 3 of the assignment routine are (i) $V(p') \leq V(p)$ and p is eligible for transmission and (ii) $V(p') \leq V(p)$, p is not eligible for transmission, $V(p'') \geq V(p)/2$ and p'' is eligible for transmission. We show that these cases are covered by Step 1: for the case (i), the value of p' is assigned during the input scheduling subphase when p is scheduled since $V(p) \geq V(p')$; for the case (ii), the value of p' is assigned during the output scheduling subphase when p'' is scheduled since $V(p'') \geq V(p)/2$. If a PGV packet is preempted, the value assigned to it is re-assigned to the preempting packet by Step 4.

Finally, we demonstrate that no packet is assigned more than 18 times its own value. Consider a packet p sent by PGV. Observe that p can be assigned at most 3 times its own value by Step 1 and at most 6 times its own value by Step 2. By the specification of Sub-Step 3.2, it does not assign any value to p if it is assigned a value by either Sub-Step 3.1 or Sub-Step 3.2. We also show that Sub-Step 3.1 does not assign any value to p if it is assigned a value by Sub-Step 3.2. That is due to the fact that by the specification of Sub-Step 3.2, if p is assigned a value by this sub-step during t_s, then p is not present in the input buffer of OPT at this time. Therefore, Sub-Step 3.1 cannot be later applied to it. We obtain that p can be assigned at most once its own value by Step 3. Hence, a packet that does not perform preemptions can be assigned at most 10 times its value.

Next we analyze Step 4. Note that this assignment is done only to packets that are actually transmitted out of the switch (i.e. they are not preempted). We say that p *transitively* preempts a packet q if either p directly preempts q or p preempts another packet that transitively preempts q. Firstly, p can preempt another packet q' in the crosspoint queue such that $V(q') \leq V(p)/2$. Observe that any preempted packet in a crosspoint queue is assigned at most once its own value by Step 1, once its own value by Step 3 and no value by Step 2. Hence, the total value that can be assigned to p by Step 4 due to transitively preempted packets when p preempts q' is bounded by twice its own value. Secondly, p can preempt another packet q'' in the output queue such that $V(q'') \leq V(p)/2$. Observe that any preempted packet in an output queue is assigned at most 3 times its own value by Step 1, twice its own value by Step 2, once its own value by Step 3, and twice its own value by Step 4. Thus, the total value that can be assigned to p by Step 4 due to transitively preempted packets when p preempts q'' is bounded by 8 times its own value. Therefore, in total no PGV packet is assigned more than 18 times its own value.

Now we are ready to prove the main theorem, which follows from Lemma 3.

Theorem 3. *The competitive ratio of the PGV policy is at most 18 for any speedup.*

5   Conclusions

As switch speeds constantly grow, centralized switch scheduling algorithms become the main performance bottleneck. In this paper we consider competitive switch policies for buffered crossbars switches with PQ buffers. The major advantage of the buffered crossbar switch architecture is that the need for centralized arbitration is eliminated and scheduling decisions can be made independently by the input and output ports.

Our main result is a 18-competitive preemptive greedy switch policy for the general case of unit size and variable value packets and arbitrary value of the switch fabric speedup. We also propose a simple greedy switch policy that achieves a competitive ratio of 4 for any value of speedup in the case of unit size and value packets. As far as we know, these are the first results on competitive analysis for the buffered crossbar switch architecture. We believe that this work advances the design of practical switch policies with provable worst-case performance guarantees for state-of-the-art switch architectures.

References

1. Albers, S., Jacobs, T.: An experimental study of new and known online packet buffering algorithms. In: Arge, L., Hoffmann, M., Welzl, E. (eds.) ESA 2007. LNCS, vol. 4698, pp. 754–765. Springer, Heidelberg (2007)
2. Albers, S., Schmidt, M.: On the Performance of Greedy Algorithms in Packet Buffering. SIAM Journal on Computing 35(2), 278–304 (2005)
3. Azar, Y., Litichevskey, M.: Maximizing throughput in multi-queue switches. Algorithmica 45(1), 69–90 (2006)
4. Azar, Y., Richter, Y.: Management of Multi-Queue Switches in QoS Networks. In: Algorithmica, vol. 43(1-2), pp. 81–96 (2005)
5. Azar, Y., Richter, Y.: An improved algorithm for CIOQ switches. ACM Transactions on Algorithms 2(2), 282–295 (2006)
6. Borodin, A., El-Yaniv, R.: Online Computation and Competitive Analysis. Cambridge University Press, Cambridge (1998)
7. Black, D., Blake, S., Carlson, M., Davies, E., Wang, Z., Weiss, W.: An Architecture for Differentiated Services, Internet RFC 2475 (December 1998)
8. Chuang, S.T., Goel, A., McKeown, N., Prabhakar, B.: Matching Output Queueing with a Combined Input Output Queued Switch. IEEE Journal on Selected Areas in Communications 17, 1030–1039 (1999)
9. Chuang, S.T., Iyer, S., McKeown, N.: Practical Algorithms for Performance Guarantees in Buffered Crossbars. In: Proc. INFOCOM 2005, vol. 2, pp. 981–991 (2005)
10. Clark, D., Fang, W.: Explicit Allocation of Best Effort Packet Delivery Service. IEEE/ACM Trans. on Networking 6(4), 362–373 (1998)
11. Dai, J., Prabhakar, B.: The throughput of data switches with and without speedup. In: Proc. IEEE INFOCOM 2000, March 2000, vol. 2, pp. 556–564 (2000)
12. Giaccone, P., Leonardi, E., Prabhakar, B., Shah, D.: Delay Performance of High-speed Packet Switches with Low Speedup. In: Proc. IEEE GLOBECOM 2002, November 2002, vol. 3, pp. 2629–2633 (2002)
13. Hahne, E.L., Kesselman, A., Mansour, Y.: Competitive Buffer Management for Shared-Memory Switches. In: Proc. SPAA, July 2001, pp. 53–58 (2001)
14. Iyer, S., Zhang, R., McKeown, N.: Routers with a Single Stage of Buffering. ACM SIGCOMM 3(4), 251–264 (2002)
15. Javidi, T., Magill, R., Hrabik, T.: A High Throughput Scheduling Algorithm for a Buffered Crossbar Switch Fabric. In: Proc. IEEE International Conference on Communications, vol. 5, pp. 1586–1591 (2001)

16. Karol, M., Hluchyj, M., Morgan, S.: Input versus Output Queuing an a Space Division Switch. IEEE Trans. Communications 35(12), 1347–1356 (1987)
17. Kesselmanm, A., Lotker, Z., Mansour, Y., Patt-Shamir, B., Schieber, B., Sviridenko, M.: Buffer Overflow Management in QoS Switches. SIAM Journal on Computing 33(3), 563–583 (2004)
18. Kesselmanm, A., Lotker, Z., Mansour, Y., Patt-Shamir, B.: Buffer Overflows of Merging Streams. In: Di Battista, G., Zwick, U. (eds.) ESA 2003. LNCS, vol. 2832, pp. 349–360. Springer, Heidelberg (2003)
19. Kesselman, A., Mansour, Y.: Harmonic Buffer Management Policy for Shared Memory Switches. Theoretical Computer Science, Special Issue on Online Algorithms, In Memoriam: Steve Seiden 324(2-3), 161–182 (2004)
20. Kesselman, A., Rosén, A.: Scheduling Policies for CIOQ Switches. Journal of Algorithms 60(1), 60–83 (2006)
21. Kesselman, A., Rosén, A.: Controlling CIOQ Switches with Priority Queuing and in Multistage Interconnection Networks. Journal of Interconnection Networks (to appear)
22. McKeown, N.: iSLIP: A Scheduling Algorithm for Input-Queued Switches. IEEE Transactions on Networking 7(2), 188–201 (1999)
23. McKeown, N., Mekkittikul, A., Anantharam, V., Walrand, J.: Achieving 100% Throughput in an Input-Queued Switch. IEEE Transactions on Communications 47(8), 1260–1267 (1999)
24. Nabeshima, M.: Performance evaluation of combined input-and crosspoint- queued switch. IEICE Trans. Commun. E83-B(3), 737–741 (2000)
25. Sleator, D., Tarjan, R.: Amortized Efficiency of List Update and Paging Rules. Communications of the ACM 28(2), 202–208 (1985)
26. Paxson, V., Floyd, S.: Wide Area Traffic: The Failure of Poisson Modeling. IEEE/ACM Transactions on Networking 3(3), 226–244 (1995)
27. Veres, A., Boda, M.: The Chaotic Nature of TCP Congestion Control. In: Proc. INFOCOM, March 2000, vol. 3, pp. 1715–1723 (2000)

Word of Mouth:
Rumor Dissemination in Social Networks

Jan Kostka, Yvonne Anne Oswald, and Roger Wattenhofer

Computer Engineering and Networks Laboratory,
ETH Zurich, Switzerland
{kostkaja, oswald, wattenhofer}@tik.ee.ethz.ch

Abstract. In this paper we examine the diffusion of competing rumors in social networks. Two players select a disjoint subset of nodes as initiators of the rumor propagation, seeking to maximize the number of persuaded nodes. We use concepts of game theory and location theory and model the selection of starting nodes for the rumors as a strategic game. We show that computing the optimal strategy for both the first and the second player is NP-complete, even in a most restricted model. Moreover we prove that determining an approximate solution for the first player is NP-complete as well. We analyze several heuristics and show that—counter-intuitively—being the first to decide is not always an advantage, namely there exist networks where the second player can convince more nodes than the first, regardless of the first player's decision.

1 Introduction

Rumors can spread astoundingly fast through social networks. Traditionally this happens by word of mouth, but with the emergence of the Internet and its possibilities new ways of rumor propagation are available. People write email, use instant messengers or publish their thoughts in a blog. Many factors influence the dissemination of rumors. It is especially important where in a network a rumor is initiated and how convincing it is. Furthermore the underlying network structure decides how fast the information can spread and how many people are reached. More generally, we can speak of diffusion of information in networks. The analysis of these diffusion processes can be useful for viral marketing, e.g. to target a few influential people to initiate marketing campaigns. A company may wish to distribute the rumor of a new product via the most influential individuals in popular social networks such as MySpace. A second company might want to introduce a competing product and has hence to select where to seed the information to be disseminated. In these scenarios it is of great interest what the expected number of persuaded nodes is, under the assumption that each competitor has a fixed budget available for its campaign.

The aim of this paper is to gain insights into the complexity of a model that captures the dissemination of competing rumors as a game where a number

A. Shvartsman and P. Felber (Eds.): SIROCCO 2008, LNCS 5058, pp. 185–196, 2008.
© Springer-Verlag Berlin Heidelberg 2008

of players can choose different starting nodes in a graph to spread messages. The payoff of each player is the number of nodes that are convinced by the corresponding rumor. We focus on one crucial aspect of such a rumor game: the choice of a set of nodes that is particularly suitable for initiating the piece of information. We show that even for the most basic model, selecting these starting nodes is NP-hard for both the first and the second player. We analyze tree and d-dimensional grid topologies as well as general graphs with adapted concepts from facility location theory. Moreover, we examine heuristics for the selection of the seed nodes and demonstrate their weaknesses. We prove that contrary to our intuition there exist graphs where the first player cannot win the rumor game, i.e., the second player is always able to convince more nodes than the first player.

2   Related Work

Recently, viral marketing experienced much encouragement by studies [12] stating that traditional marketing techniques do no longer yield the desired effect. Furthermore [12,15,16] provide evidence that people do influence each other's decision to a considerable extent. The low cost of disseminating information via new communication channels on the Internet further increases the appeal of viral marketing campaigns. Thereupon algorithmic questions related to the spread of information have come under scrutiny. Richardson and Domingos [5] as well as Kleinberg et al. [13] were among the first to study the optimization problem of selecting the most influential nodes in a social network. They assume that an initial set of people can be convinced of some piece of information, e.g., the quality of a new product or a rumor. If these people later influence their friends' decisions recursively, a cascading effect takes place and the information is distributed widely in a network. They define the *Influence Maximization Problem*, which asks to find a k-node set for which the expected number of convinced nodes at the end of the diffusion process is maximized. The authors introduced various propagation models such as the linear threshold model and the independent cascade model. Moreover, they show that determining an optimal seeding set is NP-hard, and that a natural greedy hill-climbing strategy yields provable approximation guarantees. This line of research was extended by introducing a second competitor for the most far-ranging influence. Carnes et al. [3] study the strategies of a company that wishes to invade an existing market and persuade people to buy their product. This turns the problem into a Stackelberg game [23] where in the first player (leader) chooses a strategy in the first stage, which takes into account the likely reaction of the second players (followers). In the second stage, the followers choose their own strategies having observed the Stackelberg leader decision i.e., they react to the leader's strategy. Carnes et al. use models similar to the ones proposed in [13] and show that the second player faces an NP-hard problem if aiming at selecting an optimal strategy. Furthermore, the authors prove that a greedy hill-climbing algorithm leads to a $(1 - 1/e - \epsilon)$-approximation. Around the same time, Bharathi et al. [1] introduce roughly the same model for competing rumors and they also show that there exists an efficient approximation algorithm for the second player. Moreover they present an FPTAS for the single player problem on trees.

Whereas the application of information dissemination to viral marketing campaigns is relatively new, the classic subjects of *Competitive Location Theory* and *Voting Theory* provide concepts that are related and prove very useful in this paper. Location theory studies the question where to place facilities in order to minimize the distance to their future users. One of the earliest results stems from Hotelling [11], where he examines a competitive location problem in one dimension. He analyzes the establishment of ice-cream shops along a beach where customers buy their ice-cream at the nearest shop. *Voronoi Games* [4] study the same problem in two dimensions. In these games the location set is continuous, and the consumers are assumed to be uniformly distributed. Contrary to these assumptions the dissemination of information depends on the underlying network structure, i.e., there is a discrete set of possible "locations". Closest related to the spread of rumors is the competitive location model introduced by Hakimi [9]. Here, two competitors alternately choose locations for their facilities on a network. The author assumes that the first player knows of the existence of the second player and its budget, i.e., the leader can take the possible reactions of the follower into consideration. In turn, the follower has full knowledge of the leader's chosen positions and adapts its decision accordingly. Hakimi shows that finding the leader's and the follower's position on general graphs is NP-hard. Our model differs from Hakimi's two main aspects, namely he permits locating facilities on edges, and the placing of multiple users at nodes. Voting theory [10] introduces notions such as plurality solution, Condorcet solution or Simpsons solution describing the acceptance among a set of people, some of which we will use in our analysis.

A large body of research covers the dynamics of epidemics on networks, e.g., [2,17,18,19,20] to name but a few. Many of these models are applicable to the diffusion of information for a single player, however, to the best of our knowledge no work exists on epidemics that fight each other.

3 Model and Notation

3.1 Propagation Models

The *Propagation Model* describes the dissemination of k competing rumors on an undirected graph $G(V, E)$. Initially, each node is in one of $k+1$ states. A node is in state $i \leq k$ if it believes rumor i, in state 0 if it has not heard any rumors yet. In the first step all nodes apart from the nodes in state 0, send a message containing rumor i to their neighbors, informing them about their rumor. Now, all nodes in state 0 that received one or more messages decide which rumor they believe (if any), i.e. they change their state to i if they decided to accept rumor

i, or remain in state 0, or adopt state ∞ if they reject all rumors. Nodes in state $i \in \{1 \dots k\}$ spread the rumor by forwarding a message to their neighbors. These steps are repeated recursively until no messages are transmitted any more. Observe that in this model each node transmits at most once and no node ever changes its first decision.

Depending on the process of reaching a decision after receiving one or several messages the diffusion of the rumors differs. In this paper we mostly consider the *basic model* where each node trusts the first rumor it encounters unless two or more different rumors arrive at the same time in which case the node chooses state ∞, i.e., it refuses to decide and ignores all further messages.

This model can easily be extended by varying the decision process. E.g., rumor i could be accepted and forwarded to the neighbors with probability $P_{rumor_i} = \frac{\#\text{messages } rumor_i}{\#\text{messages}}$. Thereby the decision depends on the number of messages containing rumor i versus the total number of messages received in this time slot. Moreover, edges could be oriented and a persuasiveness value could be assigned to each rumor influencing the decision. A more complex model such as the *linear threshold model* or the *independent cascade model* could be implemented. Note that our basic model is a special case of the independent cascade model. The threshold model has been introduced by Granovetter [8] and Schelling [21], who were among the first to define a model that handles the propagation of information in networks. In this model, a node u forwards a rumor i to all neighbors if the accumulated persuasiveness of the received messages i exceeds a threshold, $\sum_{m_i} psv_u(m_i) \geq t$. The independent cascade model has been proposed in the context of marketing by Goldenberg, Libai and Muller [6]. Here, a node u is given one opportunity to propagate rumor i to neighbor v with probability $p_{u,v}$. Thereafter no further attempts of node u to convince node v take place. Kempe et al. [14] show that these two models can be generalized further and ultimately are equivalent.

3.2 Strategic Rumor Game

Consider two players p_1, p_2 and a graph $G(V, E)$. Player p_1 selects a subset $V_1 \subset V$ of nodes corresponding to the set of nodes initiating rumor 1. Subsequently, p_2 selects the seeds for rumor 2, a set $V_2 \subset V$, where $V_1 \cap V_2 = \emptyset$. The rumors then propagate through the graph as specified by the propagation model. The payoff for player p_i is calculated when the propagation has terminated and equals the number of nodes that believe $rumor_i$. This model can be extended to multiple players, where each players' strategy consists of a disjoint set of nodes to initiate their rumors.

Observe that this game is related to the classic subject of competitive location theory and the equilibrium analysis of voting processes. In order to analyze our rumor game in different topologies we therefore introduce the notions *Distance Score* and *Condorcet Node*.

Definition 1. *For any two nodes* $v_i, v_j \in V$ *the number of nodes that are closer to* v_i *than to* v_j *is designated as the* distance score, $DS_i(j) = |\{v \in V : d(v, v_i) < d(v, v_j)\}|$. *A node* $v_j \in V$ *is called a* Condorcet node *if* $DS_i(j) \leq |V|/2$ *for every* $v_i \in V \setminus \{v_j\}$.

Thus a node $v_j \in V$ is called a Condorcet node if no more than one half of the nodes accept a rumor from any other node in the graph. Note that this definition differs from the original definition of a *Condorcet Point* that can be anywhere on the graph, including edges.

4   Analysis

Location theory studies the optimal distribution of facilities such that the distance to the users is minimized. In our basic model, we consider a very similar problem. Instead of two facility providers two rumors compete for users. Hakimi et al. [9] examine the facility location problem in a weighted graph, i.e., each edge is assigned a length value. The facilities are located at nodes or edges, the users are located at nodes only and multiple users are allowed per node. We adjust these concepts to our model where only one user is located at each node and the edge lengths are restricted to 1. Furthermore, the rumors cannot start on edges, i.e., the available locations are confined to the nodes.

The $(r|p)$-medianoid problem in location theory asks to locate r new facilities in the graph which compete with p existing facilities for reaching more users. Whereas the $(r|p)$-centroid problem examines how to place the p facilities when knowing that r facilities are located afterwards by a second player. We adapt these two terms for the problems faced by player 1 and player 2 in the rumor game.

Definition 2. *Player 1 solves the* $(r|p)$-centroid *problem of a graph by selecting* p *nodes to initiate rumor 1 ensuring that the number of nodes convinced by rumor 1 is maximized when player 1 knows that player 2 will choose* r *nodes.*

Definition 3. *Player 2 solves the* $(r|p)$-medianoid *problem of a graph by selecting* r *nodes to initiate rumor 2 ensuring that the number of nodes convinced by rumor 2 is maximized when player 1 has chosen* p *nodes already.*

The locational centroid and medianoid problems have been shown to be NP-complete in [9]. Our rumor game using the basic model is a restricted special case of the general facility location problem. In the following paragraphs we will prove that the computation of optimal solutions in the rumor game is of the same difficulty. To this end we need some additional notation. Let $D_G(v, Z) = \min\{d(v, z)|z \in Z\}$ for a subset $Z \subset V$, where $d(v, z)$ describes the length of a shortest path from v to z in G. Thus $D_G(v, Z)$ designates the length of the shortest path from node v to a node $z \in Z$. Let X_p be the set of the p nodes chosen by player 1 and Y_r the set of the r nodes selected by player 2. The set of

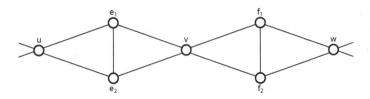

Fig. 1. Diamond structure used for the reduction of the centroid problem

nodes that are closer to a rumor published by Y_r than to the ones published by X_p is $V(Y_r|X_p) = \{v \in V | D(v, Y_r) < D_G(v, X_p)\}$. This allows us to define the part of the graph controlled by rumors placed at Y_r as $W(Y_r|X_p) = |V(Y_r|X_p)|$.

4.1 Complexity of the Centroid Problem

We demonstrate how the $(1|p)$-centroid problem can be reduced to from Vertex Cover.

Theorem 4. *The problem of finding an $(1|p)$-centroid of a graph is NP-hard.*

Proof. We prove this theorem by reducing the *Vertex Cover (VC)* problem to the $(1|p)$-centroid problem. An instance of the VC problem is a graph $G(V, E)$ and an integer $p < |V|$. We have to determine whether there is a subset $V' \subset V$ with $|V'| \leq p$ such that each edge $e \in E$ has at least one end node in V'.

Given an instance of the VC problem, we construct a graph $\bar{G}(\bar{V}, \bar{E})$ from G by replacing each edge $e_i = (u, v)$ in G by the diamond structure shown in Figure 1. Let $Y_1(X_p)$ be the node chosen by player 2 when player 1 has selected the nodes X_p. We prove our theorem by showing that there exists a set X_p of p nodes on $\bar{G}$ such that $W(Y_1(X_p)|X_p) \leq 2$ for every node $Y_1(X_p)$ on $\bar{G}$, if and only if the VC problem has a solution.

Assume V' is a solution to the VC problem in G and $|V'| = p$. Let $X_p = V'$ on $\bar{G}$. Then for any diamond joining u and v in $\bar{G}$, either u or v belongs to $V' = X_p$. It is easy to see that in this case $W(Y_1(X_p)|X_p) \leq 2$ for every node $Y_r(X_p)$ in $\bar{G}$. On the other hand suppose the set of p nodes X_p on $\bar{G}$ satisfies the requirement $W(Y_1(X_p)|X_p) \leq 2$ for every choice of node $Y_1(X_p)$ on $\bar{G}$. If on each diamond of $\bar{G}$ there exists at least one node of X_p, then we can move this node to u or $v \in V' \subset V$. It follows that each diamond has either u or v in V' and therefore V' would provide a solution to the VC problem in G. What can we say about diamonds in $\bar{G}$ joining u and v on which no node of X_p lies? Without loss of generality we may state that there has to be an adjacent diamond with at least one node of X_p, otherwise $W(Y_1(X_p)|X_p) \leq 2$ is violated. No matter whether w, f_1 or f_2 is in X_p, player 2 can select v yielding $W(Y_1(X_p)|X_p) \geq 3$. Consequently, player 1 has to choose at least one node on each diamond and the claim follows.   □

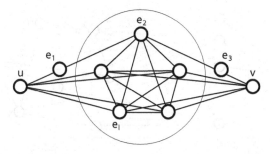

Fig. 2. Graph $\bar{G}$ used for the reduction of the approximation of the centroid problem

Note that for trees the(1|1)-centroid is always on a node in the facility location context [22]. Hence the algorithm proposed by Goldman [7] can be used to find an(1|1)-centroid on trees in time $O(n)$.

Intriguingly, even finding an approximate solution to the $(1|p)$-centroid problem is NP-hard. We define X_p^α to be an α-approximate $(1|p)$-centroid if for any $1 < \alpha \in o(n)$ it holds that $W(Y_1^{OPT}(X_p^\alpha)|X_p^\alpha) \leq \alpha W(Y_1^{OPT}(X_p^{OPT})|X_p^{OPT})$.

Theorem 5. *Computing an α-approximation of the $(1|p)$-centroid problem is NP-hard.*

Proof. This proof uses a reduction from Vertex Cover again and follows the previous proof closely. Given an instance of the VC problem, we construct a graph $\bar{G}(\bar{V}, \bar{E})$ from G by replacing each edge $e_i = (u, v)$ in G by another diamond structure shown in Figure 2. Instead of adding two nodes and five edges for every edge (u, v), we introduce a clique of $4\alpha - 2$ nodes and connect u and v to each node of the clique. Moreover we insert one node on each of the edges from u, v to one designated node of the clique. In a first step we show that $W(Y_1(X_p)|X_p) \leq 4\alpha$ for every node player 2 might pick as Y_1 if and only if VC has a solution.

Assume V' is a solution to the VC problem in G and $|V'| = p$. Let $X_p = V'$ on $\bar{G}$. Then for any diamond joining u and v in $\bar{G}$, either u or v belongs to $V' = X_p$. It is easy to see that in this case $W(Y_1(X_p)|X_p) \leq 4 \leq 4\alpha$ for every node $Y_r(X_p)$ in $\bar{G}$. On the other hand suppose X_p satisfies $W(Y_1(X_p)|X_p) \leq 4\alpha$ for every choice of node $Y_1(X_p)$ on $\bar{G}$. If on each diamond of $\bar{G}$ there exists at least one node of X_p, then we can move this node to u or $v \in V' \subset V$. It follows that each diamond has either u or v in V' and therefore V' would provide a solution to the VC problem in G.

Suppose there is a diamond without a node in X_p. In this case, it is easy to see that if $\min\{D(u, X_p), D(v, X_p)\}$ exceeds one, $W(Y_1(X_p)|X_p) \geq 4\alpha + 2$. Hence we may assume that $0 < \min\{D(u, X_p), D(v, X_p)\} \leq 1$ and we can state without loss of generality that there has to be an adjacent diamond with at least one node of X_p in distance 1 to u. No matter which of the suitable nodes is in X_p, player 2 can select u yielding $W(Y_1(X_p)|X_p) \geq 4\alpha + 1$.

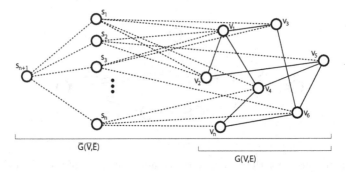

Fig. 3. Graph $\bar{G}$ used in the reduction of the medianoid problem

Consequently, player 1 has to add another node on this diamond to X_p to avoid a violation of our presumption. Thus we can easily construct a VC out of X_p. Moreover, we can prove using similar arguments that X_p exists on $\bar{G}$ such that the condition that $W(Y_1(X_p)|X_p) \leq 4$ holds for every node $Y_1(X_p)$ on $\bar{G}$ if and only if VC on G has a solution. Consequently it must hold that $W(Y_1(X_p)|X_p) \leq \alpha W(Y_1^{OPT}(X_p^{OPT})|X_p^{OPT}) \leq 4\alpha$ and the statement of the theorem follows.   □

4.2   Complexity of the Medianoid Problem

The second player has more information than the first player, however, determining the optimal set of seeding nodes for player 2 is in the same complexity class. We prove this by a reduction of the dominating set problem to the $(r|X_1)$-medianoid problem.

Theorem 6. *The problem of finding an $(r|X_1)$-medianoid of a graph is NP-hard.*

Proof. Consider an instance of the NP-complete *Dominating Set (DS)* problem, defined by a graph $G(V,E)$ and an integer $r < n$, where $n = |V|$. The answer to this problem states whether there exists a set $V' \subset V$ such that $|V'| \leq r$ and $D_G(v,V') \leq 1$ for all $v \in V$. We construct a graph $\bar{G}(\bar{V}, \bar{E})$ with node set $\bar{V} = V \cup S$, where S consists of $n+1$ nodes. Let the nodes in V be numbered from $v_1, \ldots, v_n$ and the nodes in S from $s_1, \ldots, s_{n+1}$. For each node s_i, $i \in \{1, \ldots, n\}$, we add an edge to s_{n+1}, an edge to v_i as well as an edge to every neighbor of v_i, compare Figure 3. Thus the edge set is $\bar{E} = E \cup E_s$, where $E_s = \{(s_i, v_i)|s \in S, v \in V\} \cup \{(v_{n+1}, v)|v \in S \setminus \{v_{n+1}\}\} \cup \{(s_i, v_j)|(v_i, v_j) \in E\}$. Let player 1 choose s_{n+1}. We show now that there exist r nodes in $\bar{G}$ composing Y_r such that $W(Y_r|s_{n+1}) = |V| + r$, if and only if the DS problem has a solution in G.

Assume the DS problems has a solution in G. In this case there exists $V' \subset V$ with $|V'| = r$ such that $D_G(v, V') \leq 1$ for all $v \in V$. Let Y_r contain the nodes in S corresponding to V', i.e., $Y_r = \{s_i|v_i \in V'\}$. It follows that

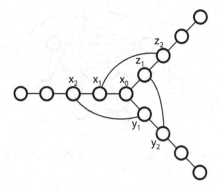

Fig. 4. Example of a graph where the first player never wins

$W(Y_r|s_{n+1}) = |V| + r$, because $\forall v \in V$ $D(v, Y_r) = 1 < d(v, s_{n+1}) = 2$. Suppose Y_r is such that $W(Y_r|s_{n+1}) = |V| + r$. For all nodes $s_i \in Y_r$ it holds that $W(s_i|s_{n+1}) \leq W(v_j|s_{n+1})$, if s_i and v_j are neighbors. This follows from the fact that on every path from a node $v \in V$ to s_{n+1} in $\bar{G}$ there is a node s_i, $i < n + 1$. By removing s_i from Y_r and adding its neighbor $v_i \in V$ to Y_r we maintain $\forall v \in V$ $D(v, Y_r) = 1 < d(v, s_{n+1}) = 2$. We repeat these steps for all nodes $s_i \in S \cap Y_r$ yielding $Y_r \subset V$. Clearly, $W(Y_r|s_{n+1}) = |V|$, letting us state for all $v \in V, D(v, Y_r) < d(v, s_{n+1}) = 2$. Thus this adapted set Y_r is a solution to DS. $\square$

Observe that the hill-climbing algorithms proposed in [3] can be adapted to provide $(1 - 1/e - \epsilon)$-approximations of the medianoid problem in polynomial time.

4.3 Advantage of the First Player

Intuitively, one would assume that the first player has an advantage over the second player because it has more choice. Hence one might think that the first player is always able to convince more nodes than the second player if it selects its seed nodes carefully. Theorem 7 proves the contrary.

Theorem 7. *In a two player rumor game where both player select one node to initiate their rumor in the graph, the first player does not always win.*

Proof. We consider an instance of the rumor game where both the first and the second player can select one node each as a seed. See Figure 4 for an example where the second player can always persuade more players than the first player regardless of the decision the first player makes. If player 1 chooses the node x_0 in the middle, the second player can select x_1 thus ensuring that 7 nodes believe rumor 2 and only 5 nodes adopt rumor 1. If player 1 decides for node x_1, player 2 can outwit the first player by choosing x_2. If player 1 designates x_2 as its seed, the second player select z_1. All other strategies are symmetric to one

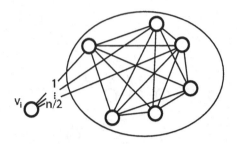

Fig. 5. Example where a Condorcet Node v_i yields a low payoff for player 1. The subgraph in the circle is a complete graph.

of the options mentioned or even less promising for player 1. Hence the second player can always convince 7 nodes whereas the first player has to content itself with 5 persuaded nodes.   □

Since there exists no Condorcet Node in the graph in Figure 4, the curious reader might wonder whether a Condorcet Node guarantees at least $n/2$ convinced nodes for the first player. This conjecture is also wrong as Figure 5 demonstrates.

4.4   Heuristics for Centroid

Having discovered that there are graphs were the second player always is more successful in distributing its rumor, we now concentrate on games where the first

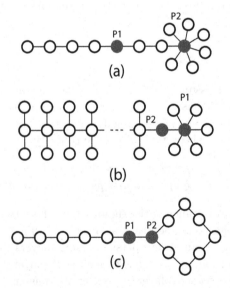

Fig. 6. Counterexamples for heuristics where player 1 wins fewer nodes than player 2 (a) Player 1 selects the node with smallest radius. (b) Player 1 selects the node with highest degree. (c) Player 1 selects the midpoint of the minimum spanning tree.

player could convince more nodes than the second player. Since determining a centroid is NP-complete we consider the following (efficiently computable) strategies the first player can pursue: choose the node with smallest radius, with largest degree or the midpoint of the minimal spanning tree. The node with minimal maximal distance to any other node is the midpoint of the spanning tree. However, for these strategies Figure 3 shows examples where they do not win. In the example shown in Figure 6(a) player 1 selects the node v_j with the smallest radius rad_{min}, i.e., the minimum over all nodes v of the greatest distance between v and any other node. In this case the second player wins more than player 1 by choosing the highest degree node v_i, if it holds $degree(v_i) > 3 \cdot rad_{min}/2$. In Figure 6(b) player 1 selects the node v_i with highest degree. If it holds $n > 2 \cdot degree(v_i)$ then player 2 wins more than half of the nodes by selecting the neighbor of v_i. When the midpoint of a spanning tree is chosen by player 1 then it is easy to see that player 2 can choose a neighbor and win more than half of the nodes, compare Figure 6(c). For all these heuristics there even exist graphs where the first player wins three nodes and the remaining nodes adopt the second rumor.

5   Conclusion

In this paper we have presented the rumor game which models the dissemination of competing information in networks. We defined a model for the game and specified how the propagation of the rumors in the network takes place. We proved that even for a restricted model computing the $(r|p)$-medianoid and $(r|p)$-centroid and its approximation is NP-complete. Moreover, we demonstrated the weaknesses of some heuristics for finding the centroid. Finally we proved the surprising fact that the first player does not always win our two-player rumor game, even when applying optimal strategies.

References

1. Bharathi, S., Kempe, D., Salek, M.: Competitive influence maximization in social networks. In: Deng, X., Graham, F.C. (eds.) WINE 2007. LNCS, vol. 4858, pp. 306–311. Springer, Heidelberg (2007)
2. Breban, R., Vardavas, R., Blower, S.: Linking population-level models with growing networks: A class of epidemic models. Physical Review E 72(4) (2005)
3. Carnes, T., Nagarajan, C., Wild, S.M., van Zuylen, A.: Maximizing influence in a competitive social network: a follower's perspective. In: Proceedings of the 9th international conference on Electronic commerce (ICEC) (2007)
4. Cheong, O., Har-Peled, S., Linial, N., Matoušek, J.: The one-round voronoi game. Discrete & Computational Geometry (2004)
5. Domingos, P., Richardson, M.: Mining the network value of customers. In: Proceedings of the 7th international conference on Knowledge discovery and data mining (KDD) (2001)

6. Goldenberg, J., Libai, B., Muller, E.: Using complex systems analysis to advance marketing theory development. Academy of Marketing Sc. Rev. (2001)
7. Goldman, A.J.: Optimal center location in simple networks. Transportation Science, 212–221 (1971)
8. Granovetter, M.: Threshold models of collective behavior. American Journal of Sociology 83(6), 1420–1443 (1979)
9. Hakimi, S.: Locations with Spatial Interactions: Competitive Locations and Games. Discrete Location Theory, 439–478 (1990)
10. Hansen, P., Thisse, J.-F., Wendell, R.E.: Equilibrium Analysis for Voting and Competitive Location Problems. Discrete Location Theory, 479–501 (1990)
11. Hotelling, H.: Stability in Competition. Economic Journal 39, 41–57 (1929)
12. Leskovec, J., Adamic, L., Huberman, B.: The dynamics of viral marketing. In: Proceedings of the 7th conference on Electronic commerce (EC) (2006)
13. Kempe, D., Kleinberg, J., Tardos, E.: Maximizing the Spread of Influence through a Social Network. In: Proceedings of the 9th international conference on knowledge discovery and data mining (KDD) (2003)
14. Kempe, D., Kleinberg, J., Tardos, É.: Influential Nodes in a Diffusion Model for Social Networks. In: Caires, L., Italiano, G.F., Monteiro, L., Palamidessi, C., Yung, M. (eds.) ICALP 2005. LNCS, vol. 3580. Springer, Heidelberg (2005)
15. Leskovec, J., Singh, A., Kleinberg, J.: Patterns of influence in a recommendation network. In: Ng, W.-K., Kitsuregawa, M., Li, J., Chang, K. (eds.) PAKDD 2006. LNCS (LNAI), vol. 3918. Springer, Heidelberg (2006)
16. Linden, G., Smith, B., York, J.: Amazon.com recommendations: Item-to-item collaborative filtering. IEEE Internet Computing 7(1), 76–80 (2003)
17. Marder, M.: Dynamics of epidemics on random networks. Physical Review E 75(6), 066103 (2007)
18. Moore, C., Newman, M.E.J.: Epidemics and percolation in small-world networks. Physical Review E 61(5), 5678–5682 (2000)
19. Newman, M.E.J.: Spread of epidemic disease on networks. Physical Review E 66(1), 016128 (2002)
20. Pastor-Satorras, R., Vespignani, A.: Epidemic dynamics and endemic states in complex networks. Physical Review E 63(6), 066117 (2001)
21. Schelling, T.: Micromotives and Macrobehavior. Norton (1978)
22. Slater, P.J.: Maximin facility location. Journal of National Bureau of Standards 79B, 107–115 (1975)
23. Stackelberg, H.V.: Marktform und Gleichgewicht. Julius Springer, Heidelberg (1934)

Non-preemptive Coordination Mechanisms for Identical Machine Scheduling Games

Konstantinos Kollias

Department of Informatics and Telecommunications, University of Athens,
Panepistimiopolis Ilisia, Athens 15784, Greece
k.kollias@di.uoa.gr*

Abstract. We study coordination mechanisms for scheduling n selfish tasks on m identical parallel machines and we focus on the price of anarchy of non-preemptive coordination mechanisms, i.e., mechanisms whose local policies do not delay or preempt tasks. We prove that the price of anarchy of every non-preemptive coordination mechanism for $m > 2$ is $\Omega(\frac{\log \log m}{\log \log \log m})$, while for $m = 2$, we prove a $\frac{7}{6}$ lower bound. Our lower bounds indicate that it is impossible to produce a non-preemptive coordination mechanism that improves on the currently best known price of anarchy for identical machine scheduling, which is $\frac{4}{3} - \frac{1}{3m}$.

1 Introduction

Computer networks are characterized by the presence of multiple autonomous users that share common system resources. These users behave selfishly and their actions result in suboptimal system performance. This situation can be modeled in the framework of *game theory* [19,20]. The users are the selfish players who choose their strategies such as to minimize their individual costs. We assume the outcome of the game will be a *Nash equilibrium* [17], which is a set of strategies, one for each player, such that no player benefits from switching to a different strategy unilaterally. In game theory it is well known that a Nash equilibrium does not necessarily constitute a socially optimal outcome. In fact, computer systems are almost certain to suffer performance degradation due to the selfishness of the users. The performance of the system is measured by a *social cost*, e.g., the maximum or the average of the players' costs. The deterioration in performance is captured by the *price of anarchy* [15,21] which is the worst case ratio of the social cost in a Nash equilibrium, to the optimal social cost.

In such settings, applying a centralized control upon the users is infeasilbe and algorithmic choices are restricted to designing the system a priori. The idea of designing system-wide rules and protocols is traditional in game theory and is termed *mechanism design* [18]. In this framework, various approaches have been proposed, such as imposing economic incentives upon players [5,7,10] or applying the Stackelberg strategy [4,12,22,23], which involves enforcing several strategies

---

* Research supported in part by IST-015964, AEOLUS.

A. Shvartsman and P. Felber (Eds.): SIROCCO 2008, LNCS 5058, pp. 197–208, 2008.

on a subset of the players. Another approach is the design of *coordination mechanisms* [6]. Coordination mechanisms are different from the forementioned approaches in that they are distributed and do not require global knowledge, which can be very important in some settings. Informally, a coordination mechanism is a set of local policies, that assign costs to the players.

Consider, for example, a *selfish scheduling* game. The game consists of n selfish tasks that wish to be executed on a system of m machines. The coordination mechanism of this game defines the local scheduling policies of the m machines. A policy of a machine decides the execution order of the tasks and imposes additional delays on some tasks, without knowing the state of the system outside the machine. The strategy of each task is a probability distribution on the m machines. The cost that each task wishes to minimize is its expected completion time. The social cost of the game is the expected makespan, *i.e.*, the expected maximum cost of all tasks. The price of anarchy of a coordination mechanism is the worst case ratio of the social cost in a Nash equilibrium, to the optimal social cost, for every possible set of selfish tasks.

We focus on *identical machine scheduling*, *i.e.*, selfish scheduling on m machines with identical speed. More specifically, we study the price of anarchy of *non-preemptive coordination mechanisms* for identical machine scheduling, which are mechanisms whose local policies simply decide the execution order of the tasks. The term "non-preemptive" for this kind of policies is due to Azar, Jain and Mirrokni [3].

1.1 Related Work

The price of anarchy of selfish scheduling was first studied in [15] by Koutsoupias and Papadimitriou. Some of their results concern m identical machines that complete all tasks simultaneously or schedule them in a random order. For both coordination mechanisms the price of anarchy was shown to be $\frac{3}{2}$, if $m = 2$, and $\Omega(\frac{\log m}{\log \log m})$, if $m > 2$. Mavronicolas and Spirakis [16] extended the result for m machines and proved that the price of anarchy is $\Theta(\frac{\log m}{\log \log m})$ for the fully mixed model, in which all tasks assign non-zero probability to all machines. Finally, Czumaj and Vocking in [9] and Koutsoupias et al in [14] independently proved that the price of anarchy of these mechanisms for $m > 2$ is $\Theta(\frac{\log m}{\log \log m})$. The forementioned results hold for all coordination mechanisms, whose m machines have identical speed and follow the same scheduling policy.

In [15] the authors also studied the related machine model in which the machines have different speeds. The price of anarchy for 2 machines, that complete all tasks simultaneously or schedule them randomly, is at least the golden ratio $\phi = \frac{1+\sqrt{5}}{2}$. In [9] the authors proved that the price of anarchy for $m > 2$ is $\Theta(\frac{\log m}{\log \log \log m})$. Another extension has been studied in [2] by Awerbuch et al. In their paper they extended the forementioned model by not allowing all players to select all machines. They proved that the price of anarchy of this restricted assignment model is $\Theta(\frac{\log m}{\log \log \log m})$.

In [6], Christodoulou et al proposed various coordination mechanisms for the identical machine model. First, they presented the *Increasing-Decreasing* mechanism. This is a non-preemptive coordination mechanism for 2 machines. The first machine's policy schedules the tasks in increasing task length order. The second machine schedules the tasks using the exact opposite order. The price of anarchy of this mechanism is $\frac{4}{3}$. They also presented the *All-Decreasing* mechanism, which is a coordination mechanism for $m \geq 2$ machines. All machines schedule the tasks in decreasing task length order and each machine j adds a delay $j\epsilon$ to all tasks, with ϵ infinitesimal. The price of anarchy of this mechanism is $\frac{4}{3} - \frac{1}{3m}$, which is the currently best known price of anarchy for this setting. Christodoulou et al [6], also introduced *truthful* coordination mechanisms. A coordination mechanism is truthful if no player can benefit from altering the task's length. Their *All-Increasing* mechanism, which is similar to the All-Decreasing mechanism but with opposite scheduling order, is truthful and has price of anarchy $2 - \frac{2}{m+1}$. In [1], Angel et al define a randomized truthful coordination mechanism for 2 machines that has better price of anarchy than the All-Increasing mechanism.

Immorlica et al in [11] studied several coordination mechanisms for 4 selfish scheduling models. They proved upper and lower bounds for the price of anarchy of these mechanisms, concerning Nash equilibria where each task assigns probability 1 to some machine. The models studied involve identical machine scheduling, related machine scheduling, the restricted assignment model and unrelated machine scheduling. In the last model, the processing time of a task on a machine is an arbitrary positive number. For some of them, they studied the convergence to a Nash equilibrium and proved it is linear. Recently, Azar et al in [3] studied coordination mechanisms for scheduling selfish tasks on unrelated machines and managed to improve the price of anarchy of this model to $\Theta(\log m)$.

1.2   Our Contributions

We study the price of anarchy of non-preemptive coordination mechanisms for identical machine scheduling. First we give lower bounds for the case of $m > 2$ machines. We prove that the price of anarchy of all non-preemptive coordination mechanisms for $m > 2$ is $\Omega(\frac{\log \log m}{\log \log \log m})$ and always at least $\frac{3}{2}$. For the case of $m = 2$ we prove a $\frac{7}{6}$ lower bound for the price of anarchy of all non-preemptive coordination mechanisms. The constant lower bounds prove that, for all m, it is impossible to produce a non-preemptive coordination mechanism with price of anarchy less than $\frac{4}{3} - \frac{1}{3m}$, which is the best known price of anarchy for this setting and is achieved by the All-Decreasing mechanism of [6].

2   The Model

An identical machine scheduling game can be fully described by a set of task lengths $w = \{w_1, w_2, \ldots, w_n\}$ and a coordination mechanism c that defines the m local scheduling policies. w_i is the length of task i, *i.e.*, the processing time of task

i on any machine. We assume that the tasks are given in increasing task length order and that tasks with the same length can be ordered increasingly in a pre-defined manner. The strategy of each task is a probability distribution over the set of machines $\{1, 2, \ldots, m\}$. Thus, if p_i^j is the probability which task i assigns to machine j, then $E = ((p_1^1, p_1^2, \ldots, p_1^m), (p_2^1, p_2^2, \ldots, p_2^m), \ldots, (p_n^1, p_n^2, \ldots, p_n^m))$ is an outcome of the game. For the game defined by a coordination mechanism c and a set of task lengths w, if E is the outcome, then $c_i^j(w; E)$ is the expected cost of task i, if the task plays purely machine j. E is a Nash equilibrium if and only if $p_i^j > 0 \Rightarrow c_i^j(w; E) \le c_i^k(w; E)$, for $k = 1, 2, \ldots, m$. We will write $sc(c; w; E)$ for the social cost of outcome E in a game defined by c and w. The social cost is the expected total makespan, $i.e.$, the expected maximum comple-tion time among all tasks. We will denote $opt(w)$ the social cost of the optimal allocation of the n tasks to the m machines. With $N(c; w)$ being the set of Nash equilibria of the game defined by c and w, we define the price of anarchy of a coordination mechanism c as:

$$PoA(c) = \max_{w} \max_{E \in N(c;w)} \frac{sc(c; w; E)}{opt(w)}.$$

2.1 Non-preemptive Coordination Mechanisms

A non-preemptive coordination mechanism consists of m non-preemptive poli-cies. The algorithm, that implements a non-preemptive policy, takes as input the tasks that are allocated to the machine and decides the order in which they will be executed. A non-preemptive policy is not allowed to delay or preempt tasks. The non-preemptive policy of a machine j can be fully described by an infinite set of task length sequences, m_j. There is exactly one sequence in m_j for every possible set of task lengths. These sequences define the order in which the received tasks will be executed. Consider, for example, the following set of sequences:

$$m_j = \{(5, 4), (1, 3, 2), (2, 3), \ldots\}.$$

The sequence $(1, 3, 2)$ defines that if machine j receives 3 tasks t_i, t_k, t_l, with lengths $w_i = 1$, $w_k = 2$, $w_l = 3$, then the tasks have completion times 1, 6 and 4 respectively. Obviously, since $(1, 3, 2) \in m_j$, all other sequences of the numbers 1, 2 and 3 are not in m_j. If a sequence contains several equal task lengths, the policy makes them distinct by ordering them in the predefined manner mentioned earlier. For example, if the machine has received 3 tasks with length 2, then the sequence that describes the execution order could be $(2_1, 2_3, 2_2)$. The task with length 2_1 is considered the smallest among the three and the task with length 2_3 is considered the largest. Thus the smallest task is executed first, the largested task is executed second and the remaining task is executed last. A non-preemptive coordination mechanism c for m machines can be written as $c = \{m_1, m_2, \ldots, m_m\}$, with m_j being the set of sequences of machine j.

3   The Price of Anarchy of Non-preemptive Coordination Mechanisms

In this section, we evaluate the performance of non-preemptive coordination mechanisms for identical machine scheduling. We do so by studying their price of anarchy. At first, we focus on coordination mechanisms for $m > 2$ machines. The following lemma gives an interesting preliminary result.

Lemma 1. *If c is a non-preemptive coordination mechanism for $m > 2$, then $PoA(c) \geq \frac{3}{2}$.*

Proof. Suppose we have a game with $n = m$ tasks of length 1, with $m > 2$ being the number of machines of a non-preemptive coordination mechanism. The set of task lengths is $w = \{w_1, w_2, \ldots, w_m\}$, with $w_i = 1$, for $i = 1, 2, \ldots, m$. As stated in section 2, it is assumed that the tasks are ordered in increasing order, even if they have equal lengths. We can see that, since $m > 2$, there always exist 2 machines that schedule 2 tasks with length 1 the same way, since the possible schedules are only 2: $(1_1, 1_2)$ and $(1_2, 1_1)$. We "mark" these machines and produce the following outcome. We find all "unmarked" machines with policies that include $(1_1, 1_2)$. Suppose we find l such machines. We select the tasks $1, 2, \ldots, l$ and allocate them, one each, to the l found machines. Now, we select the tasks $m, m - 1, \ldots, m - l - 2$ and allocate them, one each, to the rest $m - l - 2$ of the "unmarked" machines, which obviously have policies that include $(1_2, 1_1)$. Now we are left with 2 tasks that are allocated with probability $\frac{1}{2}$ to each of the "marked machines". Now we will prove that the outcome we constructed is a Nash equilibrium. All tasks in "unmarked" machines have completion time 1, which is the best possible. The task that is executed first in both "marked" machines, also has completion time 1. The final task has expected completion time $\frac{3}{2}$ in the "marked" machines. This task would have completion time 2 in

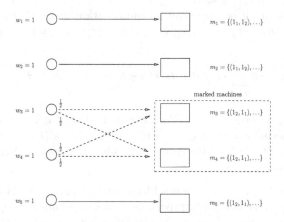

Fig. 1. The Nash equilibrium produced by our procedure for this instance, if we mark machines 3 and 4

any other machine, since all "unmarked" machines already have one task that has priority over it. The expected makespan of this Nash equilibrium is $\frac{3}{2}$, while the optimal makespan is obviously 1. This completes the proof. □

The proof of the previous lemma can be extended[1] from finding 2 machines that follow the same policy for 2 tasks of length 1, to finding k machines that follow the same policies for up to k tasks of length 1. This leads to the following theorem.

Theorem 1. *If c is a non-preemptive coordination mechanism for $m > 2$ machines, then*

$$PoA(c) = \Omega\Big(\frac{\log\log m}{\log\log\log m}\Big).$$

Proof. For a game with $n = m$ tasks of length 1, we seek to find a Nash equilibrium, similar to the one of the previous proof, but with $k > 2$ tasks being allocated with probability $\frac{1}{k}$ to k machines. For this purpose we need k machines that produce the exact same schedule for $1, 2, 3, \ldots, k-1$ and k tasks of length 1. There are $1!2!\ldots k!$ possible such policies. With $m = 1!2!\ldots k!(k-1)+1$, we are certain we can find the k machines we need. We "mark" the k machines, follow the same procedure, looking at the sequence for $1_1, 1_2$ of the "unmarked" machines, and produce a Nash equilibrium, in which each "unmarked" machine receives one task and the k tasks that are left last, are allocated with probability $\frac{1}{k}$ to the k "marked" machines. In the same way as in the previous proof, we can confirm that this is a Nash equilibrium. All tasks in "unmarked" machines have the smallest possible cost. The k tasks that play with probability $\frac{1}{k}$ all "marked" machines, have expected cost at most $1 + \frac{k-1}{k}$ in any of them and cost 2 if they play any of the "unmarked" machines. Since k tasks are allocated with probability $\frac{1}{k}$ to each one of k machines, the situation is identical to throwing k balls at random to k bins. The expected maximum number of balls in a bin is well known to be $\Theta(\frac{\log k}{\log\log k})$. Thus, this Nash equilibrium proves that the price of anarchy of all non-preemptive coordination mechanisms is $\Omega(\frac{\log k}{\log\log k})$ for $m = 1!2!\ldots k!(k-1)+1$. From $m = 1!2!\ldots k!(k-1)+1$, we get

$$\log m = \Theta(\sum_{j=1}^{k} j\log j) = \Theta(k^2\log k) \Rightarrow k^2\, logk = \Theta(\log m).$$

This gives $\log k = \Theta(\log\log m)$. We have seen that if c is a non-preemptive coordination mechanism for a game with $m = 1!2!\ldots k!(k-1)+1$ machines, then $PoA(c) = \Omega(\frac{\log k}{\log\log k})$. By combining this with $\log k = \Theta(\log\log m)$, we prove the theorem. □

Now we turn to the case of $m = 2$ machines. We will prove a lower bound for the price of anarchy of all non-preemptive coordination mechanisms by following an extensive case analysis. It begins with the following lemma.

---

[1] We would like to thank the anonymous reviewer who suggested this extension.

Lemma 2. $c = \{m_1, m_2\}$ *is a non-preemptive coordination mechanism. Suppose that both machines produce the same schedule for 2 tasks with lengths* $w_1, w_2 \in \{2, 3\}$, $w_1 \leq w_2$. *The price of anarchy of the mechanism is at least* $\frac{4}{3}$.

Proof. For a game defined by the described coordination mechanism c and the set of task lengths $w = \{w_1, w_2\}$, there always exists the Nash equilibrium $E = ((\frac{1}{2}, \frac{1}{2}), (\frac{1}{2}, \frac{1}{2}))$. Obviously $opt(w) = w_2$. The social cost of the Nash equilibrium is

$$ sc(c; w; E) = \frac{1}{2}(w_1 + w_2) + \frac{1}{2}w_2 \Rightarrow PoA(c) \geq 1 + \frac{w_1}{2w_2}, $$

which for $w_1, w_2 \in \{2, 3\}$, proves the lemma. $\qquad\square$

The forementioned lemma proves that if a mechanism $c = \{m_1, m_2\}$ is to have price of anarchy lower than $\frac{4}{3}$, then the mechanism does not have the same policy on tasks of length 2 or 3. That is, one of the policies includes $(2_1, 2_2)$ and the other includes $(2_2, 2_1)$, one includes $(3_1, 3_2)$ and the other includes $(3_2, 3_1)$, one includes $(2, 3)$ and the other includes $(3, 2)$. Without loss of generality, we will assume $(2, 3) \in m_1$ and $(3, 2) \in m_2$. We will call this property the *2-3-Mirror Property*. We need to examine all mechanisms with this property and study their price of anarchy. There are the following possible cases:

$$ m_1 = \{(2,3), (2_2, 2_1), (3_1, 3_2), \ldots\}, \quad m_2 = \{(3,2), (2_1, 2_2), (3_2, 3_1), \ldots\}. $$

$$ m_1 = \{(2,3), (2_1, 2_2), (3_1, 3_2), \ldots\}, \quad m_2 = \{(3,2), (2_2, 2_1), (3_2, 3_1), \ldots\}. $$

$$ m_1 = \{(2,3), (2_2, 2_1), (3_2, 3_1), \ldots\}, \quad m_2 = \{(3,2), (2_1, 2_2), (3_1, 3_2), \ldots\}. $$

$$ m_1 = \{(2,3), (2_1, 2_2), (3_2, 3_1), \ldots\}, \quad m_2 = \{(3,2), (2_2, 2_1), (3_1, 3_2), \ldots\}. $$

We will only examine coordination mechanisms that match the first case. We will prove that their price of anarchy is at least $\frac{7}{6}$ and show how the same proof can be applied to the other 3 cases as well. If a mechanism matches the first case, we will say that the mechanism satisfies the *Specific 2-3-Mirror Property*.

Our proof will procede with 8 lemmata which prove that for a coordination mechanism $c = \{m_1, m_2\}$, that satisfies the Specific 2-3-Mirror Property, and for every sequence of $2_1, 2_2, 3$ in m_1, the price of anarchy of c is at least $\frac{7}{6}$. The next lemma examines the case in which m_1 includes $(3, 2_2, 2_1)$. The lemmata that follow examine the remaining 5 cases.

Lemma 3. *If* $c = \{m_1, m_2\}$ *satisfies the Specific 2-3-Mirror Property and* m_1 *includes* $(3, 2_2, 2_1)$, *then* $PoA(c) \geq \frac{7}{6}$.

Proof. For the described mechanism c, suppose that when machine 2 receives 3 tasks with lengths 2, 2, 3, then the task with length 3 is executed first. Now, suppose we have a game with 4 tasks such that $w = \{2, 2, 3, 3\}$. This game has the Nash equilibrium $E = ((1, 0), (\frac{1}{2}, \frac{1}{2}), (1, 0), (0, 1))$. Examination of the costs confirms this is a Nash equilibrium. We can see that

$$ c_1^1(w; E) = \frac{1}{2}7 + \frac{1}{2}2 \text{ and } c_1^2(w; E) \geq \frac{1}{2}5 + \frac{1}{2}5, \text{ so } c_1^1(w; E) < c_1^2(w; E), $$

$$c_2^1(w; E) = c_2^2(w; E) = 5,$$

$$c_3^1(w; E) = \frac{1}{2}3 + \frac{1}{2}5 \text{ and } c_3^2(w; E) \geq \frac{1}{2}6 + \frac{1}{2}3, \text{ so } c_3^1(w; E) < c_3^2(w; E) \text{ and}$$

$$c_4^2(w; E) = 3, \text{ which is the smallest possible.}$$

Thus, we confirm that E is a Nash equilibrium. We can see that the expected social cost is $sc(c; w; E) = \frac{1}{2}7 + \frac{1}{2}5 = 6$, while $opt(w) = 5$. So, this coordination mechanism has price of anarchy at least $\frac{6}{5}$. In the rest of this paper we will present several Nash equilibria that can be cofirmed the same way. Now, suppose that when machine 2 receives 3 tasks with lengths 2, 2, 3, then the task with length 3 is executed second. For this mechanism, if $w = \{2, 2, 2, 3\}$ then $E = ((0, 1), (0, 1), (1, 0), (\frac{1}{2}, \frac{1}{2}))$ is a Nash equilibrium with expected social cost 6, while the optimal social cost is 5. This way we get a $\frac{6}{5}$ lower bound for the price of anarchy of this mechanism. Finally, we examine the case where if machine 2 receives 3 tasks with lengths 2, 2, 3, then the task with length 3 is executed third. For $w = \{2, 2, 3\}$, $E = ((\frac{2}{3}, \frac{1}{3}), (0, 1), (1, 0))$ is a Nash equilibrium with $sc(c; w; E) = \frac{14}{3}$, while $opt(w) = 4$, which proves that the price of anarchy of this mechanism is at least $\frac{7}{6}$. At this point, we have examined the described coordination mechanism with all possible sequences of $2_1, 2_2, 3$ in m_2 and proved that in all cases the price of anarchy is at least $\frac{7}{6}$. This proves the lemma.   □

Lemma 4. If $c = \{m_1, m_2\}$ satisfies the Specific 2-3-Mirror Property and m_1 includes $(3, 2_1, 2_2)$, then $PoA(c) > \frac{7}{6}$.

Proof. For a game with $w = \{2, 2, 3\}$ and c the described mechanism, there is always the Nash equilibrium $E = ((1, 0), (1, 0), (\frac{1}{4}, \frac{3}{4}))$ with expected social cost $\frac{19}{4}$. Thus, we get that the price of anarchy of the mechanism is at least $\frac{19}{16} > \frac{7}{6}$.
    □

Lemma 5. If $c = \{m_1, m_2\}$ satisfies the Specific 2-3-Mirror Property with m_1 including either $(2_1, 2_2, 3)$ or $(2_2, 2_1, 3)$, then $PoA(c) \geq \frac{6}{5}$.

Proof. For the described mechanism, suppose that when machine 2 receives 3 tasks with lengths 2, 3, 3, the task with length 2 is executed second or third. Then for a game with $w = \{2, 2, 3, 3\}$, $E = ((1, 0), (1, 0), (0, 1), (0, 1))$ is a Nash equilibrium with social cost 6. It is obvious that $opt(w) = 5$, so this mechanism has price of anarchy at least $\frac{6}{5}$. Now we need to examine the price of anarchy of the mechanism when m_2 includes $(2, 3_1, 3_2)$ or $(2, 3_2, 3_1)$. We know that m_1 includes $(2_1, 2_2, 3)$ or $(2_2, 2_1, 3)$, so we need to examine 4 cases. We will examine each one in a game with $w = \{2, 2, 3, 3\}$. If m_1 includes $(2_2, 2_1, 3)$ and m_2 includes $(2, 3_2, 3_1)$, then $E = ((\frac{3}{4}, \frac{1}{4}), (1, 0), (\frac{2}{3}, \frac{1}{3}), (0, 1))$ is a Nash equilibrium with expected social cost $\frac{13}{2}$. If m_1 includes $(2_2, 2_1, 3)$ and m_2 includes $(2, 3_1, 3_2)$, then $E = ((\frac{3}{7}, \frac{4}{7}), (1, 0), (0, 1), (\frac{2}{3}, \frac{1}{3}))$ is a Nash equilibrium with expected social cost $\frac{44}{7}$. If m_1 includes $(2_1, 2_2, 3)$ and m_2 includes $(2, 3_2, 3_1)$, then $E = ((\frac{3}{4}, \frac{1}{4}), (1, 0), (\frac{2}{5}, \frac{3}{5}), (0, 1))$ is a Nash equilibrium with expected social cost $\frac{13}{2}$. If m_1 includes $(2_1, 2_2, 3)$ and m_2 includes $(2, 3_1, 3_2)$, then $E = ((\frac{3}{7}, \frac{4}{7}), (1, 0), (0, 1), (\frac{2}{5}, \frac{3}{5}))$ is a Nash equilibrium with expected social cost $\frac{232}{35}$. Since the optimal social cost in all cases is 5, the lemma holds.   □

Lemma 6. *If* $c = \{m_1, m_2\}$ *satisfies the Specific 2-3-Mirror Property with* m_1 *including either* $(2_1, 3, 2_2)$ *or* $(2_2, 3, 2_1)$ *and* m_2 *including either* $(3, 2_2, 2_1)$ *or* $(3, 2_1, 2_2)$, *then* $PoA(c) \geq \frac{7}{6}$.

Proof. Suppose we have a game with a mechanism c, as described, and $w = \{2, 2, 2, 3, 3\}$. We will costruct an outcome that sees 2 tasks, with lengths 2 and 3, playing $(1, 0)$ and 3 tasks, with lengths 2, 2 and 3, playing $(0, 1)$. We pick one task of length 3 that would have completion time at least 6 in machine 2, if that machine received 4 tasks with lengths 2, 2, 3, 3. This task is allocated to machine 1 with cost 5, since machine 1 will receive 2 tasks with lengths 2, 3. Thus, this task has no incentive to switch strategy. The other task of length 3 is allocated to machine 2, where it is executed first. So, this task has no incentive to switch, either. The tasks of length 2, choose their strategies as follows. If m_1 includes $(2_2, 3, 2_1)$, then tasks 1 and 2 play $(0, 1)$, while task 3 plays $(1, 0)$. Otherwise, tasks 2 and 3 play $(0, 1)$, while task 1 plays $(1, 0)$. It is easy to confirm that the forementioned allocation is a Nash equilibrium with social cost 7, while the optimal social cost is 6. This proves the lemma. $\qquad\square$

Lemma 7. *If* $c = \{m_1, m_2\}$ *satisfies the Specific 2-3-Mirror Property with* m_1 *including* $(2_1, 3, 2_2)$ *and* m_2 *not including* $(3, 2_2, 2_1)$ *and* $(3, 2_1, 2_2)$, *then* $PoA(c) > \frac{7}{6}$.

Proof. For the described coordination mechanism and a game with set of task lengths $w = \{2, 2, 3\}$, $E = ((\frac{2}{7}, \frac{5}{7}), (\frac{3}{10}, \frac{7}{10}), (1, 0))$ is a Nash equilibrium with expected social cost $\frac{327}{70}$. So the price of anarchy of the mechanism is greater than $\frac{7}{6}$. $\qquad\square$

Lemma 8. *If* $c = \{m_1, m_2\}$ *satisfies the Specific 2-3-Mirror Property with* m_1 *including* $(2_1, 2_2, 2_3)$, *then* $PoA(c) \geq \frac{5}{4}$.

Proof. For the described mechanism and for a game with $w = \{2, 2, 2, 2\}$ we study the following cases. If m_2 includes $(2_3, 2_1, 2_2)$, then there exists the Nash equilibrium $E = ((\frac{1}{2}, \frac{1}{2}), (0, 1), (0, 1), (1, 0))$ with expected social cost 5. If m_2 includes $(2_3, 2_2, 2_1)$, then $E = ((1, 0), (\frac{1}{3}, \frac{2}{3}), (\frac{1}{2}, \frac{1}{2}), (0, 1))$ is a Nash equilibrium with expected social cost 5. For any other possible sequences of 2_1, 2_2, 2_3 in m_2, $E = ((0, 1), (1, 0), (\frac{1}{2}, \frac{1}{2}), (1, 0))$ is a Nash equilibrium with expected social cost 5. This proves the lemma. $\qquad\square$

Lemma 9. *If* $c = \{m_1, m_2\}$ *satisfies the Specific 2-3-Mirror Property with* m_1 *including* $(2_1, 2_3, 2_2)$, *then* $PoA(c) \geq \frac{5}{4}$.

Proof. For the described mechanism and for a game with $w = \{2, 2, 2, 2\}$ we study the following cases. If m_2 includes $(2_2, 2_1, 2_3)$, then there exists the Nash equilibrium $E = ((\frac{1}{2}, \frac{1}{2}), (0, 1), (0, 1), (1, 0))$ with expected social cost 5. If m_2 includes $(2_2, 2_3, 2_1)$, then $E = ((1, 0), (1, 0), (0, 1), (\frac{1}{2}, \frac{1}{2}))$ is a Nash equilibrium with expected social cost 5. For any other possible sequences of 2_1, 2_2, 2_3 in m_2, $E = ((0, 1), (1, 0), (1, 0), (\frac{1}{2}, \frac{1}{2}))$ is a Nash equilibrium with expected social cost 5. This proves the lemma. $\qquad\square$

Lemma 10. *If* $c = \{m_1, m_2\}$ *satisfies the Specific 2-3-Mirror Property with* m_1 *including* $(2_2, 3, 2_1)$, *and* m_2 *not including* $(3, 2_2, 2_1)$ *and* $(3, 2_1, 2_2)$, *then* $PoA(c) \geq \frac{6}{5}$.

Proof. For the described coordination mechanism, if m_2 includes $(2_1, 3, 2_2)$, then for a game with $w = \{2, 2, 2, 3\}$, $E = ((0, 1), (0, 1), (1, 0), (\frac{1}{2}, \frac{1}{2}))$ is a Nash equilibrium with expected social cost 6. Obviously, $opt(w) = 5$. If m_2 includes $(2_2, 3, 2_1)$, then for a game with $w = \{2, 2, 2, 3\}$, the Nash equilibrium $E = ((0, 1), (0, 1), (1, 0), (\frac{3}{8}, \frac{5}{8}))$ has $sc(c; w; E) = \frac{25}{4}$, while $opt(w) = 5$. If m_2 includes $(2_2, 2_1, 3)$, then for a game with set of task lengths $w = \{2, 2, 3, 3\}$, either $((0, 1), (\frac{1}{2}, \frac{1}{2}), (0, 1), (1, 0))$ or $((0, 1), (\frac{1}{2}, \frac{1}{2}), (1, 0), (0, 1))$ is a Nash equilibrium, depending on the sequence of $2, 3_1, 3_2$ that is included in m_2. In both cases the expected social cost is 6, while the optimal is 5. Finally, if m_2 includes $(2_1, 2_2, 3)$, then for a game with $w = \{2, 2, 2, 3\}$ and under the assumption that m_1 does not include $(2_1, 2_2, 2_3)$ or $(2_1, 2_3, 2_2)$, we can confirm that $E = ((0, 1), (\frac{1}{2}, \frac{1}{2}), (1, 0), (0, 1))$ is a Nash equilibrium with $sc(c; w; E) = 6$, while $opt(w) = 5$. From Lemma 8 and Lemma 9, we get that if the assumption does not hold, then the price of anarchy of the mechanism is at least $\frac{6}{5}$. This completes the proof. $\square$

At this point we have studied mechanisms that satisfy the Specific 2-3-Mirror Property, for all possible sequences of $2_1, 2_2, 3$ in m_1, and we have proved that their price of anarchy is at least $\frac{7}{6}$. So, the price of anarchy of all non-preemptive coordination mechanisms that satisfy the Specific 2-3-Mirror Property is at least $\frac{7}{6}$.

Theorem 2. *If* c *is a non-preemptive coordination mechanism for* $m = 2$, *then* $PoA(c) \geq \frac{7}{6}$.

Proof. We know that the price of anarchy of non-preemptive coordination mechanisms that satisfy the Specific 2-3-Mirror Property is at least $\frac{7}{6}$. We proved it by presenting several Nash equilibria with tasks of lengths 2 and 3. We can observe that the same proof can be applied to the other 3 cases of mechanisms that satisfy the 2-3-Mirror Property. If instead of $(2_2, 2_1) \in m_1$ and $(2_1, 2_2) \in m_2$, we have $(2_1, 2_2) \in m_1$ and $(2_2, 2_1) \in m_2$, we adjust the case analysis as follows. We reverse the "direction" of the tasks with length 2 in all sequences of the mechanism. For example, if a sequence of a policy in our case analysis is $(\ldots, 2_3, \ldots, 2_1, \ldots, 2_2, \ldots)$, we turn it into $(\ldots, 2_1, \ldots, 2_3, \ldots, 2_2, \ldots)$, or if it is $(\ldots, 2_1, \ldots, 2_2, \ldots)$, we turn it into $(\ldots, 2_2, \ldots, 2_1, \ldots)$. We also change the strategies of the tasks with length 2 as follows. The task which is placed last in the increasing order switches strategies with the task which is placed first, the task which is placed second to last switches with the task which is placed second and so on. This produces a case analysis that is symmetric to the one we followed previously and proves the same result. The same idea is applied if instead of $(3_1, 3_2) \in m_1$ and $(3_2, 3_1) \in m_2$, we have $(3_2, 3_1) \in m_1$ and $(3_1, 3_2) \in m_2$. This proves that all non-preemptive coordination mechanisms that satisfy the 2-3-Mirror Property have price of anarchy at least $\frac{7}{6}$. Combining this with Lemma 2 completes the proof. $\square$

Our constant bounds on the price of anarchy of non-preemptive coordination mechanisms yield an interesting observation, when compared with $\frac{4}{3} - \frac{1}{3m}$ which is the currently best known price of anarchy and is achieved by the preemptive All-Decreasing mechanism of [6]. For $m > 2$, the price of anarchy of every non-preemptive coordination mechanism is at least $\frac{3}{2}$, thus strictly greater than $\frac{4}{3} - \frac{1}{3m}$. For the case of $m = 2$ machines the lower bound is $\frac{7}{6}$ and is equal to $\frac{4}{3} - \frac{1}{3m}$.

Corollary 1. *For the price of anarchy of every non-preemptive coordination mechanism c, for identical machine scheduling,*

$$PoA(c) \geq \frac{4}{3} - \frac{1}{3m}.$$

Thus, we conclude that it is impossible to produce a non-preemptive coordination mechanism that improves on the currently best known price of anarchy for this setting. □

We are able to derive upper bounds for the price of anarchy of the best coordination mechanism for identical machine scheduling from previous work.

Remark 1. *The upper bound for $m = 2$ machines is given from the Increasing-Decreasing mechanism of [6] and it is $\frac{4}{3}$. For $m > 2$, it is trivial to see that if all machines schedule the tasks in decreasing task length order, then the price of anarchy is $\Theta(\frac{\log m}{\log \log m})$, so the price of anarchy of the best non-preemptive coordination mechanism for $m > 2$ is $O(\frac{\log m}{\log \log m})$.* □

4 Conclusion and Open Problems

In this paper, we gave a formal definition of non-preemptive coordination mechanisms for identical machine scheduling and studied their price of anarchy. We managed to prove lower bounds for their price of anarchy and presented upper bounds for the best price of anarchy that such a mechanism can achieve.

The proof of the lower bound for $m = 2$ machines is based on an extensive case analysis. This may seem unavoidable, but it remains open whether a more direct proof exists. Interesting open problems for future research, also include closing the gap between the upper and lower bounds for the price of anarchy of non-preemptive coordination mechanisms for identical machine scheduling. The most important problem of the area is proving upper and lower bounds on the price of anarchy of the best coordination mechanism for this setting. The known upper bound is $\frac{4}{3} - \frac{1}{3m}$ and the known lower bound is 1.

References

1. Angel, E., Bampis, E., Pascual, F.: Truthful Algorithms for Scheduling Selfish Tasks on Parallel Machines. In: Deng, X., Ye, Y. (eds.) WINE 2005. LNCS, vol. 3828. Springer, Heidelberg (2005)

2. Awerbuch, B., Azar, Y., Richter, Y.: Analysis of worst case Nash equilibria for restricted assignment (Manuscript 2002)
3. Azar, Y., Jain, K., Mirrokni, V. (Almost) Optimal Coordination Mechanisms for Unrelated Machine Scheduling. In: SODA (2008)
4. Bagchi, A.: Stackelberg differential games in economic models. Lecture Notes in Control and Information Sciences, vol. 64. Springer, Heidelberg (1984)
5. Beckmann, M., McGuire, C.B., Winstein, C.B.: Studies in the Economics of Transportation. Yale University Press (1956)
6. Christodoulou, G., Koutsoupias, E., Nanavati, A.: Coordination mechanisms. In: Díaz, J., Karhumäki, J., Lepistö, A., Sannella, D. (eds.) ICALP 2004. LNCS, vol. 3142. Springer, Heidelberg (2004)
7. Cole, R., Dodis, Y., Roughgarden, T.: How much can taxes help selfish routing? In: ACM EC (2003)
8. Czumaj, A.: Selfish routing on the Internet. In: Handbook of Scheduling: Algorithms, Models, and Performance Analysis (2004)
9. Czumaj, A., Vocking, B.: Tight bounds for worst-case equilibria. In: SODA (2002)
10. Fleischer, L., Jain, K., Mahdian, M.: Tolls for heterogeneous selfish users in multi-commodity networks and generalized congestion games. In: FOCS (2004)
11. Immorlica, N., Li, L., Mirrokni, V., Schulz, A.: Coordination mechanisms for selfish scheduling. In: Deng, X., Ye, Y. (eds.) WINE 2005. LNCS, vol. 3828. Springer, Heidelberg (2005)
12. Korilis, Y.A., Lazar, A.A., Orda, A.: Achieving network optima using Stackelberg routing strategies. IEEE/ACM Transactions on Networking (1997)
13. Koutsoupias, E.: Coordination mechanisms for congestion games. Sigact News (December 2004)
14. Koutsoupias, E., Mavronicolas, M., Spirakis, P.: Approximate equilibria and ball fusion. In: SIROCCO (2002)
15. Koutsoupias, E., Papadimitriou, C.H.: Worst-case equilibria. In: Meinel, C., Tison, S. (eds.) STACS 1999. LNCS, vol. 1563. Springer, Heidelberg (1999)
16. Mavronicolas, M., Spirakis, P.: The price of selfish routing. In: STOC (2001)
17. Nash, J.F.: Non-cooperative Games. Annals of Mathematics (1951)
18. Nisan, N., Ronen, A.: Algorithmic mechanism design. Games and Economic Behavior 35, 166–196 (2001)
19. Nisan, N., Roughgarden, T., Tardos, E., Vazirani, V.: Algorithmic Game Theory. Cambridge University Press, Cambridge (2007)
20. Osborne, M.J., Rubinstein, A.: A Course in Game Theory. The MIT Press, Cambridge (1994)
21. Papadimitriou, C.H.: Algorithms, games, and the Internet. In: STOC (2001)
22. Roughgarden, T.: Stackelberg scheduling strategies. In: STOC (2001)
23. von Stackelberg, H.: Marktform und Gleichgewicht. English translation entitled The Theory of the Market Economy. Springer, Heidelberg (1934)

Computing Approximate Nash Equilibria in Network Congestion Games[*]

Andreas Emil Feldmann[1], Heiko Röglin[2], and Berthold Vöcking[2]

[1] Institute of Theoretical Computer Science
ETH Zürich, Switzerland
andreas.feldmann@inf.ethz.ch
[2] Department of Computer Science
RWTH Aachen, Germany
{roeglin,voecking}@cs.rwth-aachen.de

Abstract. We consider the problem of computing ε-approximate Nash equilibria in network congestion games. The general problem is known to be PLS-complete for every $\varepsilon > 0$, but the reductions are based on artificial and steep delay functions with the property that already two players using the same resource cause a delay that is significantly larger than the delay for a single player.

We consider network congestion games with delay functions such as polynomials, exponential functions, and functions from queuing theory. We analyse which approximation guarantees can be achieved for such congestion games by the method of randomised rounding. Our results show that the success of this method depends on different criteria depending on the class of functions considered. For example, queuing theoretical functions admit good approximations if the equilibrium load of every resource is bounded away appropriately from its capacity.

1 Introduction

In recent years, there has been an increased interest in understanding selfish routing in large networks like the Internet. Since the Internet is operated by different economic entities with varying interests, it is natural to model these entities as selfish agents who are only interested in maximising their own benefit. *Congestion games* are a classical model for resource allocation among selfish agents. We consider the special case of *network congestion games*, in which the resources are the edges of a graph and every player wants to allocate a path between her designated source and target node. The delay of an edge increases with the number of players allocating it, and every player is interested in allocating a routing path with minimum delay.

Rosenthal [10] shows that congestion games are potential games and hence they always possess *Nash equilibria*[1], i.e. allocations of resources from which no

---

[*] This work was supported by DFG grant VO 889/2 and by the Ultra High-Speed Mobile Information and Communication Research Cluster (UMIC) established under the excellence initiative of the German government.
[1] In this paper, the term *Nash equilibrium* always refers to a pure equilibrium.

A. Shvartsman and P. Felber (Eds.): SIROCCO 2008, LNCS 5058, pp. 209–220, 2008.
© Springer-Verlag Berlin Heidelberg 2008

player wants to deviate unilaterally. Fabrikant et al. [5] show that the problem of computing a pure Nash equilibrium can be phrased as a local search problem belonging to the complexity class PLS. They show that it is already PLS-complete for the special case of network congestion games if different players are allowed to have different source and target nodes. Ackermann et al. [1] show that this result can even be extended to network congestion games with linear delay functions. This implies that there is no efficient algorithm for computing pure Nash equilibria, unless PLS ⊆ P. On the other hand, for *symmetric network congestion games*, in which all players have the same source and the same target node, Nash equilibria can be computed efficiently by solving a min-cost flow problem [5].

In many applications players incur some costs when they change their strategy. Hence, it is reasonable to assume that a player is only interested in changing her strategy if this decreases her delay significantly. This assumption leads to the notion of an *ε-approximate Nash equilibrium*, which is a state in which no player can decrease her delay by more than a factor of $1 + \varepsilon$ by unilaterally changing her strategy. For symmetric congestion games, in which the strategy spaces of the players coincide, Chien and Sinclair [3] show that ε-approximate equilibria can be computed by simulating the best response dynamics for a polynomial (in the number of players and ε^{-1}) number of steps. Unfortunately, the problem of computing an ε-approximate Nash equilibrium is still PLS-complete for every constant $\varepsilon > 0$ (and even every polynomial-time computable function ε) for general congestion games [12] and even for network congestion games[2]. The delay functions used in these reductions are quite artificial and steep with the property that already two players using the same resource cause a delay that is significantly larger than the delay for a single player. In this paper, we study natural classes of delay functions such as polynomials and functions from queuing theory, and we analyse which approximation guarantees can be achieved for these functions by the method of randomised rounding [9].

1.1   Models and Method

A network congestion game is described by a directed graph $G = (V, E)$ with m edges, a set $\mathcal{N}$ of n players, a pair $(s_i, t_i) \in V \times V$ of source and target node for each player $i \in \mathcal{N}$, and a non-decreasing delay function $d_e \colon \mathbb{R}_{\geq 0} \to \mathbb{R}_{\geq 0}$ for each edge $e \in E$. For $i \in \mathcal{N}$ we denote by $\mathcal{P}_i$ the set of all paths from node s_i to node t_i. Every player i has to choose one path P_i from the set $\mathcal{P}_i$ and to allocate all edges on this path. For a *state* $S = (P_1, \ldots, P_n) \in \mathcal{P}_1 \times \ldots \times \mathcal{P}_n$ and an edge $e \in E$, we denote by $n_e(S)$ the number of players allocating edge e in state S, i.e. $n_e(S) = |\{i \in \mathcal{N} \mid e \in P_i\}|$. The *delay* $\delta_i(S)$ to a player $i \in \mathcal{N}$ in state S is defined as equal to the delay $d_{P_i}(S) := \sum_{e \in P_i} d_e(n_e(S))$ of the chosen path P_i in S and every player wants to allocate a path with minimum delay. We say that a state S is a *Nash equilibrium* if no player can decrease her delay by changing her strategy, i.e. if state S' is obtained from S by letting one player $i \in \mathcal{N}$ change her strategy, then the delay $\delta_i(S')$ is at least as large as the delay $\delta_i(S)$. A state S is said to be an *ε-approximate Nash equilibrium* if $\delta_i(S) \leq (1 + \varepsilon) \cdot \delta_i(S')$ for

---

[2] Alexander Skopalik, personal communication.

every state S' that is obtained from S by letting one player $i \in \mathcal{N}$ change her strategy.

In order to compute approximate Nash equilibria, we use the method of randomised rounding. Therefore, we first relax the network congestion game by replacing each player by an infinite set of agents, each of which controlling an infinitesimal amount of flow. To be more precise, we transform the network congestion game into a multi-commodity flow problem and we introduce a flow demand of 1 that is to be routed from node s_i to node t_i for every player $i \in \mathcal{N}$. A flow vector $f \in \mathbb{R}_{\geq 0}^{|E|}$ induces delays on the edges. The delay of edge $e \in E$ is $d_e(f) = d_e(f_e)$, and the delay on a path P is the sum of the delays of its edges, i.e. $d_P(f) = \sum_{e \in P} d_e(f)$. A flow vector f is called a *Wardrop equilibrium* [13] if for all commodities $i \in \mathcal{N}$ and all paths $P_1, P_2 \in \mathcal{P}_i$ with $f_{P_1} > 0$ it holds that $d_{P_1}(f) \leq d_{P_2}(f)$. It is well-known that Wardrop equilibria can be computed in polynomial time using convex programming [2].

After relaxing the network congestion game and computing a Wardrop equilibrium f, we compute a decomposition of the flow f into polynomially many paths. For a commodity $i \in \mathcal{N}$ let $\mathcal{D}_i \subseteq \mathcal{P}_i$ denote the set of paths used in this decomposition, and for $P \in \mathcal{D}_i$, let f_P^i denote the flow of commodity i that is sent along path P. For fixed i the flows f_P^i can be interpreted as a probability distribution on the set $\mathcal{D}_i$. Following the method of randomised rounding, we choose, according to these probability distributions, independently for each player i a routing path from $\mathcal{D}_i \subseteq \mathcal{P}_i$. In the following, we analyse for several classes of delay functions the approximation guarantee of this approach.

1.2 Our Results

After the randomised rounding the congestion on an edge is a sum of independent Bernoulli random variables whose expectation equals the flow on that edge in the Wardrop equilibrium. By applying Chernoff bounds, we can find for each edge a small interval such that it is unlikely that any congestion takes a value outside the corresponding interval. If the delay functions are not too steep in these intervals, then the delays on the edges after the rounding are neither much smaller nor much larger than the delays in the Wardrop equilibrium, implying that the resulting state is an ε-approximate Nash equilibrium for some ε depending on the steepness of the delay functions. Due to the multiplicative definition of approximate Nash equilibria, delay functions can be steeper in intervals in which they take larger values in order to achieve the same ε.

In the literature on selfish routing, it is a common assumption that the delay functions are polynomials with nonnegative coefficients [4,11]. Hence, we start our investigations with the question which properties polynomial delay functions need to satisfy in order to guarantee that randomised rounding yields an ε-approximate Nash equilibrium with high probability. We have argued that the delay functions must not grow too fast relative to their values. For polynomials $d_e(x) = \sum_{j=0}^{g} a_j^e x^j$ with $a_j^e \geq 0$, this implies that the offset a_0^e must not be too small. If all delay functions are polynomials of some constant degree g and if, for each edge $e \in E$, the offset a_0^e satisfies

$$a_0^e \geq \frac{((1+\varepsilon)g^2 \cdot 6\ln(4m))^g}{((\sqrt{1+\varepsilon})-1)^{2g+1}} \sum_{j=1}^{g} a_j^e = \Theta\left(\frac{\ln^g m}{\varepsilon^{2g+1}}\right) \sum_{j=1}^{g} a_j^e ,$$

then an ε-approximate Nash equilibrium can be computed by randomised rounding with high probability in polynomial time. In the above asymptotic estimate ε tends to zero while m approaches infinity. If, for example, all delay functions are linear and one wants to obtain an ε-approximate equilibrium for some constant $\varepsilon > 0$, then all delay functions must have the form $d_e(x) = a_1^e x + a_0^e$ where a_0^e is sufficiently large in $\Omega(a_1^e \cdot \ln(m))$. A lower bound on a_0^e is not unrealistic since most network links have a non-negligible delay even if they are relatively uncrowded. For example, in communication networks the offset corresponds to the sum of *packet-propagation delay* and *packet-processing delay*, which should dominate the *packet-queuing delay* if an edge is not dramatically overloaded.

The second class that we study are exponential delay functions of the form $d_e(x) = \alpha_e \cdot \exp(x/\beta_e) + \gamma_e$. We show that for these functions an ε-approximation can be achieved if β_e is lower bounded by some function $f(\chi_e, m, \varepsilon)$ growing in the order of

$$f(\chi_e, m, \varepsilon) \in O\left(\frac{\ln(m) \cdot \sqrt{\chi_e}}{\varepsilon}\right) ,$$

where χ_e denotes the load of edge e in the Wardrop equilibrium and hence corresponds to the expected congestion on e after the randomised rounding. Such exponential functions grow very slowly as long as less than β_e players allocate the edge, but they start growing rapidly beyond this point. This reflects typical behaviour in practice, because one often observes that the delay on a network link grows rather slowly with its congestion until some overload point is reached after which the quality of the link deteriorates quickly. We show that it is even possible to replace the exponential function up until β_e by a polynomial. To be precise, we show that if an ε-approximate equilibrium can be computed for two functions by randomised rounding, then this is also the case for the function that takes for every input the maximum of these functions.

Finally, we consider functions that arise when using queuing theory for modelling the behaviour of network links. We consider the $M/M/1$ queuing model in which there is one queue and the inter-arrival and service times of the packets are exponentially distributed. A network of such queues constitutes a so-called Jackson network [6]. We take as delay function the limiting behaviour of the sojourn time of a packet on the network link. We interpret the congestion on an edge as the rate of the packet arrival process and we show that, in order to obtain an ε-approximate equilibrium, it suffices that the rate μ_e of the service process (corresponding to the capacity of the edge) is lower bounded by some function $g(\chi_e, m, \varepsilon)$ growing in the order of

$$g(\chi_e, m, \varepsilon) \in O\left(\chi_e + \frac{\ln(m) \cdot \sqrt{\chi_e}}{\varepsilon}\right) .$$

Our result implies that it is sufficient if the equilibrium load χ_e is bounded away from the capacity μ_e by some additive term of order only $O(\ln(m) \cdot \sqrt{\mu_e}/\varepsilon)$.

Outline. In the remainder of this paper we first state some preliminaries and illustrate our approach. After that we give a sufficient condition on the delay functions that guarantees that randomised rounding computes an ε-approximate equilibrium in polynomial time with high probability. Then we analyse which restrictions this condition imposes when applied to polynomial delay functions, exponential delay functions, and delay functions from queuing theory. Finally, we prove a theorem for combined delay functions.

2   Preliminaries

The number N_e of players that use an edge $e \in E$ after the rounding is a sum of independent Bernoulli random variables whose expectation χ_e equals the flow on e in the Wardrop equilibrium. We can use Chernoff bounds to identify, for each edge e, an interval $[l_e, u_e]$ such that it is unlikely that N_e takes a value outside this interval. We choose these intervals such that for $x \in [l_e, u_e]$ the delay $d_e(x)$ of edge e lies between $d_e(\chi_e)/\sqrt{1+\varepsilon}$ and $\sqrt{1+\varepsilon} \cdot d_e(\chi_e)$.

Lemma 1. *If for all $e \in E$ it holds that $N_e \in [l_e, u_e]$ and $d_e(\chi_e)/\sqrt{1+\varepsilon} \leq d_e(x) \leq \sqrt{1+\varepsilon} \cdot d_e(\chi_e)$ for any $x \in [l_e, u_e]$, then the resulting state is an ε-approximate Nash equilibrium.*

Proof. Let S denote the state computed by the randomised rounding. Assume that a path P_i is chosen for player i by the randomised rounding and that P_i' is a path with minimum delay after the randomised rounding. From the definition of a Wardrop equilibrium it follows that in the computed flow the delay L_i on P_i is at most as large as the delay on P_i' because flow is sent along P_i (otherwise the probability that path P_i is chosen would equal 0). Since the delay on P_i increases at most by a factor of $\sqrt{1+\varepsilon}$ and the delay on P_i' decreases at most by a factor of $\sqrt{1+\varepsilon}$ during the randomised rounding, we obtain

$$\frac{d_{P_i}(S)}{d_{P_i'}(S)} \leq \frac{\sqrt{1+\varepsilon} \cdot L_i}{L_i/\sqrt{1+\varepsilon}} = 1 + \varepsilon \ ,$$

which proves the lemma. $\square$

3   A Sufficient Condition on the Delay Functions

In this section, we present a sufficient condition on the delay functions that guarantees that an ε-approximate Nash equilibrium can be computed by randomised rounding in polynomial time. We will make use of the following Chernoff bounds.

Lemma 2 ([8]). *Let $X_1, \ldots, X_n$ be independent random variables with $\mathbf{Pr}[X_i = 1] = p_i$ and $\mathbf{Pr}[X_i = 0] = 1 - p_i$ for each $i \in \{1, \ldots, n\}$ and let the random variable X be defined as $\sum_{i=1}^n X_i$.*

- *If $\mu \geq \mathbf{E}[X]$ and $0 \leq \delta \leq 1$, then $\mathbf{Pr}[X > (1+\delta) \cdot \mu] \leq \exp\left(-\frac{\delta^2 \mu}{3}\right)$.*
- *If $\mu \leq \mathbf{E}[X]$ and $0 \leq \delta \leq 1$, then $\mathbf{Pr}[X < (1-\delta) \cdot \mu] \leq \exp\left(-\frac{\delta^2 \mu}{3}\right)$.*

We make two assumptions on the delay functions to avoid case distinctions and to keep the statement of the next theorem simple. We assume that each delay function is defined on $\mathbb{R}$ and w.l.o.g. we set $d_e(x) = d_e(0)$ for $x < 0$. Additionally, we assume that the delay function never equals zero, i.e. $d_e(x) > 0$ for all $x \in \mathbb{R}$. The latter condition is reasonable since in practice the delay of a network link never drops to zero.

For an edge e, let χ_e in the following denote the expected congestion, which equals the flow on edge e in the Wardrop equilibrium.

Theorem 3. *Using the method of randomised rounding, it is possible to compute an ε-approximate Nash equilibrium of a network congestion game with high probability in polynomial time if for each edge $e \in E$ and for all $x \in [0, \max\{6\ln(4m), \chi_e + \sqrt{3\ln(4m) \cdot \chi_e}\}]$*

$$\frac{d_e(x)}{d_e(x - \sqrt{6\ln(4m) \cdot x})} \leq \sqrt{1 + \varepsilon} \ . \tag{1}$$

Proof. Following the arguments in Section 2, we define an interval $\mathfrak{I}_e := [l_e, u_e]$ for each edge e such that, after the randomised rounding, the congestion N_e lies in this interval with probability at least $1 - \frac{1}{2m}$ and such that $d_e(u_e)/d_e(\chi_e) \leq \sqrt{1 + \varepsilon}$ and $d_e(\chi_e)/d_e(l_e) \leq \sqrt{1 + \varepsilon}$. Given these properties, one can easily see that after the randomised rounding and with probability at least $1/2$, $N_e \in \mathfrak{I}_e$ for all edges e. If this event occurs, then the resulting state is an ε-approximate Nash equilibrium due to Lemma 1. Since the failure probability is at most $1/2$, repeating the randomised rounding, say, n times independently yields an exponentially small failure probability.

Since we assume $d_e(x) = d_e(0)$ for $x < 0$, we can assume that Inequality (1) holds for all $x \in \mathbb{R}_{\leq 0}$. For an edge e, we set

$$\mathfrak{I}_e = [l_e, u_e] = \left[\chi_e - \sqrt{3\ln(4m) \cdot \chi_e}, \max\{6\ln(4m), \chi_e + \sqrt{3\ln(4m) \cdot \chi_e}\}\right] \ .$$

Since $l_e \geq \chi_e - \sqrt{6\ln(4m) \cdot \chi_e}$, Inequality (1) and the monotonicity of d_e imply

$$\frac{d_e(\chi_e)}{d_e(l_e)} \leq \frac{d_e(\chi_e)}{d_e(\chi_e - \sqrt{6\ln(4m) \cdot \chi_e})} \leq \sqrt{1 + \varepsilon} \ .$$

If $\chi_e \geq 3\ln(4m)$, then $u_e = \chi_e + \sqrt{3\ln(4m) \cdot \chi_e}$ and

$$\frac{d_e(u_e)}{d_e(\chi_e)} \leq \frac{d_e(\chi_e + \sqrt{3\ln(4m) \cdot \chi_e})}{d_e(\chi_e)} \leq \sqrt{1 + \varepsilon} \ ,$$

where the last inequality follows from substituting x by $x + \sqrt{3\ln(4m) \cdot x}$ in Inequality (1) and by using the monotonicity of the delay function d_e. If $\chi_e \leq 3\ln(4m)$, then $u_e = 6\ln(4m)$ and

$$\frac{d_e(u_e)}{d_e(\chi_e)} \leq \frac{d_e(6\ln(4m))}{d_e(0)} \leq \sqrt{1 + \varepsilon} \ ,$$

where the last inequality follows directly from (1) by setting $x = 6\ln(4m)$. Altogether, this implies that we have achieved the desired properties that $d_e(u_e)/d_e(\chi_e) \leq \sqrt{1+\varepsilon}$ and $d_e(\chi_e)/d_e(l_e) \leq \sqrt{1+\varepsilon}$.

It remains to analyse the probability with which the congestion N_e of an edge e takes on a value in the interval $\mathfrak{I}_e$ defined above. Since the congestion N_e is the sum of independent Bernoulli random variables, we can apply the Chernoff bound stated in Lemma 2, yielding

$$
\begin{aligned}
\mathbf{Pr}\left[N_e < l_e\right] &= \mathbf{Pr}\left[N_e < \left(1 - \sqrt{\frac{3\ln(4m)}{\chi_e}}\right)\chi_e\right] \\
&\leq \exp\left(-\frac{1}{3}\left(\sqrt{\frac{3\ln(4m)}{\chi_e}}\right)^2 \chi_e\right) = \frac{1}{4m} .
\end{aligned}
$$

If $\chi_e \geq 3\ln(4m)$, then $u_e = \chi_e + \sqrt{3\ln(4m) \cdot \chi_e}$, for which we obtain

$$
\begin{aligned}
\mathbf{Pr}\left[N_e > u_e\right] &= \mathbf{Pr}\left[N_e > \left(1 + \sqrt{\frac{3\ln(4m)}{\chi_e}}\right)\chi_e\right] \\
&\leq \exp\left(-\frac{1}{3}\left(\sqrt{\frac{3\ln(4m)}{\chi_e}}\right)^2 \chi_e\right) = \frac{1}{4m} .
\end{aligned}
$$

If $\chi_e \leq 3\ln(4m)$, then $u_e = 6\ln(4m)$, for which we obtain

$$
\begin{aligned}
\mathbf{Pr}\left[N_e > u_e\right] &= \mathbf{Pr}\left[N_e > (1+1) \cdot 3\ln(4m)\right] \\
&\leq \exp\left(-\frac{1}{3} \cdot 3\ln(4m)\right) = \frac{1}{4m} .
\end{aligned}
$$

Altogether, this implies that $\mathbf{Pr}\left[N_e \notin \mathfrak{I}_e\right] \leq \mathbf{Pr}\left[N_e < u_e\right] + \mathbf{Pr}\left[N_e > l_e\right] \leq 1/2m$, as desired. □

4   Analysis of Classes of Delay Functions

In this section we analyse which conditions Theorem 3 imposes when applied to polynomial delay functions, exponential delay functions, and delay functions from queuing theory.

4.1   Polynomial Delay Functions

We consider polynomial delay functions with nonnegative coefficients and constant degree g. That is, the delay function has the form $d(x) = \sum_{j=0}^{g} a_j x^j$, where $a_j \geq 0$ for $j \in \{0, \ldots, g-1\}$ and $a_g > 0$. Since the coefficients are nonnegative, the function d is non-decreasing. To fulfil the assumption that the delay function never equals zero, we also assume that $a_0 > 0$.

Theorem 4. *A polynomial delay function d with degree g and nonnegative co-efficients satisfies Condition* (1) *in Theorem 3 for all $x \in \mathbb{R}_{\geq 0}$ if*

$$a_0 \geq \frac{((1+\varepsilon)g^2 \cdot 6\ln(4m))^g}{(\sqrt{1+\varepsilon}-1)^{2g+1}} \sum_{j=1}^{g} a_j = \Theta\left(\frac{\ln^g m}{\varepsilon^{2g+1}}\right) \sum_{j=1}^{g} a_j^e \ . \tag{2}$$

Proof. To establish the theorem we show that (2) implies Inequality (1) from Theorem 3 for any $x \geq 0$. In the following, we assume $g > 0$ because for constant functions Inequality (1) is trivially satisfied.

In order to show that Inequality (1) is satisfied we use two upper bounds on the function

$$f(x) = \frac{d(x)}{d(x - \sqrt{6\ln(4m) \cdot x})} \ ,$$

of which one is monotonically increasing and the other is monotonically decreasing. We show that the upper bounds are chosen such that their minimum is bounded from above by $\sqrt{1+\varepsilon}$ for every $x \geq 0$. Since d is non-decreasing and we assumed that $d(x)$ equals $d(0)$ for any $x \leq 0$, we obtain, for every $x \geq 0$,

$$f(x) = \frac{d(x)}{d(x - \sqrt{6\ln(4m) \cdot x})} \leq \frac{d(x)}{d(0)} = \frac{1}{a_0}d(x) \ . \tag{3}$$

The second upper bound on $f(x)$ is presented in the following lemma.

Lemma 5. *For all $x > g^2 \cdot 6\ln(4m)$,*

$$f(x) \leq \frac{1}{1 - g\sqrt{\frac{6\ln(4m)}{x}}} \ .$$

Proof. Since the second derivative of any polynomial with nonnegative coefficients is greater or equal to 0, the delay function is convex. The fact that the first order Taylor approximation of a convex function is always a global underestimator yields, for $x \geq 0$,

$$d(x - \sqrt{6\ln(4m) \cdot x}) \geq d(x) - \sqrt{6\ln(4m) \cdot x} \cdot d'(x) \ . \tag{4}$$

The lower bound in (4) is positive for $x > g^2 \cdot 6\ln(4m)$ because

$$d(x) - \sqrt{6\ln(4m) \cdot x} \cdot d'(x) = \frac{d(x)}{\sqrt{x}}\left(\sqrt{x} - \sqrt{6\ln(4m)} \cdot \frac{xd'(x)}{d(x)}\right)$$

$$\geq \frac{d(x)}{\sqrt{x}}\left(\sqrt{x} - \sqrt{g^2 \cdot 6\ln(4m)}\right) > 0 \ ,$$

where the second to the last inequality follows because $xd'(x)/d(x)$ is the so-called *elasticity* of d, which can readily be seen to be upper bounded by g for polynomials with degree g and nonnegative coefficients. Hence, for $x > g^2 \cdot 6\ln(4m)$, we obtainc

$$\frac{d(x)}{d(x - \sqrt{6\ln(4m) \cdot x})} \leq \frac{d(x)}{d(x) - \sqrt{6\ln(4m) \cdot x} \cdot d'(x)}$$

$$= \frac{1}{1 - \frac{xd'(x)}{d(x)}\sqrt{\frac{6\ln(4m)}{x}}}$$

$$\leq \frac{1}{1 - g\sqrt{\frac{6\ln(4m)}{x}}} \quad ,$$

which concludes the proof of the lemma. □

Let $x_\varepsilon = \frac{(1+\varepsilon)g^2 \cdot 6\ln(4m)}{((\sqrt{1+\varepsilon})-1)^2}$. We show that, for $x \leq x_\varepsilon$, the upper bound in (3) yields $f(x) \leq \sqrt{1+\varepsilon}$ and that, for $x \geq x_\varepsilon$, Lemma 5 implies $f(x) \leq \sqrt{1+\varepsilon}$. Since the upper bound in (3) is non-decreasing, it suffices to observe that $d(x_\varepsilon)/a_0 \leq \sqrt{1+\varepsilon}$, which follows from

$$\frac{1}{a_0}d(x_\varepsilon) = \frac{1}{a_0}\sum_{j=0}^{g} a_j x_\varepsilon^j = 1 + \frac{1}{a_0}\sum_{j=1}^{g} a_j x_\varepsilon^j \leq 1 + \frac{x_\varepsilon^g}{a_0}\sum_{j=1}^{g} a_j \leq \sqrt{1+\varepsilon} \ ,$$

where the last inequality follows from (2) and we used the fact that $x_\varepsilon \geq 1$ if $g \geq 1$. Since the upper bound on $f(x)$ given in Lemma 5 is non-increasing and

$$\frac{1}{1 - g\sqrt{\frac{6\ln(4m)}{x_\varepsilon}}} = \sqrt{1+\varepsilon} \ ,$$

the theorem follows. □

4.2 Exponential Delay Functions

The lower bound on a_0 in Theorem 4 is determined by the fact that we allow any input value from the domain $\mathbb{R}_{\geq 0}$, which is a natural assumption. However, this means that the bound is too restrictive in the case that the congestion is large, since then also the interval in which the congestion falls is located at some point far to the right of the abscissa. From the fact that the upper bound given in Lemma 5 is decreasing, we can see that in these intervals only smaller values than needed in order to guarantee the approximation factor ε are reached. Since this seems to be a typical characteristic of polynomials, this raises the question whether Theorem 3 can also be applied to delay functions that grow superpolynomially from a certain point on. The next theorem gives an affirmative answer.

Theorem 6. *A delay function d of the form*

$$d(x) = \alpha \cdot \exp\left(\frac{x}{\beta}\right) + \gamma$$

satisfies Condition (1) in Theorem 3 in some interval $[0, u]$ if $\alpha > 0$, $\gamma \geq 0$, and $\beta \geq 2\sqrt{u \cdot 6\ln(4m)}/\ln(1+\varepsilon)$.

Proof. We have to show that all functions of the suggested form comply with
(1). This follows because, for $x \in [0, u]$,

$$\frac{d(x)}{d(x - \sqrt{6\ln(4m) \cdot x})} \leq \exp\left(\frac{\sqrt{6\ln(4m) \cdot x}}{\beta}\right) \leq \exp\left(\frac{\ln(1+\varepsilon)}{2}\right) = \sqrt{1+\varepsilon} \ ,$$

which concludes the proof. □

In Theorem 3 the upper bound u_e is set to $\max\{6\ln(4m), \chi_e + \sqrt{3\ln(4m) \cdot \chi_e}\}$.
When substituting u accordingly in the lower bound on β we obtain a bound
$f(\chi_e, m, \varepsilon)$ growing in the order of

$$f(\chi_e, m, \varepsilon) \in O\left(\frac{\ln(m) \cdot \sqrt{\chi_e}}{\varepsilon}\right) \ .$$

4.3   Delay Functions from Queuing Theory

In Kendall's notation [7], we consider the $M/M/1$ queuing model. This means
that the queue is processed in a first-come first-served manner, the inter-arrival
times at the queue as well as the service times of the packets are exponentially
distributed, and each network link can process only one packet at each point in
time.

The following basic theorem from queuing theory describes the limiting behaviour of the sojourn time of a packet on a network link. This is the time that
packet k, where k tends to infinity, spends on that link in total until it arrives
at the next node, i.e. it includes the waiting time in the queue plus the service
time of the packet. In the $M/M/1$ queuing model, the arrival of jobs is a Poisson process whose rate is denoted by λ, and the processing of jobs is a Poisson
process whose rate is denoted by μ. A basic assumption that has to be fulfilled
in order for the theorem to hold is that the occupation rate $\rho = \frac{\lambda}{\mu}$ is strictly
smaller than 1. Otherwise there would be more arrivals than the link can handle,
which would result in an unbounded growth of the queue.

Theorem 7 ([7]). *In an $M/M/1$ queuing system with arrival rate λ, service
rate μ, and in which $\rho < 1$ the limiting behaviour of the expected sojourn time is*

$$E[S] = \frac{1}{\mu - \lambda}.$$

In the following theorem we interpret the congestion as the arrival rate λ and
we assume that the considered link has a certain service rate μ.

Theorem 8. *A delay function d of the form*

$$d(x) = \begin{cases} \frac{1}{\mu - x} & \text{if } x < \mu \\ \infty & \text{if } x \geq \mu \end{cases}$$

satisfies Condition (1) in Theorem 3 in some interval $[0, u]$ if

$$\mu \geq u + \frac{\sqrt{6\ln(4m)u}}{(\sqrt{1+\varepsilon}) - 1} \ .$$

Proof. Since $x \leq u$ and $\mu \geq u$ we can use the finite part of $d(x)$ to obtain

$$\frac{d(x)}{d(x - \sqrt{6\ln(4m) \cdot x})} = 1 + \frac{\sqrt{6\ln(4m) \cdot x}}{\mu - x} \leq 1 + \frac{\sqrt{6\ln(4m) \cdot u}}{\mu - u} \leq \sqrt{1 + \varepsilon} .$$

The first inequality follows from the fact that the function is monotonically increasing in x and $x \leq u$, while the second inequality follows directly from the lower bound on μ. $\qquad\square$

When setting u to u_e, analogous to the case of the exponential functions, we obtain a lower bound $g(\chi_e, m, \varepsilon)$ on μ growing in the order of

$$g(\chi_e, m, \varepsilon) \in O\left(\chi_e + \frac{\ln(m) \cdot \sqrt{\chi_e}}{\varepsilon}\right) .$$

5 Combined Delay Functions

In the previous section we applied Theorem 3 to several classes of functions. In this section we prove a general theorem showing that if two delay functions satisfy (1) in Theorem 3, then also their maximum satisfies this property. This allows us to combine different types of delay functions. One weak point of Theorem 6 concerning exponential functions is that it works only for exponential functions that grow slowly up until $2\sqrt{u \cdot 6\ln(4m)}/\ln(1+\varepsilon)$. The following theorem allows us to combine such an exponential function with a polynomial that satisfies Theorem 4. If we take the maximum over these two functions, we obtain a function that grows polynomially until some point and exponentially thereafter.

Theorem 9. *Let p and q denote two delay functions that satisfy Condition (1) in Theorem 3 in some interval $x \in [0, u]$. Then the function*

$$d(x) = \max\{p(x), q(x)\}$$

also satisfies this condition for $x \in [0, u]$.

Proof. Let $x \in [0, u]$, $x' = x - \sqrt{6\ln(4m) \cdot x}$, and without loss of generality assume $d(x) = p(x)$. Since p satisfies (1), we know that $p(x)/p(x') \leq \sqrt{1 + \varepsilon}$. Hence, if $d(x') = p(x')$, then (1) follows immediately. If, however, $d(x') = q(x')$ then $q(x') \geq p(x')$ and by the definition of $d(x)$ we obtain

$$\frac{d(x)}{d(x')} = \frac{p(x)}{q(x')} \leq \frac{p(x)}{p(x')} \leq \sqrt{1 + \varepsilon} .$$

The last inequality holds again because p satisfies (1). $\qquad\square$

6 Conclusions

In this paper, we have considered network congestion games with delay functions from several different classes. We have identified properties that delay functions from these classes have to satisfy in order to guarantee that an approximate Nash equilibrium can be computed by randomised rounding in polynomial time. Additionally, we have presented a method of combining these delay functions.

It remains an interesting open question to explore the limits of approximability further and to close the gap between the PLS-completeness results and the positive results presented in this paper. This could be done by either proving PLS-completeness of computing approximate equilibria for more natural delay functions or by extending the positive results to larger classes of functions. We believe that other techniques than randomised rounding are required for the latter.

References

1. Ackermann, H., Röglin, H., Vöcking, B.: On the impact of combinatorial structure on congestion games. In: Proc. of the 47th Ann. IEEE Symp. on Foundations of Computer Science (FOCS), pp. 613–622 (2006)
2. Beckmann, M., McGuire, C.B., Winsten, C.B.: Studies in the Economics of Transportation. Yale University Press (1956)
3. Chien, S., Sinclair, A.: Convergence to approximate Nash equilibria in congestion games. In: Proc. of the 18th Ann. ACM–SIAM Symp. on Discrete Algorithms (SODA), pp. 169–178 (2007)
4. Christodoulou, G., Koutsoupias, E.: The price of anarchy of finite congestion games. In: Proc. of the 37th Ann. ACM Symp. on Theory of Computing (STOC), pp. 67–73 (2005)
5. Fabrikant, A., Papadimitriou, C., Talwar, K.: The complexity of pure Nash equilibria. In: Proc. of the 36th Ann. ACM Symp. on Theory of Computing (STOC), pp. 604–612 (2004)
6. Jackson, J.: Jobshop-like queueing systems. Management Science 10(1), 131–142 (1963)
7. Kendall, D.G.: Some problems in the theory of queues. Royal Statistic Society 13(2), 151–182 (1951)
8. Mitzenmacher, M., Upfal, E.: Probability and Computing: Randomized Algorithms and Probabilistic Analysis. Cambridge University Press, Cambridge (2005)
9. Raghavan, P., Thompson, C.D.: Randomized rounding: a technique for provably good algorithms and algorithmic proofs. Combinatorica 7, 365–374 (1987)
10. Rosenthal, R.W.: A class of games possessing pure-strategy Nash equilibria. Int. Journal of Game Theory 2, 65–67 (1973)
11. Roughgarden, T., Tardos, É.: How bad is selfish routing? Journal of the ACM 49(2), 236–259 (2002)
12. Skopalik, A., Vöcking, B.: Inapproximability of pure Nash equilibria. In: Proc. of the 40th Ann. ACM Symp. on Theory of Computing (STOC) (to appear, 2008)
13. Wardrop, J.G.: Some theoretical aspects of road traffic research. In: Proc. of the Institute of Civil Engineers, Pt. II, vol. 1, pp. 325–378 (1952)

On the Performance of Beauquier and Debas' Self-stabilizing Algorithm for Mutual Exclusion

Viacheslav Chernoy[1], Mordechai Shalom[2], and Shmuel Zaks[1]

[1] Department of Computer Science, Technion, Haifa, Israel
vchernoy@tx.technion.ac.il, zaks@cs.technion.ac.il
[2] TelHai Academic College, Upper Galilee, 12210, Israel
cmshalom@telhai.ac.il

Abstract. In [Dij74] Dijkstra introduced the notion of self-stabilizing algorithms and presented an algorithm with three states for the problem of mutual exclusion on a ring of processors. In [BD95] a similar three state algorithm with an upper bound of $5\frac{3}{4}n^2 + O(n)$ and a lower bound of $\frac{1}{8}n^2 - O(n)$ were presented for its stabilization time. For this later algorithm we prove an upper bound of $1\frac{1}{2}n^2 + O(n)$, and show a lower bound of $n^2 - O(n)$.

1 Introduction

The notion of self stabilization was introduced by Dijkstra in [Dij74]. He considers a system, consisting of a set of processors, and each running a program of the form: **if** *condition* **then** *statement*. A processor is termed *privileged* if its condition is satisfied. A *scheduler* chooses any privileged processor which then executes its statement (*i.e.,* makes a move); if there are several privileged processor, the scheduler chooses any of them. Such a scheduler is termed *centralized.* A scheduler that chooses any subset of the privileged processors, which are then making their moves simultaneously, is termed *distributed.* Thus, starting from any initial configuration, we get sequences of moves (termed *executions*). The scheduler thus determines all possible executions of the system. A specific subset of the configurations is termed *legitimate.* The system is *self-stabilizing* if any possible execution will eventually get – that is, after a finite number of moves – only to legitimate configurations. The number of moves from any initial configuration until the system stabilizes is often referred to as *stabilization time* (see, *e.g.,* [BJM06, CG02, NKM06, TTK00]).

Dijkstra studied in [Dij74] the fundamental problem of mutual exclusion, for which the subset of legitimate configurations includes the configurations in which exactly one processor is privileged. n processors are arranged in a ring, so that each processor can communicate with its two neighbors using a shared memory, and where not all processors use the same program. Three algorithms were presented – without correctness or complexity proofs – in which each processor could be in one of $k > n$, four and three states, respectively. Most attention thereafter was devoted to the third algorithm – to which we refer as algorithm

A. Shvartsman and P. Felber (Eds.): SIROCCO 2008, LNCS 5058, pp. 221–233, 2008.
© Springer-Verlag Berlin Heidelberg 2008

$\mathcal{D}$ in this paper – which is rather non-intuitive, and for which Dijkstra presented in [Dij86] a proof of correctness (another proof was given in [Kes88], and a proof of correctness under a distributed scheduler was presented in [BGM89]). Though while dealing with proofs of correctness one can sometimes get also complexity results, this was not the case with this proof of [Dij86]. In [CSZ07] an upper bound of $5\frac{3}{4}n^2$ was provided for the stabilization time of algorithm $\mathcal{D}$. In [BD95] Beauquier and Debas introduced a similar three state algorithm, to which we refer in this paper as algorithm $\mathcal{BD}$, and they show an upper bound of $5\frac{3}{4}n^2 + O(n)$ for its stabilization time. They also prove an $\Omega(n^2)$ lower bound for the stabilization time of a family of algorithms, which includes both algorithm $\mathcal{D}$ and algorithm $\mathcal{BD}$; for algorithm $\mathcal{BD}$ their proof implies a lower bound of $\frac{1}{8}n^2 - O(n)$.

2   Our Contribution

In this work we improve the analysis of algorithm $\mathcal{BD}$. We show an upper bound of $1\frac{1}{2}n^2 + O(n)$ for its worst case stabilization time, and, a lower bound of $n^2 - O(n)$. For the upper bound we first present a proof which is similar to the one of [BD95]. This proof uses the more conventional tool of potential function that is used in the literature of self-stabilizing algorithms to deal mainly with the issue of correctness (see, e.g., [Dol00]). In our case the use of this tool is not straightforward, since the potential function can also increase by some of the moves. We use this tool to achieve a complexity result; namely, an upper bound of $2\frac{3}{5}n^2 + O(n)$. We then use amortized analysis. This more refined technique enables us to achieve an upper bound of $1\frac{1}{2}n^2 + O(n)$. Since by [CSZ08] algorithm $\mathcal{D}$ has a lower bound of $\frac{11}{6}n^2$, this implies that algorithm $\mathcal{BD}$ has a better worst case performance than algorithm $\mathcal{D}$.

In Section 3 we present algorithm $\mathcal{D}$, discuss some of its properties, and then present algorithm $\mathcal{BD}$. The lower bound for algorithm $\mathcal{BD}$ is discussed in Section 4. In Section 5 we present some properties of the algorithm, which are used in the upper bound analysis; the discussion using potential function is presented in Section 6, and the one using amortized analysis is presented in Section 7. We summarize with some remarks in Section 8. Some of the proofs are omitted or sketched in this Extended Abstract.

3   Algorithm $\mathcal{BD}$

Since algorithm $\mathcal{BD}$ is based on algorithm $\mathcal{D}$, we first present algorithm $\mathcal{D}$. We assume n processors $p_0, p_1, \ldots, p_{n-1}$ arranged in a ring; that is, processor p_i is adjacent to $p_{(i-1) \bmod n}$ and $p_{(i+1) \bmod n}$, for every i. Processor p_i has a local state $x_i \in \{0, 1, 2\}$. Two processors – namely, p_0 and p_{n-1} – run special programs, while all intermediate processors p_i, $1 \le i \le n - 2$, run the same program. The programs of the processors are as follows:

Algorithm $\mathcal{D}$

Program for processor p_0:
IF $x_0 + 1 = x_1$ **THEN**
  $x_0 := x_0 + 2$
END.

Program for processor p_i, $1 \le i \le n - 2$:
IF $(x_{i-1} - 1 = x_i)$ **OR** $(x_i = x_{i+1} - 1)$ **THEN**
  $x_i := x_i + 1$
END.

Program for processor p_{n-1}:
IF $(x_{n-2} = x_{n-1} = x_0)$ **OR** $(x_{n-2} = x_{n-1} + 1 = x_0)$ **THEN**
  $x_{n-1} := x_{n-2} + 1$
END.

The legitimate configurations for this problem are those in which exactly one processor is privileged. The configuration $x_0 = \cdots = x_{n-1}$ and $x_0 = \cdots = x_i \ne x_{i+1} = \cdots = x_{n-1}$ are legitimate. In [Dij86] it was shown that algorithm $\mathcal{D}$ self stabilizes under a centralized scheduler; namely, starting from any initial configuration the system achieves mutual exclusion. We mention the notation that was used in the proof and needed for our discussion. Given an initial configuration $x_0, x_1, \ldots, x_{n-1}$, and placing the processors on a line, consider each pair of neighbors p_{i-1} and p_i, for $i = 1, \ldots, n - 1$ (note p_{n-1} and p_0 are not considered to be neighbors). In this work we denote *left arrow* and *right arrow*, introduced in [Dij86], by '$<$' and '$>$'. Notation $x_{i-1} < x_i$ means $x_i = (x_{i-1} + 1) \bmod 3$ and $x_{i-1} > x_i$ means $x_i = (x_{i-1} - 1) \bmod 3$. Thus, for each two neighboring processors with states x_{i-1} and x_i, either $x_{i-1} = x_i$, or $x_{i-1} < x_i$, or $x_{i-1} > x_i$. For a given configuration $C = x_0, x_1, \ldots, x_{n-1}$, Dijkstra introduces the function

$$f(C) = \#left\ arrows + 2\#right\ arrows . \qquad (1)$$

Example 1. For $n = 7$, a possible configuration C is $x_0 = 1$, $x_1 = 1$, $x_2 = 0$, $x_3 = 1$, $x_4 = 2$, $x_5 = 2$, $x_6 = 0$. This configuration is denoted as $1 = 1 > 0 < 1 < 2 = 2 < 0$. Then, we have $f(C) = 3 + 2 \times 1 = 4$.

There are eight possible types of moves of the system: one possible move for processor p_0, five moves for any intermediate processor p_i, $0 < i < n - 1$, and two moves for p_{n-1}. These possibilities are summarized in Table 1.

In this table C_1 and C_2 denote the configurations before and after the move, respectively, and $\Delta f = f(C_2) - f(C_1)$. In the table we show only the local parts of these configurations. For example, in the first row, p_0 is privileged; therefore in C_1 we have $x_0 < x_1$, and in C_2 $x_0 > x_1$, and since one left arrow is replaced by a right arrow, $\Delta f = f(C_2) - f(C_1) = 1$.

We now present algorithm $\mathcal{BD}$. It is similar to algorithm $\mathcal{D}$ with the following changes: moves of type 0 depend on processor p_{n-1} (and not only on p_1), moves 6 and 7 (of p_{n-1}) are more complicated, and only one new arrow may be created by a move of type 7_2 (see Table 2). There is no change in the moves of intermediate processors p_i.

Table 1. Dijkstra's algorithm

Type	Proc.	C_1	C_2	Δf
0	p_0	$x_0 < x_1$	$x_0 > x_1$	$+1$
1	p_i	$x_{i-1} > x_i = x_{i+1}$	$x_{i-1} = x_i > x_{i+1}$	0
2	p_i	$x_{i-1} = x_i < x_{i+1}$	$x_{i-1} < x_i = x_{i+1}$	0
3	p_i	$x_{i-1} > x_i < x_{i+1}$	$x_{i-1} = x_i = x_{i+1}$	-3
4	p_i	$x_{i-1} > x_i > x_{i+1}$	$x_{i-1} = x_i < x_{i+1}$	-3
5	p_i	$x_{i-1} < x_i < x_{i+1}$	$x_{i-1} > x_i = x_{i+1}$	0
6	p_{n-1}	$x_{n-2} > x_{n-1} < x_0$	$x_{n-2} < x_{n-1}$	-1
7	p_{n-1}	$x_{n-2} = x_{n-1} = x_0$	$x_{n-2} < x_{n-1}$	$+1$

Algorithm $\mathcal{BD}$

Program for processor p_0:
IF $(x_0 + 1 = x_1)$ **AND** $(x_0 = x_{n-1})$ **THEN**
$\quad x_0 := x_0 - 1$
END.

Program for processor p_i, $1 \le i \le n - 2$:
IF $(x_i + 1 = x_{i-1})$ **OR** $(x_i + 1 = x_{i+1})$ **THEN**
$\quad x_i := x_i + 1$
END.

Program for processor p_{n-1}:
IF $(x_{n-1} = x_0 + 2)$ **AND** $(x_{n-1} \neq x_{n-2})$ **THEN**
$\quad x_{n-1} := x_{n-1} + 2$
ELSIF $x_{n-1} = x_0$ **THEN**
$\quad x_{n-1} := x_{n-1} + 1$
END.

For describing a relation between states of processors p_0 and p_{n-1}, we define the function $\hat{f}$ for any configuration C as follows:

$$\hat{f}(C) = f(C) \bmod 3 . \tag{2}$$

Table 2. Beauquier and Debas' algorithm

Type	Proc.	C_1	C_2	$\Delta\hat{f}$	Δf	Δh
0	p_0	$x_0 < x_1$, $\hat{f} = 1$	$x_0 > x_1$	$+1$	$+1$	$n - 2$
1	p_i	$x_{i-1} > x_i = x_{i+1}$	$x_{i-1} = x_i > x_{i+1}$	0	0	-1
2	p_i	$x_{i-1} = x_i < x_{i+1}$	$x_{i-1} < x_i = x_{i+1}$	0	0	-1
3	p_i	$x_{i-1} > x_i < x_{i+1}$	$x_{i-1} = x_i = x_{i+1}$	0	-3	$-(n + 1)$
4	p_i	$x_{i-1} > x_i > x_{i+1}$	$x_{i-1} = x_i < x_{i+1}$	0	-3	$3i - 2n + 2 \le n - 4$
5	p_i	$x_{i-1} < x_i < x_{i+1}$	$x_{i-1} > x_i = x_{i+1}$	0	0	$n - 3i - 1 \le n - 4$
6_1	p_{n-1}	$x_{n-2} > x_{n-1}$, $\hat{f} = 2$	$x_{n-2} < x_{n-1}$	-1	-1	$n - 2$
6_2	p_{n-1}	$x_{n-2} < x_{n-1}$, $\hat{f} = 2$	$x_{n-2} = x_{n-1}$	-1	-1	$-(n - 1)$
7_1	p_{n-1}	$x_{n-2} < x_{n-1}$, $\hat{f} = 0$	$x_{n-2} > x_{n-1}$	$+1$	$+1$	$-(n - 2)$
7_2	p_{n-1}	$x_{n-2} = x_{n-1}$, $\hat{f} = 0$	$x_{n-2} < x_{n-1}$	$+1$	$+1$	$n - 1$
7_3	p_{n-1}	$x_{n-2} > x_{n-1}$, $\hat{f} = 0$	$x_{n-2} = x_{n-1}$	$+1$	-2	-1

By (1) and the definition of arrows we get $\hat{f}(C) \equiv (x_{n-1} - x_0) \pmod 3$.

The possible types of moves of $\mathcal{BD}$ are summarized in Table 2.

In this table we also include the changes of the function $\hat{f}$ (note that $\hat{f}(C) = 0$ iff $x_{n-1} = x_0$) and the function h (that will be introduced in Section 6) implied by each move.

Intuitively, every arrow is related to a token, which is transferred until achieving p_0 or p_{n-1} or colliding with another token. As opposed to algorithm $\mathcal{D}$, algorithm $\mathcal{BD}$ may generate at most one new token (move 7_2) during any execution, which results in its having a better stabilization time.

4   Lower Bound

In this section we introduce the lower bound for algorithm $\mathcal{BD}$. We denote configurations by regular expressions over $\{<, >, =\}$. For example, $[<^3==<>>]$ and $[<^3=^2<>^2]$ are possible notations for the configuration $x_0 < x_1 < x_2 < x_3 = x_4 = x_5 < x_6 > x_7 > x_8$. Note that this notation does not loose relevant information, since the behavior of the algorithm is dictated by the arrows (see Table 2). We now present the lower bound:

Theorem 1. *The stabilization time of algorithm $\mathcal{BD}$ is at least $(n-4)^2$.*

Proof. Assume $n = 3k + 1$. For any $0 \le i \le k$, let $C_i := [=^{3i}<^{3k-3i}]$. In particular, C_0 is $[<^{3k}]$ and C_{k-1} is $[=^{3k-3}<^3]$. We show an execution with $2 \times 3i + 1 + 2 \times (3i+1) + 1 + 2 \times (3i+2) + 1 = 18i + 9$ moves, starting from C_i and ending at C_{i+1} (the example contains only moves of types 1, 2, 4 and 5).

$$\left[=^{3i}<^{3k-3i}\right], \text{ or:}$$
$$\left[=^{3i}<<<<<^{3k-3i-4}\right], \quad \text{after } 2 \times 3i \text{ steps of type 2:}$$
$$\left[<<=^{3i}<<<^{3k-3i-4}\right], \quad \text{after } 1 \text{ step of type 5:}$$
$$\left[>==^{3i}<<<^{3k-3i-4}\right], \quad \text{after } 2 \times (3i+1) \text{ steps of type 2:}$$
$$\left[><<==^{3i}<^{3k-3i-4}\right], \quad \text{after } 1 \text{ step of type 5:}$$
$$\left[>>===^{3i}<^{3k-3i-4}\right], \quad \text{after } 2 \times (3i+2) \text{ steps of type 1:}$$
$$\left[===^{3i}>><^{3k-3i-4}\right], \quad \text{after } 1 \text{ step of type 4:}$$
$$\left[===^{3i}=<<^{3k-3i-4}\right], \quad \text{or: } \left[=^{3(i+1)}<^{3k-3(i+1)}\right]$$

Then, starting from C_0 we get an execution that reaches C_{k-1} in $\sum_{i=0}^{k-2}(18i + 9) = 9(k-1)^2$ moves. We substitute $k = \frac{1}{3}(n-1)$ to get $(n-4)^2$. □

5   Properties of Algorithm $\mathcal{BD}$

In this section we derive some properties of algorithm $\mathcal{BD}$. They refine the ones in [BD95], and enable us to improve the analysis of the upper bound in the next two sections. By inspection of Table 2 we have:

Observation 1. *For any configuration C:*

1. *Any move of processor p_i, $1 \leq i \leq n-2$, does not change the function $\hat{f}$, i.e., $\Delta \hat{f} = 0$.*
2. *p_0 is privileged according to case 0 **iff** $\hat{f}(C) = 1$ and $x_0 < x_1$.*
3. *p_{n-1} is privileged according to case 7 **iff** $\hat{f}(C) = 0$.*
4. *p_{n-1} is privileged according to case 6 **iff** $\hat{f}(C) = 2$ and $x_{n-2} \neq x_{n-1}$.*
5. *After processor p_0 makes a move (case 0), we reach a configuration C such that $\hat{f}(C) = 2$.*
6. *After processor p_{n-1} makes a move (cases 6 or 7), we reach a configuration C such that $\hat{f}(C) = 1$.*

By [BD95] it follows that any execution of algorithm $\mathcal{BD}$ self-stabilizes. Given any execution or a segment of execution e before stabilization, $t_i(e)$ (or simply t_i) denotes the number of type i moves in e. Note that i may also indicate a sub-type, e.g. the number of moves of type 6_1 is t_{6_1} (see Table 2). We denote by a the number of arrows in the initial configuration of e. Moves of types 3, 4, 5, 6_2 and 7_3 are termed *collisions*.

Intuitively, an execution with the maximal number of moves will not contain moves of type 3, since such a collision decreases the number of arrows by 2 while other collisions (i.e., moves of type 4,5) decrease it only by 1. The following key lemma allows us to focus on executions e for which $t_3(e) = 0$, and is the basis for the amortized analysis in Section 7:

Lemma 1. *For every execution e, there is an execution e' containing no moves of type 3, such that $|e'| \geq |e|$.*

Proof (sketch). The proof is by induction on $t_3(e)$. At each inductive step we replace a segment of e by another segment with similar length.

Base: $t_3(e) = 0$. In this case $e' = e$ satisfies the claim.

Step: If $t_3(e) > 0$, then there is at least one type 3 move in e. Let p_i be the processor that made this move. Immediately after this move p_i is disabled. As there are no deadlocks, p_i will be re-enabled again. This happens at the first time there is an $>$ (resp. $<$) arrow at the left (resp. right) of p_i. This may happen in either of type $1, 2, 7_2$ moves. For the discussion we use the following notations. When describing executions or segments of them, we denote one move of type t by $\xrightarrow{(t)}$, and a series of x moves $\xrightarrow{x}$. In addition, when describing configurations we use regular expressions as defined in Section 4 and we use the wildcard character $'?'$ to denote any one of $\{=, <, >\}$.

- **A type 1 or type 2 move:** We describe the case of a type 1 move. The other case is symmetric. In this case, we have in e a segment of the form:

$$\left[?^i > <?^{n-3-i} \right] \xrightarrow{(3)} \left[?^i == ?^{n-3-i} \right] \xrightarrow{x} \left[?^{i-1} > == ?^{n-3-i} \right] \xrightarrow{1} \left[?^{i-1} => = ?^{n-3-i} \right]$$

Consider the execution e'' which is obtained by replacing the above segment by the following one:

$$\left[?^i > <?^{n-3-i} \right] \xrightarrow{x} \left[?^{i-1} >> <?^{n-3-i} \right] \xrightarrow{(4)} \left[?^{i-1} = << ?^{n-3-i} \right] \xrightarrow{(5)} \left[?^{i-1} => = ?^{n-3-i} \right]$$

The length of both segments is $x + 2$, thus $|e''| = |e|$. Clearly $t_3(e'') < t_3(e)$). By the induction hypothesis there is an execution e' such that $t_3(e') = 0$ and $|e'| \geq |e''| = |e|$, as required.

- **A type 7_2 move:** In this case, we have in e a segment of the following form:

$$\left[?^{n-3} ><\right] \overset{(3)}{\to} \left[?^{n-3} ==\right] \overset{x}{\to} \hat{f} = 0 \left[?^{n-3} ==\right] \overset{(7_2)}{\to} \hat{f} = 1 \left[?^{n-3} =<\right]$$

Consider the execution e'' which is obtained by replacing the above segment by the following one:

$$\left[?^{n-3} ><\right] \overset{x}{\to} \hat{f} = 0, \left[?^{n-3} ><\right] \overset{(7_1)}{\to} \hat{f} = 1, \left[?^{n-3} >>\right] \overset{(4)}{\to} \hat{f} = 1, \left[?^{n-3} =<\right]$$

We proceed exactly as in the first case.

$\square$

In particular for any worst case execution e there is an execution e' with the same number of moves. Therefore we assume without loss of generality that a worst case execution e does not contain type 3 moves.

The following lemma presents the main properties of the algorithm.

Lemma 2. *Let e be an execution until stabilization, such that $t_3(e) = 0$.*

1. *A move of types 7 may occur at most once. This can happen only before any move of type 0 or 6.*
2. *Between any two successive moves of type 0, there is exactly one move of type 6.*
3. *Between any two successive moves of type 6, there is exactly one move of type 0.*
4. *Between any two successive moves of type 0, there is at least one move of type 4.*
5. *Between any two successive moves of type 6_1, there is at least one move of type 5 or 6_2.*
6. *$a + t_{7_2} - t_{7_3} - t_4 - t_5 - t_{6_2} \geq 0$.*

Proof (sketch).

- Cases 1, 2 and 3 follow from Observation 1.
- Case 4: the arrow '$>$' that is created in the first type 0 move must disappear, in order to allow to an arrow '$<$' to reach the left end and to initiate the next type 0 move.
- Case 5 is similar to case 4.
- Case 6 holds since the number of arrows is always non-negative.

$\square$

We summarize all constrains in the following system:

$$\begin{cases} t_{7_1} + t_{7_2} + t_{7_3} & \leq 1 \\ t_0 & = t_{6_1} + t_{6_2} \\ t_0 & \leq t_4 \\ t_{6_1} & \leq t_5 + t_{6_2} \\ t_4 + t_5 + t_{6_2} + t_{7_3} & \leq a + t_{7_2} \\ a & \leq n \end{cases}$$

From the above we derive:

Lemma 3. *The following conditions hold:*

1. $t_4 + t_5 + t_{6_2} + t_{7_3} \le a + 1$.
2. $t_0 + t_{6_1} - t_{6_2} + t_4 + t_5 \le 2(a + 1)$.
3. $t_0 + t_4 + t_5 + t_{6_1} + t_{6_2} + t_{7_1} + t_{7_2} + t_{7_3} = O(n)$.

Proof.

1. $t_4 + t_5 + t_{6_2} + t_{7_3} \le a + t_{7_2} \le a + 1$.
2. $t_0 + t_{6_1} - t_{6_2} + t_4 + t_5 \le 2(t_4 + t_5) \le 2(a + 1)$.
3. $t_0 + t_4 + t_5 + t_{6_1} + t_{6_2} + t_{7_1} + t_{7_2} + t_{7_3} \le 2(t_4 + t_5 + t_{6_2}) + 1 \le 2a + 3 = O(n)$.

$\square$

Using the inequalities of Lemma 3, we now proceed in two ways to derive the upper bound. We first use the tool of potential function (Section 6), and we then use amortized analysis which enables us to track the route followed by individual arrows, and thus achieve a tighter bound (Section 7).

6 Upper Bound Using a Potential Function

We now introduce the function h. This function decreases by 1 during each move of types 1, 2 or 7_3 and decreases by $(n+1)$, $(n-1)$ and $(n-2)$ during each move of types 3, 6_2, 7_1 correspondingly. Unfortunately, moves of other types increase the function. By combining results of the previous section and the properties of h we manage to derive the upper bound on the number of moves to reach stabilization. We note that in [BD95] the same technique of potential function is used to derive the upper bound. The difference between the analyzes is that we use a simpler potential function, and a more refined analysis, to get a better upper bound (and, in addition, we use the technique of amortized analysis (in the next section) that allows to get an even better bound).

Given a configuration $C = x_0, x_1, \ldots, x_{n-1}$, we define the function $h(C)$ as follows:

$$h(C) = \sum_{\substack{1 \le i \le n-1 \\ x_{i-1} < x_i}} i + \sum_{\substack{1 \le i \le n-1 \\ x_{i-1} > x_i}} (n - i) \qquad (3)$$

The changes of the function h in each of the eleven possible types of moves are summarized in Table 2.

Example 2. Simple properties of h:

- $h\left(\left[=^{n-1}\right]\right) = 0$.
- $h\left(\left[=^{i-1}<=^{n-1-i}\right]\right) = i$.
- $h\left(\left[=^{n-i-1}>=^{i-1}\right]\right) = i$.
- $h\left(\left[<=^{n-3}>\right]\right) = 2$.

$-h\left(\left[<^{n-1}\right]\right) = \sum_{i=1}^{n-1} i = \frac{1}{2}n(n-1).$

$-h\left(\left[>^{\lfloor\frac{n-1}{2}\rfloor}<^{\lceil\frac{n-1}{2}\rceil}\right]\right) = \frac{3}{4}n^2 - n + O(1).$

These changes can be obtained by using the examples or directly from the definition of h. For example, for a move of type 0 we get that $\Delta h = (n-1)-(1) = n-2$, and for a move of type 3 $\Delta h = (0) - ((i+1) + (n-i)) = -(n+1)$.

Lemma 4. *For any configuration C, $0 \leq h(C) \leq \frac{3}{4}n^2$.*

Proof. Omitted.   □

Theorem 2. *Starting from any initial configuration, the stabilization time of algorithm $\mathcal{BD}$ is bounded by $2\frac{3}{4}n^2 + O(n)$.*

Proof. We denote by Δh_i the changes of h in a move of type i. Let C be the initial configuration of e. Considering the last (legitimate) configuration C' of e, by Lemma 4 we get $h(C') = h(C) + \sum_i(t_i \cdot \Delta h_i) \geq 0$. Therefore $t_1 + t_2 \leq h(C) + \sum_{i\neq 1,2}(t_i \cdot \Delta h_i) \leq h(C) + (t_0 - t_3 + t_4 + t_5 + t_{6_1} - t_{6_2} + t_7)(n-1)$. Recalling $t_7 \leq 1$ and applying Lemmas 3 (part 2) and 4, we get $t_1 + t_2 \leq \frac{3}{4}n^2 + 2n(n-1) + O(n) = 2\frac{3}{4}n^2 + O(n)$. By Lemma 3 (part 3), the number of moves of other types is $O(n)$. This completes the proof.   □

7   Upper Bound Using Amortized Analysis

In this section we present the upper bound of $1\frac{1}{2}n^2 + O(n)$. Our proof consists of the following steps: we first explore the types of collisions that could occur in an execution (Lemma 6), then we bound the weight of each type of collision (Lemma 7), and finally summing up these weights for all collisions we get the upper bound (Theorem 3).

Assume we are given an execution with no moves of type 3 until stabilization, whose existence is guaranteed by Lemma 1. We start by introducing the term *life-cycle* of an arrow. Informally, a life-cycle of an arrow is the sequence of moves from the moment it appears in this execution until the moment it disappears. The life-cycle of an arrow appearing in the initial configuration starts at that configuration. We say that a move of type 7_2 creates an arrow and in this way starts its life-cycle. A move of type 6_1 (resp. 7_1) destroys one arrow ending its life-cycle. A move of type 4 (resp. 5) destroys two arrows, ending their life-cycles, and creates a new arrow, thus starting its life-cycle. Other moves only change the direction of the arrow and do not terminate their life-cycle.

Next we introduce the term *mark*. If an arrow is created by a move 7_2 (resp. 4, 5), it is marked by 7 (resp. 4,5): '$<_7$' (resp '$<_4$', '$>_5$'). If an arrow makes a move of type 0 (resp. 6_1, 7_1) it gets an additional mark 0 (resp. 6, 7).

That allow us to introduce *the type of an arrow* – according to marks the arrow collected during its life. We define *the weight of an arrow* to be the number of moves of types 1 and 2 the arrow makes during its life-cycle.

Example 3. An arrow of type $<_7$ starts its life-cycle being created by a move of type 7_2, then it makes some, possibly 0, number of moves of type 2 and is destroyed. The weight of any arrow of this type is bounded by n

Example 4. An arrow of type $>_{40}$ starts its life-cycle being created by a move of type 4, then it reaches processor p_0 and makes a move of type 0. After making some, possibly 0, moves of type 1, it is destroyed. Its weight is bounded by $2n$.

Example 5. An arrow of type $<_{76}$ starts its life-cycle in initial configuration in position $n-1$. Then processor p_{n-1} makes a move of types 7_1 follows by 6_1. Then the arrow possibly makes some moves of type 2. Clearly, at most one arrow of this type can be in an execution. Its weight is bounded by n.

The various types of arrows are summarized in Table 3. Collisions of cases 1-4 and 11-12 (resp. 5-10; 13-16; 17) are moves of type 4 (resp. 5; 6_2; 7_3). Amount '*' means that the exact number of such arrows (collisions) are unknown.

Table 3. Types of arrows

Case	Arrow	Weight	Amount
1	$>\ >_5$	n	*
2	$>_0$	$2n$	1
3	$>_{40}$	$2n$	*
4	$<\ <_4$	n	*
5	$<_7$	n	1
6	$<_6$	$2n$	1
7	$<_{56}$	$2n$	*
8	$>_7\ >_{47}$	0	1
9	$<_{76}\ <_{476}$	n	1

Table 4. Types of collisions

Case	Collision	Weight	Amount
1	$>>\ >>_5\ >_5>\ >_5>_5$	$2n$	*
2	$>_0>$	$2n$	1
3	$>_0>_5$	$3n$	1
4	$>_{40}>\ >_{40}>_5$	$3n$	*
5	$<<\ <_4<\ <<_4\ <_4<_4$	$2n$	*
6	$<<_6$	$2n$	1
7	$<_4<_6$	$3n$	1
8	$<<_{56}\ <_4<_{56}$	$3n$	*
9	$<<_7\ <_4<_7$	$2n$	1
10	$<<_{76}\ <_4<_{76}\ <<_{476}\ <_4<_{476}$	$2n$	1
11	$>>_7\ >>_{47}\ >_5>_7\ >_5>_{47}$	n	1
12	$>_0>_7\ >_0>_{47}\ >_{40}>_7\ >_{40}>_{47}$	$2n$	1
13	$<\ <_7$	0	1
14	$<_4$	0	*
15	$<_6$	n	1
16	$<_{56}$	n	*
17	$>\ >_5$	n	1

We now conclude the following:

Lemma 5. *For any execution e not achieving self-stabilization, the following true:*

- *Only arrows of types $>$, $>_5$, $>_0$, $>_{40}$, $>_7$, $>_{47}$, $<$, $<_4$, $<_7$, $<_6$, $<_{56}$, $<_{76}$, $<_{476}$ can appear in e.*
- *The weight of an arrow of any type is bounded by $2n$.*
- *The number of arrows of types $>_0$, $>_7$, $>_{47}$, $<_7$, $<_6$, $<_{76}$, $<_{476}$ is $O(1)$.*

Proof. Omitted.   □

Consider a collision (a move of types 4, 5, 6_2, 7_3). We define *the weight of a collision* to be the sum of weights of the arrows destroyed in the collision. Note that a move of types 4 and 5 destroys two arrows while a move of types 6_2 and 7_3 destroys only one. Clearly, the weight of any collision is bounded by $4n$. In Lemma 7 we provide the better bound on the weight of a collision.

The types of arrows destroyed in the collision define *the type of the collision*. As we interested in $O(n^2)$ bound we may ignore $O(1)$ collisions of any types. Therefor, without loss of generality, we consider collisions of types $>$, $>_5$, $>_{40}$, $<$, $<_4$, $<_{56}$ only (the number of other collisions is $O(1)$). For details see Table 4.

Lemma 6. *The only collisions of types* $>>$, $>>_5$, $>_5>$, $>_5>_5$, $>_{40}>$, $>_{40}>_5$, $<<$, $<<_4$, $<_4<$, $<_4<_4$, $<_{56}<$, $<_4>_{56}$, $<$, $<_4$, $<_{56}$, $>$, $>_5$ *can occur in an execution.*

Proof. Omitted.   □

Clearly, the number of moves of both types 1 and 2 made by all arrows until stabilization is equal to the sum of weights of all collisions occurred in the execution. Our purpose is to estimate the last. Consider the ith collision, for some $i \geq 1$. Denote by $a(i)$ the number of arrows in the configuration in which the collision i occurs. Recall that a denotes the number of arrows in the initial configuration. Denote by $t_7(i) \in \{0,1\}$ the number of moves of type 7_2 made before the collision i occurs. Clearly, $a - i + t_7(i) = a(i)$ (pay attention: $t_3 = 0$). The following key lemma is the main tool for estimating the weight of a collision.

Lemma 7. *Given an execution until self-stabilization. Consider the ith collision in the execution. The weight of the collision is bounded by* $\max\{3n, 3(n - a + i)\}$.

Proof (sketch). The initial configuration has a arrows and $n - a$ empty places where arrows are allowed to move. After i collisions, the number of empty place is $n - a(i) = n - a + i - t_7(i)$. Assuming first that $t_7(i) = 0$ and noting that only one of two arrows, participating in the collision, could change the direction during its life-cycle, we conclude that the weight of the collision is bounded by $3(n - a(i)) = 3(n - a + i)$.

Assuming $t_7(i) = 1$, we note that the same bound is true for this case too.   □

Using the last lemma we compute the tighter bound on the number of moves until stabilization.

Theorem 3. *The number of moves until stabilization is bounded by* $1\frac{1}{2}n^2 + O(n)$.

Proof (sketch). Note that Lemma 3 bounds by $O(n)$ the number of moves of all types except of types 1 and 2. The number of these moves we bound as follows. Applying Lemmas 7 and 3 we get

$$t_1 + t_2 \leq \sum_{i=1}^{t_4+t_5+t_{6_2}+t_{7_3}} \min\{3(n-a+i), 3n\} \leq \sum_{i=1}^{a+1} \min\{3(n-a+i), 3n\}.$$

Simplifying we derive

$$\leq \sum_{i=1}^{a} 3(n - a + i) + 3n = 3an - 1\frac{1}{2}a^2 + 1\frac{1}{2}a + 3n.$$

Since $0 < a < n$ it follows that $t_1 + t_2 \leq 1\frac{1}{2}n^2 + O(n)$.   □

8   Conclusion and Remarks

In this paper we proved an upper bound of $1\frac{1}{2}n^2 + O(n)$ and a lower bound of $n^2 - O(n)$ for algorithm $\mathcal{BD}$, both of which improving the results presented in [BD95] for a centralized scheduler. In the full version of the paper we will discuss the extension of the results to the distributed scheduler case.

We also note that in Section 7 and specifically in Lemma 7, the analysis might be refined using more information from Table 4 – specifically, note that the number of collisions having weight at most $2n$ is not negligible – thus resulting in a better upper bound.

References

[BD95]   Beauquier, J., Debas, O.: An optimal self-stabilizing algorithm for mutual exclusion on bidirectional non uniform rings. In: Proceedings of the Second Workshop on Self-Stabilizing Systems, pp. 17.1–17.13 (1995)

[BGM89]   Burns, J.E., Gouda, M.G., Miller, R.E.: On relaxing interleaving assumptions. In: Proceedings of the MCC Workshop on Self-Stabilizing Systems, MCC Technical Report No. STP-379-89 (1989)

[BJM06]   Beauquier, J., Johnen, C., Messika, S.: Brief announcement: Computing automatically the stabilization time against the worst and the best schedules. In: Dolev, S. (ed.) DISC 2006. LNCS, vol. 4167, pp. 543–547. Springer, Heidelberg (2006)

[CG02]   Cobb, J.A., Gouda, M.G.: Stabilization of general loop-free routing. Journal of Parallel and Distributed Computing 62(5), 922–944 (2002)

[CSZ07]   Chernoy, V., Shalom, M., Zaks, S.: On the performance of Dijkstra's third self-stabilizing algorithm for mutual exclusion. In: 9th International Symposium on Stabilization, Safety, and Security of Distributed Systems (SSS), Paris, November 2007, pp. 114–123 (2007)

[CSZ08]   Chernoy, V., Shalom, M., Zaks, S.: A self-stabilizing algorithm with tight bounds for mutual exclusion on a ring (submitted for publication) (2008)

[Dij74]   Dijkstra, E.W.: Self stabilizing systems in spite of distributed control. Communications of the Association of the Computing Machinery 17(11), 643–644 (1974)

[Dij86]   Dijkstra, E.W.: A belated proof of self-stabilization. Distributed Computing 1, 5–6 (1986)

[Dol00]   Dolev, S.: Self-Stabilization. MIT Press, Cambridge (2000)

[Kes88]   Kessels, J.L.W.: An exercise in proving self-stabilization with a variant function. Information Processing Letters 29, 39–42 (1988)

[NKM06] Nakaminami, Y., Kakugawa, H., Masuzawa, T.: An advanced performance analysis of self-stabilizing protocols: stabilization time with transient faults during convergence. In: 20th International Parallel and Distributed Processing Symposium (IPDPS 2006), Rhodes Island, Greece, April 25-29 (2006)

[TTK00] Tsuchiya, T., Tokuda, Y., Kikuno, T.: Computing the stabilization times of self-stabilizing systems. IEICE Transactions on Fundamentals of Electronic Communications and Computer Sciences E83A(11), 2245–2252 (2000)

Self-stabilizing Cuts in Synchronous Networks

Thomas Sauerwald[1,*] and Dirk Sudholt[2,**]

[1] Dept. of CS, University of Paderborn, Paderborn, Germany
sauerwal@upb.de
[2] Dept. of CS, Dortmund University of Technology, Dortmund, Germany
dirk.sudholt@cs.uni-dortmund.de

Abstract. Consider a synchronized distributed system where each node can only observe the state of its neighbors. Such a system is called self-stabilizing if it reaches a stable global state in a finite number of rounds. Allowing two different states for each node induces a cut in the network graph. In each round, every node decides whether it is (locally) satisfied with the current cut. Afterwards all unsatisfied nodes change sides independently with a fixed probability p. Using different notions of satisfaction enables the computation of maximal and minimal cuts, respectively. We analyze the expected time until such cuts are reached on several graph classes and consider the impact of the parameter p and the initial cut.

1 Introduction

1.1 Motivation

In the language of distributed computing a system is called self-stabilizing if it reaches a global state with some desired property in finite time, regardless of the initialization. This implies that the system is able to stabilize even in the presence of faults [2,4]. Such self-stabilizing processes have been investigated for various graph problems like maximal matchings [11,15], independent sets [8], and domination [6]. A lot of research effort has been spent on self-stabilizing vertex coloring algorithms [7,9,12,13,14], motivated by code assignment problems in wireless networks.

In this work we consider self-stabilizing algorithms for maximal and minimal cuts in a synchronized distributed system. The network is given by an undirected graph $G = (V, E)$. As we do not make use of IDs for the nodes, we assume that the network is anonymous. However, we assume that there is a central clock synchronization. In each round every node has one out of two possible states, which induces a cut of the network. In every round every node decides whether it is satisfied with the current cut, judging from a local perspective, i.e., the state

---

* Supported by the German Science Foundation (DFG) Research Training Group GK-693 of the Paderborn Institute for Scientific Computation (PaSCo).
** Supported by the German Science Foundation (DFG) as a part of the Collaborative Research Center "Computational Intelligence" (SFB 531).

A. Shvartsman and P. Felber (Eds.): SIROCCO 2008, LNCS 5058, pp. 234–246, 2008.

of its neighbors. Unsatisfied nodes strive to (locally) improve the cut by changing sides. In order to break symmetries, we investigate a randomized algorithm where in each round every unsatisfied node changes sides with a fixed probability p.

By different notions of satisfaction different types of cuts can be produced. We say that a node is *max-satisfied* if at least half of its neighbors are on the other side of the cut. If all nodes are max-satisfied, the current cut cannot be increased by flipping a single node. Hence the current cut is *maximal*, i. e., locally optimal w. r. t. the cut size (as opposed to *maximum* cuts representing global optima). From a global perspective, the system may be viewed as a self-stabilizing algorithm for maximal cuts.

The system may also be regarded from a local perspective. For example, the problem can be seen as a relaxed code assignment problem where nodes are forced to use different codes to communicate. In a cut where all nodes are max-satisfied every node can communicate with a majority of neighbors, even if only two codes are available. There are also connections to game theory where the nodes represent players competing for services. If some players asking for the same service are close to each other (are connected by an edge), then the benefit of this service has to be split among all these players.

On the other hand, a node is *min-satisfied* if at least half of its neighbors are on the same side of the cut. This notion of satisfaction results in *minimal* cuts (as opposed to *minimum* cuts). Finding a minimum cut in a graph is an important task in computer science with applications to clustering, chip design, and network reliability. In our distributed and anonymous setting, however, we are content with minimal cuts.

Using the above-mentioned two notions of satisfaction, we show that the system self-stabilizes and then focus on the expected time until a stable cut is obtained. We prove for both satisfaction models that planar graphs stabilize in linear time for appropriate constant values of p. The choice of p is crucial since using constant p on dense graphs results in exponential stabilization times for the max-satisfaction model, with high probability. Finally, we investigate classes of sparse graphs like rings, torus graphs, and hypercubes. On rings the expected stabilization time is logarithmic for constant p. For some torus graphs, the choice of the initial cut decides between linear and logarithmic expected stabilization times.

1.2 Related Work

Our work is related to the design of distributed approximation algorithms [5] since our algorithm approximates maximum and minimum cuts. This is especially interesting as Elkin [5] concludes in his survey that the distributed approximability of maximum and minimum cut is still unsolved. However, the focus on this work is different; due to the restrictions in our distributed model we only settle for maximal and minimal cuts, i. e., local optima.

Gradinariu and Tixeuil [9] investigated a self-stabilizing coloring algorithms that is similar to our model. In their work, a node agrees with its neighborhood if it is colored with the maximal color value that is not used by any of its neighbors.

In their distributed setting a node that disagrees with its neighborhood changes its color with probability $1/2$. It is shown that this strategy stabilizes with a $(B + 1)$-coloring in expected time $O((B - 1) \log n)$ where B is a bound on the maximal degree and n is the number of nodes. This work loosely relates to our work as every 2-coloring represents a maximum cut. However, as typically $B + 1 > 2$ colors may be used, vertex coloring and cut problems are quite different.

1.3   Our Results

After presenting necessary definitions in Section 2, we start with general upper bounds for the expected stabilization time in both min-satisfaction and max-satisfaction models in Section 3. In particular, we derive an upper bound $O(n/p)$ for all planar graphs with n nodes if $p \leq 1/12$. This bound suggests to choose p large, but for dense graphs this may lead to exponential stabilization times. Section 4 presents such examples for the max-satisfaction process on the complete graph K_n and dense random graphs in the $\mathcal{G}(n, 1/2)$-model. On K_n the expected stabilization time is exponential for $p = 1/2$, but polynomial if $p = O((\log n)/n)$ (and $p \geq n^{-O(1)}$). For sparse graphs the choice of p is less important. As shown in Section 5, rings stabilize in expected time $O((\log n)/p)$ if $p = 1 - \Omega(1)$. Moreover, the investigation of torus graphs shows that the initialization can be crucial. With a worst-case initialization torus graphs stabilize in expected time $\Omega(n/p)$, while random initialization yields a bound of $O((\log n)/p^2)$ on certain torus graphs. Section 6 finishes with conclusions and remarks on future work. Due to space limitations proofs from Section 4 are omitted. An extended version with these proofs is available as technical report [16].

2   Definitions

Let $G = (V, E)$ be an undirected graph. For $U, W \subseteq V$ let $E(U, W)$ be the set of all edges between U and W and $E(U) = E(U, U)$. For $v \in V$ let $\deg(v)$ denote the degree of v. Let $\Delta(G) = \max_{v \in V} \deg(v)$ be the maximum degree in G and $a(G) = \max_{U \subseteq V, |U| > 1} \left\lceil \frac{|E(U)|}{|U| - 1} \right\rceil$ be the (edge) arboricity of G (see [1]). We use $a(G)$ as a measure of local density in the graph and observe that $a(G)$ is small iff G is "nowhere dense." The number of nodes is always denoted by n.

At each point of time all nodes are either in state 0 or in state 1. In round t let $V_t(1) \subseteq V$ denote the set of nodes in state 1; $V_t(0) = V \backslash V_t(1)$ is the corresponding complementing set. We synonymously use the term coloring and say that a node v is c-colored if $v \in V_t(c)$, $c \in \{0, 1\}$. In this case we denote $\deg_t^+(v) = |E(\{v\}, V_t(1 - c))|$ and $\deg_t^-(v) = \deg(v) - \deg_t^+(v)$. We define two notions of satisfaction mentioned before.

Definition 1. *A node v is max-satisfied at time t if $\deg_t^+(v) \geq \deg_t^-(v)$. A node v is min-satisfied at time t if $\deg_t^+(v) \leq \deg_t^-(v)$.*

Fixing one notion of satisfaction, let V_t^{sat} denote the set of all nodes that are satisfied at time t and $V_t^{\mathrm{unsat}} := V \setminus V_t^{\mathrm{sat}}$ denote the set of unsatisfied nodes. Given $0 < p < 1$, the self-stabilizing cut algorithm is formally defined as follows.

SELF-STABILIZING CUT ALGORITHM

1: In round t execute the following rule simultaneously for all nodes v:
2:     if $v \in V_t^{\mathrm{unsat}}$ then
3:         invert state of v for round $t + 1$ with probability p.

A cut where all nodes are satisfied is called *stable*. The stabilization time is defined as the first round with a stable cut. We are interested in the expected stabilization time, where the initial cut may be chosen uniformly at random or by an adversary. In the latter case, we speak of the worst-case expected stabilization time.

Observe that for bipartite graphs one can easily switch between the two models of satisfaction. Given a bipartition $V = U \cup W$ of the graph $G = (V, E)$, flipping (inverting) all nodes in U turns every cut edge into a non-cut edge and vice versa. Thereby, the meaning of $\deg_t^+(v)$ and $\deg_t^-(v)$ is exchanged and a node becomes min-satisfied iff it has been max-satisfied before. In particular, a stable cut for one model becomes a stable cut for the other model after this transformation.

More precise, let the function h on the state space $\{0,1\}^n$ be such a transformation, then the following holds. Consider the algorithm applied to both models. If the max-satisfaction model starts in state x_0 and the min-satisfaction model starts in state $y_0 = h(x_0)$, then at any point of time t for any state x_t the probability that the max-satisfaction model is in state x_t equals the probability that the min-satisfaction model is in state $y_t = h(x_t)$. This symmetrical behavior implies that the random stabilization times for the two models have the same probability distribution. It therefore suffices to focus on one model when dealing with bipartite graphs.

In the max-satisfaction model, shortly max-model, a stable configuration represents a maximal cut, i.e., a cut that cannot be enlarged by changing a single node. This is because a local improvement implies an unsatisfied node. The same holds for the min-model and minimal cuts. In a non-distributed setting one may easily obtain maximal and minimal cuts by local search, simply changing a single unsatisfied node in each round. The number of cut edges is then strictly increasing over time, implying that at most $|E|$ iterations are needed in order to find a maximal or minimal cut. The self-stabilizing cut algorithm can simulate an iteration of local search if exactly one specific unsatisfied node is flipped, which happens with probability $p \cdot (1-p)^{|V_t^{\mathrm{unsat}}|-1} > 0$. Hence, there is a positive probability that the algorithm simulates a whole run of local search ending with a stable cut.

Proposition 1. *In both the max-model and the min-model, the self-stabilizing cut algorithm stabilizes in finite time with probability 1.*

In the following, we will present more precise results, i. e., we prove bounds between logarithmic, polynomial, and exponential orders for different graph classes. As we are especially interested in the impact of the parameter p, we state our results w. r. t. n and p.

3   A General Upper Bound

In this section we derive general upper bounds for both the max-model and the min-model. Thereby, we exploit that under certain conditions there is a probabilistic tendency to increase the cut size in the max-model and to decrease the cut size in the min-model, respectively.

Theorem 1. *On any graph $G = (V, E)$, if $p \leq 1/(4a(G))$, the expected stabilization time for both the max-model and the min-model is bounded from above by $2|E|/p$.*

Proof. It suffices to consider the max-model as the arguments for the min-model are symmetric. Let $\mathcal{P}_t = (V_t(0), V_t(1))$ and let $f(\mathcal{P}_t)$ be the number of cut edges in $\mathcal{P}_t$. Consider one round of the algorithm and let V_t^{flip} be the set of nodes changing sides (flipping) in round t. If v is the only node to be flipped in round t, this operation increases the cut size by $\deg_t^-(v) - \deg_t^+(v) \geq 1$. If V_t^{flip} is an independent set, the total increase of the cut size is $\sum_{v \in V_t^{\text{flip}}}(\deg_t^-(v) - \deg_t^+(v)) \geq |V_t^{\text{flip}}|$. However, if two changing nodes share an edge, this edge is counted wrongly for both nodes. This implies

$$f(\mathcal{P}_{t+1}) - f(\mathcal{P}_t) \geq \sum_{v \in V_t^{\text{flip}}} (\deg_t^-(v) - \deg_t^+(v)) - 2|E(V_t^{\text{flip}})|$$

$$\geq |V_t^{\text{flip}}| - 2|E(V_t^{\text{flip}})|.$$

The expected gain in cut size is at least

$$\mathbf{E}\left(f(\mathcal{P}_{t+1}) - f(\mathcal{P}_t)\right) \geq p|V_t^{\text{unsat}}| - 2p^2|E(V_t^{\text{unsat}})|.$$

Observe $|E(V_t^{\text{unsat}})| \leq a(G) \cdot (|V_t^{\text{unsat}}| - 1) < a(G) \cdot |V_t^{\text{unsat}}|$ by definition of $a(G)$. Along with the assumption $p \leq 1/(4a(G))$, we arrive at

$$\mathbf{E}\left(f(\mathcal{P}_{t+1}) - f(\mathcal{P}_t)\right) \geq p|V_t^{\text{unsat}}| - 2p^2 \cdot a(G) \cdot |V_t^{\text{unsat}}| \geq p/2 \cdot |V_t^{\text{unsat}}|.$$

As long as the current cut is not stable, $|V_t^{\text{unsat}}| \geq 1$, hence the expected increase in cut size is at least $p/2$.

We now use drift analysis arguments from He and Yao [10, Lemma 1]. Consider a Markov chain with states $X_0, X_1, \ldots$ for domain $\mathbb{R}_0^+$. Let $\alpha, \delta > 0$ and assume we are interested in the first time until the Markov chain first reaches a value at least α. If δ is a lower bound for the expected increase in one step, i. e., $\mathbf{E}(X_{t+1} - X_t \mid X_t) \geq \delta$ for $X_t < \alpha$, the expected first hitting time for a value at least α is at most α/δ. Symmetrically, if $\mathbf{E}(X_t - X_{t+1} \mid X_t) \geq \delta$ for $X_t > 0$, the expected time to reach value 0 starting with α is at most α/δ.

We apply these statements to the random cut size and finish our considerations prematurely if a maximal cut is reached. Hence, the expected time until a cut of size $|E|$ is reached or a maximal cut is found beforehand is bounded by $|E|/(p/2) = 2|E|/p$. □

Section 5 contains examples where this bound is asymptotically tight. Note that the simple strategy of choosing $p = 1/(2n)$ is oblivious of the graph at hand and, nevertheless, yields a polynomial bound of $4|E|n$ rounds. This also proves that the expected stabilization time can be polynomial for *any* graph if the parameter p is chosen appropriately.

From Theorem 1 one can easily derive a handy upper bound for all planar graphs. The arboricity of a planar graph is known to be at most 3. A proof follows by contradiction. If there is a set $U \subseteq V$ with $|U| > 1$ such that $a(G) \geq \frac{|E(U)|}{|U|-1} > 3$, this implies $|E(U)| > 3|U| - 3$. However, this contradicts the fact that the number of edges in a planar graph with k nodes is at most $3k - 6$ (see, e.g., [3]). Therefore $a(G) \leq 3$ holds if G is planar.

Corollary 1. *On any planar graph $G = (V, E)$, if $p \leq 1/12$, the expected stabilization time for the max-model and the min-model is bounded by $2|E|/p \leq 6n/p$.*

4 Dense Graphs

The upper bounds from the previous section grow with $1/p$, suggesting to always choose p large. In this section, however, we prove for the max-model that in dense graphs large values for p may result in exponentially large stabilization times. The complete graph K_n is the simplest dense graph. For even n, a cut is maximal (and maximum in this case) if $|V_t(0)| = n/2$. However, if p is chosen too large, it may happen that too many nodes change sides simultaneously and a majority of 0-nodes is turned into a similarly large majority of 1-nodes, and so forth. This may result in a non-stable equilibrium that is hard to overcome. The following result shows that for large p the max-model needs exponential time to stabilize. Due to space limitations, proofs for the following theorems are placed in an extended version of this work [16].

Theorem 2. *Consider the complete graph K_n, n even, with $n^{-1/3} \leq p \leq 1/2$ and an arbitrary, non-stable initialization. Then the stabilization time of the max-model is at least $\frac{1}{2}\exp(\frac{np^3}{192})$ with probability $1 - o(1)$.*

On the other hand, the effect of too many flipping nodes decreases with decreasing p. The following result shows that if $p = O((\log n)/n)$ (and, of course, $p \geq n^{-O(1)}$) the expected stabilization time is polynomial.

Theorem 3. *Consider the complete graph K_n, n even, with an arbitrary initialization. Then the expected stabilization time of the max-model is bounded above by $1/p \cdot (1 - p)^{-n/2}$.*

Negative results for an unlucky initialization can also be shown for random graphs of a probability space $\mathcal{G}(n, p')$ defined as follows. The random graph consists of n nodes and between any pair of nodes, an edge occurs independently with probability p'. The case $p' = 1/2$ is especially interesting as $G \in \mathcal{G}(n, 1/2)$ is a uniform sample among all graphs with n nodes.

Theorem 4. *Consider a graph G in $\mathcal{G}(n, 1/2)$, n even, and assume that initially $\frac{20}{32}n \leq |V_0(0)| \leq \frac{23}{32}n$. Then the stabilization time of the max-model with $p = \frac{1}{2}$ is $\exp(\Omega(n))$ with probability $1 - \exp(-\Omega(n))$ (w. r. t. the randomized construction of G and the randomized self-stabilizing cut algorithm).*

5   Ring Graphs, Torus Graphs, and Hypercubes

We now consider commonly used network topologies like ring graphs (and other graphs with maximum degree 2), torus graphs, and hypercubes.

5.1   Ring Graphs

Consider a graph $G = (V, E)$ with maximum degree 2. Theorem 1 yields an upper bound $O(n/p)$ if $p \leq 1/8$. We improve upon this result exploiting that on these topologies satisfied nodes cannot become unsatisfied again.

Definition 2. *A set of nodes $S \subseteq V$ is called* stable *w. r. t. the current cut $\mathcal{P}_t$ if all nodes in S are satisfied and will remain so in all future rounds with probability 1. A node v is called* stable *if it is contained in a stable set; otherwise, v is called* unstable.

Isolated nodes are trivially stable, hence we assume that G does not contain isolated nodes. Then in the max-model (min-model) a node v is satisfied iff it has at least one neighbor w on the other side of the cut (on the same side of the cut). This condition also implies that w is satisfied. Even stronger, v and w will remain satisfied forever since the edge $\{v, w\}$ will never be touched again. Therefore, on graphs with maximum degree 2 all satisfied nodes are stable.

Theorem 5. *The expected stabilization time for the max-model and the min-model on a graph $G = (V, E)$ with $\Delta(G) \leq 2$ is $O((\log n)/(p(1-p)))$.*

Proof. Consider a node v that is unsatisfied in round t and the random decision whether to flip v or not. At least one decision makes v satisfied in round $t + 1$. The "right" random decision for v is made with probability at least $q := \min\{p, 1-p\}$. In expectation $q|V_t^{\text{unsat}}|$ nodes become satisfied (and therefore stable), hence $\mathbf{E}\left(|V_{t+1}^{\text{unsat}}| \mid |V_t^{\text{unsat}}|\right) \leq (1-q) \cdot |V_t^{\text{unsat}}|$ for any $V_t^{\text{unsat}} \subseteq V$. Using the law $\mathbf{E}\left(|V_{t+1}^{\text{unsat}}|\right) = \mathbf{E}\left(\mathbf{E}\left(|V_{t+1}^{\text{unsat}}|\right) \mid |V_t^{\text{unsat}}|\right)$ and a trivial induction yields $\mathbf{E}\left(|V_t^{\text{unsat}}|\right) \leq (1-q)^t \cdot |V_0^{\text{unsat}}| \leq (1-q)^t \cdot n$.

Choosing $T := \left\lceil \log_{(1-q)} \frac{1}{2n} \right\rceil$ yields $\mathbf{E}\left(|V_T^{\text{unsat}}|\right) \leq 1/2$. By Markov's inequality $\mathbf{Pr}\left(|V_T^{\text{unsat}}| \geq 1\right) \leq 1/2$. Hence after T rounds all nodes are satisfied with probability at least $1/2$, regardless of the initial cut. If this is not the case, we consider another period of T rounds and repeat the argumentation. The expected number of periods is at most 2, hence the expected stabilization time is bounded by

$$2T \leq 2 \left(\log_{(1-q)} \frac{1}{2n} \right) + 2 = \frac{2\ln(2n)}{\ln \left(\frac{1}{1-q} \right)} + 2 \leq \frac{2\ln(2n)}{q} + 2 = O\left(\frac{\log n}{q} \right)$$

where the second inequality follows from $1/(1-x) \geq e^x$ for $x < 1$. The theorem follows since $q = \Theta(p(1-p))$. $\qquad\square$

5.2   Torus Graphs

We denote by $G_{r \times s} = (V, E)$ for $r, s \geq 4$ both even a two-dimensional torus graph, defined by

$$V = \{(x, y) \mid 0 \leq x \leq r - 1, 0 \leq y \leq s - 1\} \text{ and}$$
$$E = \{(x_1, y_1), (x_2, y_2) \mid (x_2 = x_1 \wedge y_2 = (y_1 + 1) \bmod s) \vee$$
$$(x_2 = (x_1 + 1) \bmod r \wedge y_2 = y_1)\}.$$

$G_{r \times s}$ thus consists of r rows and s columns (see Figure 1). Note that due to the assumptions on r and s all torus graphs are bipartite and regular as all nodes have degree 4. Recall that the max-model can be transferred into an equivalent min-model by inverting states of all nodes in one set of the bipartition. The visualization is easier for the min-model where large monochromatic areas in the torus are "good." Hence we will argue with the min-model in the following; however, all results also hold for the max-model.

In the min-model we can derive an intuitive characterization of stable nodes, referring to states synonymously as colors. A sufficient condition for a c-colored node v to be stable is that v belongs to a cycle of c-colored nodes. Moreover, v is stable if it belongs to a path connecting two such cycles. The following lemma shows that these two conditions are also necessary for stability.

Lemma 1. *Consider the min-model for $G_{r \times s}$. A c-colored node v, $c \in \{0, 1\}$, is stable iff v belongs to a cycle of c-colored nodes or v is on a path of c-colored nodes connecting two such cycles.*

Proof. Consider the subgraph $G_c = (V_c, E_c)$ induced by all c-colored nodes. On a cycle $C \subseteq V_c$ all $u \in C$ are satisfied, hence they will remain so forever. Consider a path $P \subseteq V_c$ connecting two cycles $C_1, C_2 \subseteq V_c$. As all nodes in $C_1 \cup C_2$ remain satisfied, all $u \in P$ remain satisfied as well.

On the other hand, if v is neither on a cycle nor on a path connecting two cycles, then v cannot be stable. Assume that v is satisfied since otherwise the claim is trivial. Let S be the union of all cycles in V_c, then $v \in V_c \setminus S$. Let T be the connected component of $V_c \setminus S$ that contains v. As T does not contain cycles, T is a tree. Consider v as the root of T, then v has at least two subtrees in T

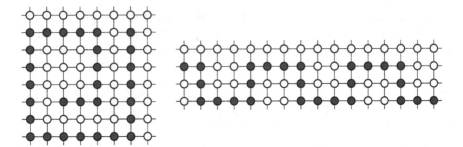

Fig. 1. Torus graphs $G_{8\times8}$ (left) and $G_{4\times16}$ (right). The coloring shows worst-case initial cuts in the min-model, where only the end nodes of the black paths are unsatisfied. All white nodes are stable by Lemma 1.

since v is satisfied. As v does not lie on a path connecting two cycles, at most one of v's subtrees is connected to S. In a subtree not connected to S every leaf is unsatisfied. If the next subsequent rounds only flip leaves of T, all subtrees of v (except one, if v is connected to S) are gradually eliminated, leaving v unsatisfied. We conclude that v cannot be stable.   □

We first consider the worst-case expected stabilization time. It is easy to see that we can color the nodes in $G_{r\times s}$ such that all 1-nodes form a path of length $\Omega(n)$ where every 1-node is adjacent to at most two other 1-nodes and all 0-nodes are stable. Figure 1 gives two examples. In such a cut only the two ends of the path are unsatisfied. As long as the path has length at least 2, this property is preserved since flipping an end node renders its neighbor on the path unsatisfied. The algorithm is thus forced to flip the nodes on this path one after another, starting from both ends simultaneously. It is then easy to prove the following lower bound.

Theorem 6. *The worst-case expected stabilization time for both the max-model and the min-model on $G_{r\times s}$ is $\Omega(n/p)$.*

An upper bound can be shown using that unsatisfied nodes have a good chance to become part of a cycle of equally colored nodes.

Lemma 2. *Consider the torus graph $G_{r\times s}$. If the current cut contains an unsatisfied node v, the probability that v becomes stable within the next two rounds is at least $p^2(1-p)^5$.*

Proof. W. l. o. g. v is 1-colored and we consider the min-model. We name nodes around v according to their direction from v and identify nodes with their corresponding colors. First consider the case $\deg^+(v) = 0$, implying $v_N = v_E = v_S = v_W = 0$. If any node from $\{v_{NW}, v_{NE}, v_{SE}, v_{SW}\}$ is 0-colored, say v_{NW}, flipping v and not flipping v_N, v_W, and v_{NW} creates a cycle. As v_{NW} is satisfied, the probability for such an event is at least $p(1-p)^2$. Now, assume $v_{NW} = v_{NE} = v_{SE} = v_{SW} = 1$. Then flipping v_N and v_E creates a cycle of 1-nodes. The probability for this to happen is at least p^2.

Let $\deg^+(v) = 1$ and w.l.o.g. assume that v_N is 1-colored. If v_{SW} or v_{SE} is 0-colored, a 0-cycle is created with probability at least $p(1-p)^2$. Hence, assume $v_{SW} = v_{SE} = 1$. If v_{NW} or v_{NE} is 1-colored, say v_{NW}, then v_W is unsatisfied and flipping it and not flipping v_{NW} creates a 1-cycle, with probability $p(1-p)$. The only remaining case is $v_{NW} = v_{NE} = 0$. If the next round flips v and doesn't flip v_W, v_{NW}, v_E, and v_{NE}, then v_N becomes unsatisfied in the following round. Flipping v_N and not flipping v_{NW} creates a cycle. The probability for these two rounds to be successful is at least $p^2(1-p)^5$.   □

The expected time to decrease the number of unstable nodes is bounded above by $1/(p^2(1-p)^5) = O(1/p^2)$ if, say, $p \leq 1/2$, hence the following theorem is immediate.

Theorem 7. *The worst-case expected stabilization time for both the max-model and the min-model on $G_{r \times s}$ is $O(n/p^2)$ if $p \leq 1/2$.*

We believe that with random initialization the expected stabilization time is much smaller. It is very unlikely that random initialization creates long paths of unstable 1-nodes. However, such paths of length $\Theta(\log n)$ are still quite likely. Using the same arguments leading to Theorem 6, a lower bound of $\Omega((\log n)/p)$ can be shown. An upper bound is more difficult. We present a bound that is of order $O((\log n)/p^2)$ if the number of rows (or, symmetrically, the number of columns) is constant (and $p \leq 1/2$).

Theorem 8. *After random initialization, the expected stabilization time for both the max-model and the min-model on $G_{r \times s}$ is $O((\log n) \cdot 2^r/p^2)$ if $p \leq 1/2$.*

Proof. Let $L_i := \{(x,i) \mid 0 \leq x \leq r-1\}$, $1 \leq i \leq s$, be the nodes in the i-th column in the graph and note $|L_i| = r$. The probability that all nodes in L_i are initialized zero (or initialized one) is exactly 2^{-r+1}. In this case, L_i is a stable set. The probability that there is no stable set among the consecutive columns $L_i, L_{i+1}, \ldots, L_{i+\gamma-1}$, where $\gamma = 2 \cdot 2^{r-1} \cdot \ln n$ for a fixed i is

$$\left(1 - 2^{-r+1}\right)^\gamma = \left(1 - 2^{-r+1}\right)^{2 \cdot 2^{r-1} \cdot \ln n} \leq n^{-2}.$$

Dividing the torus into blocks containing γ consecutive columns each, the probability that each block contains at least one stable column is at least $1 - n^{-1}$. Assume that every block contains a stable column and denote by S the set of stable nodes after initialization. Then $G \backslash S$ consists of connected components, each of which consists of at most $2r\gamma$ nodes. Consider one component C. If two subsequent rounds turn an unsatisfied node in C into a stable node, we speak of a success. Unless C is stable, there is at least one unsatisfied node in C and by Lemma 2 the success probability in two rounds is at least $q := p^2(1-p)^5$. We now argue that with high probability C becomes stable within $2T$ rounds, $T := 4r\gamma/q$. Imagine a sequence of coin flips where each coin shows heads with probability q. By the Chernoff bound the probability that less than $2r\gamma$ out of T coins show heads is at most

$$\exp(-qT/8) = \exp(-r\gamma/2) \leq n^{-2}$$

as $r \geq 2$. As $|C| \leq 2r\gamma$, the probability that C does not become stable within $2T$ rounds is at most n^{-2}. Taking the union bound over at most n components, the whole graph is stable after $2T$ rounds with probability at least $1 - n^{-1}$.

The unconditional probability that the bound $2T$ holds is at least $1 - 2n^{-1}$. In case there is a block without stable column or in case the system has not stabilized after $2T$ rounds, we use the upper bound $O(n/p^2)$ by Theorem 7 to estimate the remaining stabilization time. As this is only necessary with probability at most $2n^{-1}$, the unconditional expected stabilization time is bounded by $2T + O(1/p^2) = O((\log n) \cdot 2^r/p^2)$.   □

The bound from Theorem 8 depends crucially on r. However, we do not believe that the stabilization time is significantly affected by the aspect ratio of the torus. Instead, we conjecture that an upper bound $O((\log n)^k/p^k)$ for some $k = O(1)$ holds for all torus graphs.

5.3   Hypercubes

Recall that the node set of a d-dimensional hypercube is given by $\{0,1\}^d$ and edges are between nodes which differ in exactly one coordinate. We are interested in the worst-case expected stabilization time on hypercubes. For torus graphs we identified paths of unstable 1-nodes that delay the stabilization process. As nodes in the d-dimensional hypercube have larger degree if $d > 4$, we identify larger structures of unstable nodes.

Theorem 9. *The worst-case expected stabilization time for both the max-model and the min-model on a d-dimensional hypercube with $n = 2^d$ nodes, $d \geq 4$ even, is $\Omega(n^{1/2} + 1/p)$.*

Proof. As the hypercube is bipartite, it suffices to argue for the min-model. Given a graph $G' = (V', E')$, a *snake-in-box* in G' is a sequence of connected nodes $s'_1, \ldots, s'_\ell$ such that $\{s'_i, s'_j\} \in E'$ implies $j = i \pm 1$ (identifying $s'_{\ell+1}$ with s'_1 and s'_0 with s'_ℓ). It is known how to construct a snake-in-box with length $5/24 \cdot 2^d - 44$ in the d-dimensional hypercube [17]. Let $s_1, \ldots, s_\ell$ be a snake-in-box in the $(d/2)$-dimensional hypercube with $\ell \geq 5/24 \cdot 2^{d/2} - 44$ and let $S = \{s_1, \ldots, s_{\ell-1}\}$. Let $v[i] \in \{0,1\}$ denote the value of the i-th coordinate of v and define an initial cut as follows:

$$v \in V_0(1) \iff (v[1]v[2]\ldots v[d/2] \in S) \wedge (v[d-1]v[d] = 00).$$

Each 0-node with $v[d-1]v[d] = 00$ is satisfied since flipping one of the last $d/2$ bits results in a 0-neighbor. All other 0-nodes are satisfied since flipping one of the first $d - 2 \geq d/2$ bits leads to a 0-neighbor. We conclude that all 0-nodes are satisfied and, therefore, stable. Dividing all 1-nodes into *layers*, layer i contains all 1-nodes v with $v[1]\ldots v[d/2] = s_i$. For a 1-node v flipping a bit at position $i \in \{d/2+1, \ldots, d-2\}$ results in a 1-neighbor. Due to the snake-in-box property of S, v has at most two additional 1-neighbors obtained by flipping single bits among the first $d/2$ positions. More precise, after initialization all 1-nodes in layers 1 and $\ell - 1$ are unsatisfied with a 1-degree (i. e. number of 1-neighbors) of $d/2 - 1$ while every other 1-node has 1-degree $d/2$ and thus is satisfied. If such an unsatisfied node flips, all its 1-neighbors with 1-degree $d/2$ become unsatisfied.

A layer is called *satisfied* w. r. t. the current cut if it only contains satisfied 1-nodes. Observe that in every round all satisfied layers are connected in the subgraph of all 1-nodes. We focus on the outermost satisfied layers and define as potential the minimum difference $\alpha - \beta$ for $\alpha \leq \beta$ such that for every satisfied layer i we have $\alpha < i < \beta$. Layers α and β therefore "surround" all satisfied layers. The initial potential equals $\ell - 2$ and a potential of 0 is necessary for a stable cut. Layers α and β both contain unsatisfied 1-nodes and a round flipping one of these nodes decreases β or $-\alpha$ by 1, respectively. The probability of decreasing the potential by 1 or 2 in one round is at most $\delta := \min\{1, 2^{d/2-1} \cdot p\}$, taking the union bound over at most $2^{d/2-1}$ unsatisfied 1-nodes in layers α and β. The expected waiting time for such an event is bounded below by $1/\delta$, hence the expected time until the potential has decreased to 0 is bounded below by $1/\delta \cdot (\ell - 2)/2 = \Omega(n^{1/2} + 1/p)$. $\qquad\square$

6 Conclusions and Future Work

We investigated a self-stabilizing algorithm for maximal and minimal cuts in a restricted distributed environment. The time until the system stabilizes depends on the model of satisfaction, the underlying network, the parameter p, and the initial cut. Surprisingly, the expected stabilization time can range from logarithmic to exponential values. While sparse graphs such as planar graphs, rings, and torus graphs stabilize in expected time $O(n/p^{O(1)})$ (or even in logarithmic time) for max- and min-models, on many dense graphs the stabilization time for the max-model is exponential with high probability if p is constant. Moreover, we have seen for certain torus graphs that there is an exponential gap between random and worst-case initialization.

Several open questions remain, for example a tight bound on the expected stabilization time for all torus graphs and hypercubes with random initialization. Our models use a fixed probability p for flipping unsatisfied nodes. One may also think of other, local strategies, for example flipping an unsatisfied node v with probability proportional to $1/\deg(v)$ or depending on the degrees of v's neighbors.

Acknowledgment. We would like to thank Martin Gairing for helpful comments on an earlier version of this paper.

References

1. Chen, B., Matsumoto, M., Wang, J., Zhang, Z., Zhang, J.: A short proof of Nash-Williams' theorem for the arboricity of a graph. Graphs and Combinatorics 10(1), 27–28 (1994)
2. Dasgupta, A., Ghosh, S., Tixeuil, S.: Selfish stabilization. In: Stabilization, Safety, and Security of Distributed Systems (2006)

3. Diestel, R.: Graph Theory. Springer, Heidelberg (2005)
4. Dijkstra, E.W.: Self-stabilizing systems in spite of distributed control. Communications of the ACM 17(11), 643–644 (1974)
5. Elkin, M.: Distributed approximation: a survey. SIGACT News 35(4), 40–57 (2004)
6. Gairing, M., Goddard, W., Hedetniemi, S.T., Kristiansen, P., McRae, A.A.: Distance-two information in self-stabilizing algorithms. Parallel Processing Letters 14(3-4), 387–398 (2004)
7. Ghosh, S., Karaata, M.H.: A self-stabilizing algorithm for coloring planar graphs. Distributed Computing 7(1), 55–59 (1993)
8. Goddard, W., Hedetniemi, S.T., Jacobs, D.P., Srimani, P.K.: Self-stabilizing protocols for maximal matching and maximal independent sets for ad hoc networks. In: 17th International Parallel and Distributed Processing Symposium (IPDPS 2003), p. 162. IEEE Computer Society, Los Alamitos (2003)
9. Gradinariu, M., Tixeuil, S.: Self-stabilizing vertex coloration and arbitrary graphs. In: Procedings of the 4th International Conference on Principles of Distributed Systems, OPODIS 2000, pp. 55–70 (2000)
10. He, J., Yao, X.: A study of drift analysis for estimating computation time of evolutionary algorithms. Natural Computing 3(1), 21–35 (2004)
11. Hedetniemi, S.T., Jacobs, D.P., Srimani, P.K.: Maximal matching stabilizes in time $O(m)$. Information Processing Letters 80(5), 221–223 (2001)
12. Hedetniemi, S.T., Jacobs, D.P., Srimani, P.K.: Linear time self-stabilizing colorings. Information Processing Letters 87(5), 251–255 (2003)
13. Huang, S.-T., Hung, S.-S., Tzeng, C.-H.: Self-stabilizing coloration in anonymous planar networks. Information Processing Letters 95(1), 307–312 (2005)
14. Kosowski, A., Kuszner, Ł.: Self-stabilizing algorithms for graph coloring with improved performance guarantees. In: Rutkowski, L., Tadeusiewicz, R., Zadeh, L.A., Żurada, J.M. (eds.) ICAISC 2006. LNCS (LNAI), vol. 4029, pp. 1150–1159. Springer, Heidelberg (2006)
15. Manne, F., Mjelde, M., Pilard, L., Tixeuil, S.: A new self-stabilizing maximal matching algorithm. In: Prencipe, G., Zaks, S. (eds.) SIROCCO 2007. LNCS, vol. 4474, pp. 96–108. Springer, Heidelberg (2007)
16. Sauerwald, T., Sudholt, D.: Self-stabilizing cuts in synchronous networks. Technical Report CI-244/08, Collaborative Research Center 531, Technische Universität Dortmund (2008)
17. Tovey, C.A.: Local improvement on discrete structures. In: Local search in combinatorial optimization, pp. 57–89. Princeton University Press, Princeton (1997)

Quiescence of Self-stabilizing Gossiping among Mobile Agents in Graphs

Toshimitsu Masuzawa[1,*] and Sébastien Tixeuil[2,**]

[1] Osaka University, Japan
[2] Université Pierre-et-Marie-Curie - Paris 6, France

Abstract. This paper considers gossiping among mobile agents in graphs: agents move on the graph and have to disseminate their initial information to every other agent. We focus on self-stabilizing solutions for the gossip problem, where agents may start from arbitrary locations in arbitrary states. Self-stabilization requires (some of the) participating agents to keep moving forever, hinting at maximizing the number of agents that could be allowed to stop moving eventually.

This paper formalizes the *self-stabilizing agent gossip problem*, introduces the *quiescence number* (i.e., the maximum number of eventually stopping agents) of self-stabilizing solutions and investigates the quiescence number with respect to several assumptions related to agent anonymity, synchrony, link duplex capacity, and whiteboard capacity.

1 Introduction

Distributed systems involving mobile entities called *agents* or *robots* recently attracted a widespread attention as they enable adaptive and flexible solutions to several problems. Intuitively, agents[1] are mobile entities operating in a network that is modeled by a graph; agents have limited computing capabilities and are able to move from a node to one of its neighbors. The *gossip problem* among mobile agents was introduced by Suzuki et al. [9] as one of the most fundamental schemes supporting cooperation among mobile agents. The problem requires each agent to disseminate the initially given information to all other agents. Suzuki et al. [9] investigated the problem of minimizing the number of agent moves for the gossip problem in *fault-free* networks, and presented asymptotically optimal distributed solutions on several network topologies.

With the advent of large-scale networks that involve a total number of components in the order of the million, the fault (and attack) tolerance capabilities become at least as important as resource minimization. In this paper, we consider the gossip problem in networks where both agents and nodes can be hit by

---

* This work is supported in part by MEXT: Global COE (Center of Excellence) Program and JSPS: Grant-in-Aid for Scientific Research ((B)19300017).
** This author is supported in part by the FRACAS ARC project from INRIA and by the SOGEA projects of the ACI "Sécurité et Informatique" of the French Ministry of Research. Additional support from the INRIA research-team Grand Large.
[1] Agents and robots can be used interchangeably in this paper.

A. Shvartsman and P. Felber (Eds.): SIROCCO 2008, LNCS 5058, pp. 247–261, 2008.
© Springer-Verlag Berlin Heidelberg 2008

unpredictable faults or attacks. More precisely, we consider that transient faults arbitrarily corrupt the agent states and the node states (or the whiteboard contents), and devise algorithmic solutions to recover from this catastrophic situation. The faults and attacks are *transient* in the sense that there exists a point from which they don't appear any more and agents can work correctly according to their programs. In practice, it is sufficient that the faults and attacks are sporadic enough for the network to provide useful services most of the time. Our solutions are based on the paradigm of *self-stabilization* [4], an elegant approach to forward recovery from transient faults and attacks as well as initializing large-scale systems. Informally, a self-stabilizing system is able to recover from any transient fault in finite time, without restricting the nature or the span of those faults.

Related works. Mobile (software) agents on graphs were studied in the context of self-stabilization, e.g., in [2,6,7,8], but the implicit model is completely different from ours. In the aforementioned works, agents are software entities that are exchanged through messages between processes (that are located in the nodes of the network), and thus can be destroyed, duplicated, and created at will. The studied problems include stabilizing a network by means of a single non-stabilizing agent in [2,7], regulating the number of superfluous agents in [6], and ensuring regular traversals of k agents in [8].

The agent rendez-vous problem is closely related to the agent gossip problem and has been thoroughly investigated in previous works (e.g., [1]). The agent rendez-vous problem requires that all agents initially scattered in a network should meet at a single node not determined in advance. Thus, any solution for the rendez-vous problem is also a solution for the agent gossip problem: agents can exchange their initial information at the meeting point. However, the gossip among agents can be achieved without rendez-vous of all the agents. Suzuki et al. [9] shows that the gossip problem requires less (with respect to the number of agent moves) than the rendez-vous problem in some fault-free cases.

In this paper, we follow the model previously used in [3], that studies necessary and sufficient condition for the problems of naming and electing agents in a network that is subject to transient faults. The model assumes that the number of agents is fixed during any execution of the algorithm, but the agents can start from any arbitrary location in the network and in any arbitrary initial state. Agents can communicate with other agents only if they are currently located on the same node, or make use of so-called *whiteboards* - public memory variables located at each node. Of course, whiteboards may initially hold arbitrary contents due to a transient fault or attack.

Our contribution. The contribution of this paper is twofold:

1. We introduce the *quiescence number* of self-stabilizing agent-based solutions to quantify communication efficiency after convergence. Self-stabilizing agent-based solutions inherently require (some of the) participating agents to *keep moving* forever. This hints at maximizing the number of agents that

Table 1. Quiescence numbers of the k-gossip problem

Distinct agents

	synchronous model		asynchronous model	
whiteboards	half-duplex	full-duplex	half-duplex	full-duplex
FW	$k - 1$ (Th. 1 & 3)		0 (Th. 6)	
CW	$k - 1$ (Th. 1 & 3)		0 (Th. 7)	-1 (Th. 5)
NW	-1 (Th. 4)		-1 (Th. 4)	-1 (Th. 5)

Anonymous agents

	synchronous model		asynchronous model	
whiteboards	half-duplex	full-duplex	half-duplex	full-duplex
FW	≥ 0 (Th. 10)		0 (Th. 9)	
CW	-1 (Th. 8)		-1 (Th. 8)	
NW	-1 (Th. 4)		-1 (Th. 4)	-1 (Th. 5)

could be allowed to stop moving after some point in every execution. The quiescence number denotes the maximum possible number of stopping agents[2] 2. We study the quiescence number of self-stabilizing k-gossiping (that denotes the gossiping among k agents). The quiescence numbers we obtain are summarized in Table 1, where "-1" represents impossibility of 0-quiescence (that is, the problem is impossible to solve in a self-stabilizing way, even if agents are all allowed to move forever). We consider the quiescence number under various assumptions about synchrony (synchronous/asynchronous), node whiteboards (**FW/CW/NW**), edge capacity (half-duplex/full-duplex) and anonymity of the agents. The details of the assumptions are presented in the next section.

Outline. In Section 2, we present the computing model with various assumptions we consider in this paper. We also introduce the gossip problem and define the quiescence number of the gossip problem. Section 3 provides impossibility/possibility results in the model where each agent has a unique *id*. Section 4 briefly considers the quiescence numbers in the model of anonymous agents. Concluding remarks are presented in Section 5.

2   Preliminaries

Model. The network is modeled as a connected graph $G = (V, E)$, where V is a set of nodes, and E is a set of edges. We assume that nodes are *anonymous*, that is, no node has a unique *id* and all the nodes with the same degree are identical. We also assume that nodes have local distinct labels for incident links, however no assumption is made about the labels. Each node also maintains a so-called *whiteboard* which agents can read from and write to. Those whiteboards may store a finite yet unbounded amount of information.

---

[2] Minimizing communication after convergence in conventional self-stabilizing solutions has been largely investigated with *silent* [5] protocols.

Agents (or *robots*) are entities that move between neighboring nodes in the network. Each agent is modeled by a *deterministic* state machine. An agent staying at a node may change its state, leave some information on the whiteboard of the node, and move to one of the node's neighbors based on the following information: *(i)* the current state of the agent, *(ii)* the current states of other agents located at the same node, *(iii)* the local link labels of the current node (and possibly the label of the incoming link used by the agent to reach the node), and *(iv)* the contents of the whiteboard at the node. In other words, the only way for two agents to communicate is by being hosted by the same node or by using node whiteboards.

In this paper, we consider several variants of the model, which fall into several categories:

1. **Agent anonymity:** we consider two variants, distinct agents and anonymous agents. Each agent has a unique identifier taken from a arbitrarily large namespace in the *distinct* agent model, and all agents are anonymous and identical in the *anonymous* agent model.

2. **Synchrony:** we consider two variants, synchronous model and asynchronous model. In the *synchronous* model, all the agents are synchronized by rounds in the lock-step fashion. Every agent executes its action at every round and can move to a neighboring node. When two (or more) agents are at the same node, all the agents execute their actions in one round but in sequence. In the *asynchronous* model, there is no bound on the number of moves that an agent can make between any two moves of another agent. However, we assume that each agent is eventually allowed to execute its action. When two (or more) agents are at the same node, they execute their actions sequentially. However, agents located at different nodes may execute their actions concurrently.

3. **Link duplex capacity:** we consider two variants, full-duplex links and half-duplex links. A link is *full-duplex* if two agents located at neighboring nodes can exchange their position at the same time, crossing the same link in opposite directions without meeting each other. A link is *half-duplex* if only one direction can be used at a given time[3].

4. **Whiteboard capacity:** we consider three distinct hypothesis for information stored in the nodes' whiteboards. In the **NW** (No Whiteboard) model, no information can be stored in the whiteboard. In the **CW** (Control Whiteboard) model, only control information can be stored in the whiteboard. In the **FW** (Full Whiteboard) model, any information can be stored in the whiteboard (including gossip information, defined later in this section).

Of course, there is a strict inclusion of the hypotheses, and a solution that requires only, e.g., the **NW** or the **CW** classes will work with the less restricted classes (**CW** and **FW**, and **FW**, respectively). Conversely if an impossibility result is shown for less restricted classes, e.g., **CW** and **FW**, it remains valid in the more restricted classes (**NW**, and **NW** and **CW**, respectively).

---

[3] If two agents at the different ends of a half-duplex link try to migrate along the link simultaneously, only one of them succeeds to migrate.

The first set of hypotheses (or the agent anonymity) divides between Sections 3 and 4. In each section, the remaining hypotheses (synchrony, link duplex capacity, and whiteboard capacity) are denoted by a tuple. For example, "($Synch$, **FW**, $half$)-model" denotes the synchronous model with **FW** whiteboards and half-duplex links. The wildcard "$*$" in the triplet denotes all possibilities for the category. For example, "($*$, **FW**, $half$)-model" denotes both the ($Synch$, **FW**, $half$)-model and the ($Asynch$, **FW**, $half$)-model.

Gossip problem specification. We consider the *gossip problem* among agents: agents are given some initial information (called *gossip information*), and the goal of a protocol solving the problem is that each agent disseminates its gossip information to every other agent in the system. Each agent can transfer the gossip information to another agent by meeting it at a node or by leaving the gossip information in the whiteboard of a node. In the latter case, a **FW** whiteboard is required. The gossip information can be *relayed* by other agents, that is, any agent that has already obtained the gossip information of another agent can transfer all collected gossip information to other agents and can store it in the whiteboard of a node.

In this paper, we consider *self-stabilizing* solutions for the gossip problem. The solutions guarantee that every agent *eventually* knows the gossip information of all the agents in the system even when the system is started from an *arbitrary* configuration. So, agents may start from arbitrary locations in arbitrary states, and the nodes' whiteboards may initially contain arbitrary information. We assume that k agents are present in the network at any time, yet k is unknown to the agents. We also assume that the network topology or size are unknown to the agents. In the sequel, the k-*gossip problem* denotes the problem of gossiping among k agents.

Quiescence number. We introduce the *quiescence* of solutions for the k-gossip problem to describe the fact that some agents, although executing local code, stop moving at some point of any execution.

Definition 1. *A distributed algorithm for mobile agents is l-quiescent (for some integer l) if any execution reaches a configuration after which l (or more) agents remain still forever.*

Definition 2. *The* quiescence number *of a problem is the maximum integer l such that a l-quiescent algorithm exists for the problem. For convenience, the quiescence number is considered to be -1 if there exists no 0-quiescent algorithm (i.e., the problem is not solvable).*

Suzuki et al. [9] considered the **CW** and **FW** whiteboard models, and showed that the difference does not impact the move complexity of *non-stabilizing* solutions for the gossip problem. In this paper, we clarify some differences among the **NW**, the **CW** and the **FW** whiteboard models with respect to the quiescence number of *self-stabilizing* solutions for the gossip problem.

3   Self-stabilizing k-Gossiping among Distinct Agents

Our first result observes that the self-stabilization property of a k-gossiping protocol implies that at least one mobile agent must keep moving forever in the system.

Theorem 1. *There exists no k-quiescent self-stabilizing solution to the k-gossip problem in the $(*, *, *)$-model.*

Proof sketch. For contradiction, assume that a k-quiescent self-stabilizing algorithm exists. In any network with any set of k agents, the algorithm eventually reaches a *terminal* configuration of agents, i.e., a configuration from which all the agents never move thereafter.

Consider two n-sized networks N_1 and N_2 (with $n > k$) and mutually disjoint sets of k agents scattered on each of the networks. Each of the networks reaches a terminal configuration where each agent has collected the gossip information of all the other agents in the network it resides in. Notice that agents in different networks have collected the different gossip information. Then we construct a $2n$-sized network from the networks as follows: choose a node with no agent on it from each network (there exists such a node since $n > k$), and connect the networks by joining the two nodes. For the $2n$-sized network, the initial states of the nodes and the initial states (including the locations) of $2k$ agents are borrowed from the terminal configurations. As agents do not have the knowledge of the actual numbers of agents and nodes in the system nor the network topology, none of them is able to distinguish between the two systems, the n-sized networks and the $2n$-sized network. Thus, all the agents never move thereafter. Since agents may only communicate by meeting other agents at the same node or by using whiteboards, the k agents from N_1 are never able to communicate with any agent from N_2, hence the result.

The above discussion is valid independently of the assumptions concerning the synchrony, the link duplex capacity, or the whiteboard capacity. □

Notice that Theorem 1 does not hold if agents know the number k of existing agents. With the assumption of known k, it could be possible for agents to stop moving when k agents are located at a same node, i.e., k-quiescence may be attainable if the *rendez-vous* of k agents is possible. We now show that in the asynchronous case, no self-stabilizing algorithm can ensure that at least one agent does not move forever.

Theorem 2. *There exists no l-quiescent self-stabilizing solution, for any l ($1 \leq l \leq k - 1$), to the k-gossip problem in $(Asynch, *, *)$-model.*

Proof sketch. Assume that for every network, there exists a 1-quiescent self-stabilizing solution, that is, in any execution on any network with any set of k agents, there exists an agent that does not move after a certain configuration. The agent is only aware of the states of agents at the same node and the contents of the whiteboard at the node, and this information is sufficient to make the agent quiescent. In particular, the solution must work in a network that is regular (i.e., all nodes have the same degree) and non trivial. Consider k mutually disjoint sets of k agents and executions of each of the k sets on the regular network. By collecting the quiescent agents and the nodes they reside in, we can construct a configuration in which every agent is quiescent. Since we consider asynchronous systems, a quiescent agent cannot start moving unless another agent reaches the node. As a result, the agents never meet with each other and the gossiping cannot be achieved.   □

While Theorem 2 precludes l-quiescence in *asynchronous* models for any l ($1 \leq l \leq k - 1$), the impossibility result does not hold for *synchronous* systems. Actually, in synchronous arbitrary networks, we present in Algorithms 3.1, 3.2, 3.3, and 3.4 a positive result: a $(k-1)$-quiescent self-stabilizing solution to the k-gossip problem with **CW** whiteboards. The algorithm is based on the observation that gossiping can easily be achieved when a single agent repeatedly traverses the network: the agent alternates indefinitely a traversal to collect information and a traversal to distribute information. In our scheme, each agent may move according to a depth-first-traversal (DFT) in the network, and eventually an agent with minimal identifier (among all agents) keeps traversing forever, while other agents eventually stop. Since the network is synchronous, a stopped agent at node u waits for the traversal of the minimal identifier agent a bounded period of time, then starts moving if no such agent visits u within the bound.

Each node v has variables $InLink_v$ and $OutLink_v$ in its whiteboard to store information about the DFT of each agent i. We assume for simplicity that v locally labels each incident link with an integer a ($0 \leq a \leq \Delta_v - 1$) where Δ_v is degree of v, and $v[a]$ denotes the neighbor of v connected by the link labeled a. Variables $InLink_v$ and $OutLink_v$ have the following properties:

- A tuple (i, a) ($0 \leq a \leq \Delta_v - 1$) in variable $InLink_v$ of node v implies that agent i visited v first from $v[a]$ (i.e., $v[a]$ is the parent of v in the depth-first-tree). A tuple $(i, \perp)$ in $InLink_v$ implies that i did not visit v yet, or that i completed the DFT part starting from v (and returned to the parent of v in the depth-first-tree). For the starting node of the DFT, $(i, \perp)$ is always stored in $InLink_v$. We assume that only a single tuple of each agent i can be stored in $InLink_v$ (this can be enforced having $InLink_v$ implemented through an associative memory) and we consider that the absence of any tuple involving i denotes that $(i, \perp)$ is actually present.
- A tuple (i, a) ($0 \leq a \leq \Delta_v - 1$) in variable $OutLink_v$ of node v implies that agent i left v for $v[a]$ but did not return from $v[a]$ (i.e., i is in the DFT starting from $v[a]$). A tuple $(i, \perp)$ in $OutLink_v$ implies that i did not visit v yet, or that i completed the DFT part starting from v (and returned to the parent of v in the depth-first-tree). We assume the same additional constraints as for $InLink_v$.

In a legitimate configuration, tuples related to agent i in $InLink_v$ and $OutLink_v$ of all nodes induce a path from the starting node to the currently visited node. However, in an arbitrary initial configuration, $InLink_v$ and $OutLink_v$ may contain arbitrary tuples for agent i (several incomplete paths, cycles, no starting node, etc.). We circumvent this problem by having each agent executing DFTs repeatedly. In order to distinguish the current DFT from the previous one, each agent i maintains a boolean flag t_bit_i that is flipped when a new DFT is initiated. Each node v also maintains a variable T_table_v to store t_bit_i from the last visit of agent i in the form of a tuple (i, bit). For simplicity, we consider that $(i, true)$ is in T_table_v if no tuple of i is contained in T_table_v.

We now describe the mechanism to stop the remaining $k - 1$ agents. We assume that each node v maintains variables $MinID_v$, $WaitT_v$, and $Waiting_v$ in its whiteboard. The minimum id among all agents having visited v is stored in $MinID_v$, and the (computed) time required to complete a DFT is stored in $WaitT_v$. The completion time of a DFT is measured by the count-up timer $Timer_v$ of v as follows. Agent p with the minimum id repeatedly makes DFTs. When visiting v for the first time at each DFT, p sets the count-up timer of v to $WaitT_v$ and resets the timer. Eventually, p completes each DFT in $2m$ rounds, where m is the number of edges in the network, and $WaitT_v = 2m$ remains true thereafter. When visiting v, an agent p' finds a smaller id in $MinID_v$ and stays at v until the timer value of v reaches $WaitT_v$. Since p eventually completes each DFT in $2m$ rounds, each agent other than p eventually remains at a node v (v's timer is reset regularly enough to never expire).

Lemma 1. *Starting from any initial configuration, in every execution of Algorithms 3.1, 3.2, 3.3, and 3.4, eventually the agent with the minimum identifier repeatedly depth-first-traverses the network.*

Proof sketch. Let p be the agent with the minimum id (among all the agents in the system). When p visits node v, if $p \leq MinID_v$ then $MinID_v = p$ is executed. Otherwise (i.e., when $MinID_v$ stores an identifier that is not the id of any existing agent), p suspends its DFT and waits for timeout at v (p is appended into $Waiting_v$). Since no agent with the *fake id* exists in the network, $read(Timer_v) \geq WaitT_v$ eventually holds (in function $timeout\_check\_and\_execute_v$). When this is the case, $MinID_v = \min\{j \mid j \in Waiting_v\}(= p)$ is executed and p resumes the suspended DFT. Once $MinID_v$ is changed to p, $MinID_v$ never stores an id smaller than p again.

Now consider a DFT initiated by agent p with $t\_bit_p = b$ ($b \in \{true, false\}$). In a legitimate configuration, p initiates a DFT from a node v satisfying $(p, \perp) \in InLink_v$. However, in the initial configuration, $(p, a) \in InLink_v$ may hold for some a ($0 \leq a \leq \Delta_v - 1$) where v is the node p is initially located at. We first show that p eventually terminates the DFT starting from such an initial configuration and initiates a new DFT with $t\_bit_p = \neg b$. When p with $t\_bit_p = b$ visits a node u in a forward move, p changes its tuple in T_table_u to (p, b) if $(p, b) \notin T\_table_u$. Otherwise, p backtracks. Since (p, b) in T_table_u never changes to $(p, \neg b)$ as long as p continues the DFT with $t\_bit_p = b$, p can make at most m forward moves in the DFT. On the other hand, agent p backtracks from u to $u[a]$ only when $(p, a) \in InLink_u$ holds. When backtracking from u to $u[a]$, p changes its tuple in $InLink_u$ to $(p, \perp)$. Thus, p can make at most n backtracking moves in the DFT. Consequently, p eventually terminates the DFT even when it starts the DFT from a node v with $(p, \perp) \notin InLink_v$.

Algorithm 3.1. Protocol (Part 1: constants, variables and timers)

constants of agent i
 i: id of i;
constants of node v
 deg_v: degree of v;
local variables of agent i
 t_bit_i: bool;
 // an alternating bit to distinguish current and previous traversals
local variables of node v
 T_table_v : set of tuples (id, t_bit);
 // (id, t_bit) implies the latest visit of agent id was done with t_bit
 $InLink_v$: set of tuples $(id, port)$;
 // $(id, port)$ implies agent id first came from $v[port]$ in the current traversal
 // For each id, only the tuple updated last is stored
 // $(id, \perp)$ is stored if v is the initial node of the traversal
 // $(id, \perp)$ is considered to be stored if no $(id, *)$ is present
 $OutLink_v$: set of tuples $(id, port)$;
 // $(id, port)$ implies agent id went out from v to $v[port]$ last time it visited v
 // For each id, only the tuple updated last is stored
 $MinID_v$: agent id;
 // the minimum id of the agents that have visited v
 $WaitT_v$: int;
 // The amount of time agents with the non-minimum id should wait
 $Waiting_v$: set of agents;
 // The set of agents waiting for timeout at v
timers of node v
 $Timer_v$: count-up timer;
 The timer value is automatically increased by one at every round
functions on the local timer of node v
 $reset(Timer_v)$: Reset the timer value to 0
 $read(Timer_v)$: Return the timer value

Now consider a DFT initiated by agent i with $t\_bit_i = b$ at node v with $(i, \perp) \in InLink_v$. Let $G' = (V', E')$ be a connected component containing v of $G^{\neg b} = (V^{\neg b}, E^{\neg b})$ where $V^{\neg b} = \{u \in V \mid (i, \neg b) \in T\_table_u$ when i initiates the $DFT\}$ and $E^{\neg b} = (V^{\neg b} \times V^{\neg b}) \cap E$. Since the algorithm can be viewed as a distributed version of a sequential DFT, it means i makes a DFT in G' and its outgoing edges (if they exist). When the DFT completes, the tuple of i stored in T_table_u changes to (i, b) at each u in V', while the tuple of i stored in T_table_w remains unchanged at w ($\notin V'$) during the DFT. Thus, if G' is not the whole network, the connected component G'' (similarly defined as G' for the next DFT with $t\_bit_i = \neg b$) contains at least one more node than G'. Since the network is finite, eventually i makes DFTs repeatedly over the whole network. □

Theorem 3. *The protocol defined by Algorithms 3.1, 3.2, 3.3, and 3.4 is a $(k-1)$-quiescent self-stabilizing solution to the k-gossip problem in arbitrary networks in the $(Synch, \boldsymbol{CW}, *)$-model.*

Algorithm 3.2. Protocol (Part 2: Main behavior)

```
Behavior of node v at each round
  for each arriving agent i do
    visit_v(i);
  timeout_check_and_execute_v;
```

Proof sketch. Let p be the agent with the minimum id. From Lemma 1, eventually p makes DFTs repeatedly over the whole network. Once p completes the DFT, $MinID_v$ never becomes smaller than p at any node v.

Now consider p's DFT of the whole network that is initiated at a configuration satisfying $MinID_v \geq p$ at every node v. Then, p repeatedly performs DFTs without waiting at any node, and p completes each DFT in $2m$ rounds. This implies that timeout never occurs at any node starting from the second DFT. Any agent q other than p suspends its DFT when visiting any node u. Agent q can return to its suspended traverse only when timeout occurs at u. However, since timeout never occurs at u, q never returns to its suspended traverse and remains at u forever.   □

To complete our results for the synchronous case, let us observe that in the $(Synch, \mathbf{CW}, *)$ and $(Synch, \mathbf{FW}, *)$ models, the quiescence number of the k-gossip problem among distinct agents is $k - 1$ (by Theorems 1 and 3). There remains the case of **NW** whiteboards, unfortunately the following theorem show that when the memory of agents is bounded (the bound may depend on the network size n), the k-gossip problem among distinct agents is not solvable.

Theorem 4. *The quiescence number of the* $(*, \mathbf{NW}, *)$-*model is* -1 *for the k-gossip problem among distinct agents, when state space of each agent is bounded (but may depend on the network size n).*

Proof sketch. We prove the impossibility for synchronous ring networks. We assume for the purpose of contradiction that a 0-quiescent solution exists, and that each agent has at most s states. Notice that s is not necessarily a constant and may depend on the network size.

We consider system executions where each agent starts its execution from a predetermined state. Since no information can be stored in the whiteboards (model **NW**), the behavior of an agent depends solely on its own state and id (the network being regular). When an agent executes an action, it changes its state then (potentially) moves (clockwise or counterclockwise). Since each agent has at most s states, it repeats a cyclic execution of at most length s unless the agent meets another agent. Since only three kinds of moves are possible, there exists at most 3^{s+1} moving patterns in the cyclic behavior of length s or less. Now we consider a sufficiently large domain of agent identifiers (e.g., $k \times 3^{s+1}$). All possible agents are partitioned into at most 3^{s+1} groups depending on their moving patterns, and thus, some group contains k or more agents. Now consider k agents in the group of size k or more, that are placed regularly in different nodes

Algorithm 3.3. Protocol (Part 3: Behavior when agent i reaches v from $v[a]$)

```
function visit_v(i);
    // Executed when agent i visits node v from v[a] (a may be initially corrupted)
    if ((i,t_bit_i) ∉ T_table_v) { // first visit of i at v in the current traversal
        add (i,t_bit_i) to T_table_v; add (i,a) to InLink_v;
        if (i ≤ MinID_v) {
            MinID_v = i;
            WaitT_v = read(Timer_v);
            reset(Timer_v); // Timer is reset to start measuring the traversal time
            if (deg_v ≥ 2) {
                add (i,next_v(a)) to OutLink_v; // next_v(a) = (a + 1) mod deg_v
                migrate to v[next_v(a)];
            }
            else { // deg_v = 1 then backtrack to v[a]
                add (i,⊥) to InLink_v;
                migrate to v[a];
            }
        }
        else // i > MinID_v
            add i to Waiting_v;
    }
    else if ((i,a) ∉ OutLink_v)
        // i previously visited v in the current traversal, i backtracks to v[a].
        migrate to v[a];
    // The followings are the cases when i backtracks to v from v[a].
    else if ((next_v(a) == 0) and ((i,⊥) ∈ InLink_v)) {
        // v is the initial node of i's traversal and i completes the current traversal
        if (i ≤ MinID_v) {
            MinID_v = i; WaitT_v = read(Timer_v); reset(Timer_v);
            // Initiate a new traversal
            t_bit_i = ¬t_bit_i;
            add (i,t_bit_i) to T_table_v; add (i,0) to OutLink_v;
            migrate to v[0];
        }
        else // i > MinID_v
            add i to Waiting_v;
    }
    else if ((i,next_v(a)) ∈ InLink_v) {
        // v is not the initial node of i's traversal,
        // i completes the current traversal from v
        add (i,⊥) to InLink_v; add (i,⊥) to OutLink_v;
        migrate to v[next_v(a)]; // i backtracks
    }
    else { // i did not complete the current traversal from v
        add (i,next_v(a)) to OutLink_v;
        migrate to v[next_v(a)];
    }
```

Algorithm 3.4. Protocol (Part 4: Behavior when Timeout occurs)

```
function timeout_check_and_execute_v;
    if read(Timer_v) ≥ WaitT_v; { // Timeout occurs
        MinID_v = min{j | j ∈ Waiting_v};
        Let i be such that MinID_v = i;
        Waiting_v = Waiting_v − {i};
        reset(Timer_v); // Timer is reset to start measuring the traversal time
        if ((i, a) ∈ InLink_v for some a (0 ≤ a ≤ deg_v − 1)) {
            // v is not the initial node of i's traversal
            Let a be such that (i, a) ∈ InLink_v;
            if (deg_v ≥ 2) {
                add (i, next_v(a)) to OutLink_v;
                i migrates to v[next_v(a)];
            }
            else { // deg_v = 1 then backtrack to v[a]
                add (i, ⊥) to InLink_v;
                i migrates to v[a];
            }
        else { // v is the initial node of i's traversal
            // Initiate a new traversal
            t_bit_i = ¬t_bit_i;
            add (i, t_bit_i) to T_table_v;
            add (i, 0) to OutLink_v;
            migrate to v[0];
        }
    }
}
```

in the initial configuration of the nodes. Since agents in the group makes the same moving pattern in the cycle, the agents repeat the cyclic action without meeting each other in the synchronous execution. In the models with the whiteboards **NW**, the gossiping cannot be achieved without meetings of agents, which is a contradiction.   □

Note that the impossibility result holds even though the agents all start from a well known predefined initial state. Thus, if the initial location of agents is not controlled, even non-stabilizing solution are impossible to design. For *asynchronous* models, the remaining question is about the possibility of 0-quiescence.

Theorem 5. *The quiescence number of* $(Asynch, \mathbf{CW}, full)$ *and* $(Asynch, \mathbf{NW}, full)$ *model is* -1 *for the k-gossip problem among distinct agents.*

Proof sketch. We show that there exists no 0-quiescent self-stabilizing solution to the k-gossip problem in $(Asynch, \mathbf{CW}, full)$-model. Let us assume for the purpose of contradiction that there exists a 0-quiescent self-stabilizing solution. All k agents must keep moving in the 0-quiescent solution since 1-quiescence is impossible from Theorem 2.

Now consider a particular agent p. In the asynchronous system with *full-duplex* links, there exists an execution such that p never meets any other agent: before p reaches a node u, all the agents staying at u leave u. (Such an execution is possible for every node because the system is asynchronous and all agents must keep moving at every activation.) Notice that full-duplex links allow the agents to leave u without meeting p: scheduling allows to have all agents exiting u by the same link used by p to arrive at u to be moving concurrently with p. It follows that in the execution, agent p cannot disseminate its own information (agents have to meet one another in **CW** model). Hence the result.   □

Theorem 6. *The quiescence number of $(Asynch, \boldsymbol{FW}, *)$-model is 0 for the k-gossip problem among distinct agents.*

Proof sketch. Theorems 2 shows that 1-quiescence is impossible. Thus, it is sufficient to present a 0-quiescent self-stabilizing solution in $(Asynch, \boldsymbol{FW}, *)$-model.

Consider the following protocol outline. Every agent repeatedly performs DFTs of the network. When an agent visits a node, it stores its gossip information in the whiteboard and collects the gossip information stored in the whiteboard. After a DFT has been completed by every agent, all whiteboards contain the gossip information of all the agents, and every agent can obtain all the gossip information by performing an additional DFT.

The self-stabilizing DFT can be realized in the same way as the protocol presented in Theorem 3: each agent simply behaves as the agent with the minimum id of the protocol, yet does not need to wait at any node.   □

Theorem 7. *The quiescence number of $(Asynch, \boldsymbol{CW}, half)$-model is 0 for the k-gossip problem among distinct agents.*

Proof sketch. Theorems 2 shows that 1-quiescence is impossible to attain. Thus, it is sufficient to present a 0-quiescent self-stabilizing solution in $(Asynch, \boldsymbol{CW}, half)$-model.

Consider the following protocol outline. Every agent repeatedly performs DFTs of the network while recording at every traversed node the last targeted neighboring node. By the recorded information, other agents can trace a particular agent. When an agent visits a node and finds a smaller id than its own, it starts tracing the agent with the smaller id. Eventually all the agent other than agent p (that has minimal id) continue tracing p, that in turns perform a DFT forever. Since we assume the half-duplex edges, agents cannot miss one another on a link, and agents perform the same DFT and the agent with minimal id. Then a similar argument as in [3] implies that all agents other than p meet p infinitely often. Thus, by means of agent p, every agent can disseminate its gossip information to all other agents.

The self-stabilizing DFT can be realized in the same way as the protocol presented in Theorem 3. The only difference is in the way to detect the *fake ids*. In the protocol of Theorem 3, fake ids are detected by a timeout mechanism. Here, each agent records at each node the distance from the starting node in the depth-first-tree. In any trace labeled with a fake id, the tracing agent eventually detects contradiction in the distances and then decides that the traced *id* is a fake one. Agent p detecting a fake *id* erases the false records on the path p traced.   □

4   Self-stabilizing k-Gossip among Anonymous Agents

Distinct agents being a stronger assumption than anonymous agents, all the impossibility results for distinct agents also hold for anonymous agents. In this section, we consider only the model variations that the impossibility results for distinct agents do not cover. The following impossibility results can be derived from the impossibility results on the rendezvous among anonymous agents [1].

Theorem 8. *The quiescence number of* $(*, \boldsymbol{CW}, *)$-*model is* -1 *for the k-gossip problem among anonymous agents.*

Proof sketch. Consider a synchronous ring network where all the whiteboards of nodes contain the same initial information. Assume that all the agents are in the same state in the initial configuration. In the synchronous system, all the agents move exactly the same and they never meet each other, and thus, the gossiping cannot be completed.   □

Theorem 9. *The quiescence number of* $(Asynch, \boldsymbol{FW}, *)$-*model is* 0 *for the k-gossip problem among anonymous agents.*

Proof sketch. From Theorem 2, it is sufficient to present a 0-quiescent self-stabilizing solution to the gossip problem in $(Asynch, \boldsymbol{FW}, *)$-model.

Since the whiteboards $\boldsymbol{FW}$ is available, the k-gossiping can be completed if every agent repeatedly traverses the network. However, an anonymous agent cannot record at a node that it has visited the node since its record cannot be distinguished from that of others. Thus, anonymous agents cannot execute the DFT like the ones in Theorem 3. Instead, each agent can traverse all the paths of a given length, say ℓ, using the link labels (i.e., traverse all the paths in the lexicographic order of the label sequences). When completing the traverse of the paths of length ℓ, the agent starts traversing the paths of length $\ell + 1$. By repeating the traverses with incrementing the length, eventually the agent can traverse the whole network.   □

For the *synchronous* anonymous agents, Theorem 9 guarantees that the quiescence number is at least 0. On the other hand, the impossibility of k-quiescence for *synchronous distinct* agents (Theorem 1) leads to the following theorem.

Theorem 10. *The quiescence number of* $(Synch, \boldsymbol{FW}, *)$-*model is not larger than* $k - 1$ *and not smaller than* 0 *for the k-gossip problem among anonymous agents.*   □

5 Conclusion

This paper introduced the notion of quiescence for mobile agent protocols in a self-stabilizing setting. This notion complements the notion of silence [5] used in "classical" self-stabilizing protocols. While k-quiescence of k-gossiping among distinct agents is easily attainable in *non-stabilizing* solutions (assuming **FW** and **CW** whiteboards) [9], this paper shows that *self-stabilization* prevents k-quiescent solutions in any considered model, and even 0-quiescent solutions in some particular models. Thus, our paper shed new light on the inherent difference between non-stabilizing and self-stabilizing solutions of agent-based systems.

We would like to point out interesting open questions:

1. What is the exact quiescence number of the $(Synch, \mathbf{FW}, *)$-model for the k-gossip problem among *anonymous* agents? (besides being not smaller than 0 and not larger than $k - 1$)
2. What is the connection between the quiescence number and the topology ?
3. Does there exist a non-trivial *non-stabilizing* problem with quiescence number lower than k ?

References

1. Barriére, L., Flocchini, P., Fraigniaud, P., Santoro, N.: Rendezvous and election of mobile agents: impact of sense of direction. Theory of Computing Systems 40(2), 143–162 (2007)
2. Beauquier, J., Herault, T., Schiller, E.: Easy stabilization with an agent. In: Proceedings of the 5th Workshop on Self-Stabilizing Systems (WSS), pp. 35–51 (2001)
3. Blin, L., Gradinariu Potop-Butucaru, M., Tixeuil, S.: On the self-stabilization of mobile robots in graphs. In: Proceedings of OPODIS, pp. 301–314 (2007)
4. Dolev, S.: Self-stabilization. MIT Press, Cambridge (2000)
5. Dolev, S., Gouda, M.G., Schneider, M.: Memory requirements for silent stabilization. Acta Inf. 36(6), 447–462 (1999)
6. Dolev, S., Schiller, E., Welch, J.: Random walk for self-stabilizing group communication in ad-hoc networks. In: Proceedings of SRDS, pp. 70–79 (2002)
7. Ghosh, S.: Agents, distributed algorithms, and stabilization. In: Proceedings of International Computing and Combinatorics Conference (COCOON), pp. 242–251 (2000)
8. Herman, T., Masuzawa, T.: Self-stabilizing agent traversal. In: Proceedings of the 5th Workshop on Self-Stabilizing Systems (WSS), pp. 152–166 (2001)
9. Suzuki, T., Izumi, T., Ooshita, F., Kakugawa, H., Masuzawa, T.: Move-optimal gossiping among mobile agents. Theoretical Computer Science 393(1–3), 90–101 (2008)

Gathering with Minimum Delay in Tree Sensor Networks

Jean-Claude Bermond[1,*], Luisa Gargano[2,**], and Adele A. Rescigno[2]

[1] MASCOTTE, joint project CNRS-INRIA-UNSA, Sophia-Antipolis, France
[2] Dip. di Informatica ed Applicazioni, Universitá di Salerno, 84084 Fisciano, Italy

Abstract. Data gathering is a fundamental operation in wireless sensor networks in which data packets generated at sensor nodes are to be collected at a base station. In this paper we suppose that each sensor is equipped with an half–duplex interface; hence, a node cannot receive and transmit at the same time. Moreover, each node is equipped with omnidirectional antennas allowing the transmission over distance R. The network is a multi-hop wireless network and the time is slotted so that one–hop transmission of one data item consumes one time slot. We model the network with a graph where the vertices represent the nodes and two nodes are connected if they are in the transmission/interference range of each other. Due to interferences a collision happens at a node if two or more of its neighbors try to transmit at the same time. Furthermore we suppose that an intermediate node should forward a message as soon as it receives it. We give an optimal collision free gathering schedule for tree networks whenever each node has at least one data packet to send.

1 Introduction

A wireless sensor network is a multi-hop wireless network formed by a large number of low-cost sensor nodes, each equipped with a sensor, a processor, a radio, and a battery. Due to the many advantages they offer – e.g. low cost, small size, and wireless data transfer – wireless sensor networks become attractive to a vast variety of applications like space exploration, battlefield surveillance, environment observation, and health monitoring.

A basic activity in a sensor network is the systematic gathering of the sensed data at a base station for further processing. A key challenge in such operation is due to the physical limits of the sensor nodes, which have limited power and unreplenishable batteries. It is then important to bound the energy consumption of data dissemination [10,18,24]. However, an other important factor to consider in data gathering applications is the *latency* of the information dissemination process. Indeed, the data collected by a node can frequently change thus making essential that they are received by the base station as soon as it is possible without being delayed by collisions [26].

* Partially supported by the CRC CORSO with France Telecom, by the European FET project AEOLUS.
** Work partially done while visiting INRIA at Sophia-Antipolis.

A. Shvartsman and P. Felber (Eds.): SIROCCO 2008, LNCS 5058, pp. 262–276, 2008.
© Springer-Verlag Berlin Heidelberg 2008

Another application, which motivates this work, concerns the use in telecommunications networks a problem asked by FRANCE TELECOM about "how to provide Internet connection to a village" (see [6]). Here we are given a set of communication devices placed in houses in a village (for instance, network interfaces that connect computers to the Internet). They require access to a gateway (for instance, a satellite antenna) to send and receive data through a multi-hop wireless network. Therefore, this problem is the same as data collection in sensor network. Here the main objective is to minimize the delay.

In this paper, we will study optimal–time data gathering in tree networks.

1.1   Network Model

We adopt the network model considered in [3,12,17]. In this model each node is equipped with an half–duplex interface, hence,

  i) *a node cannot receive and transmit at the same time.*

Moreover, each node is equipped with omni directional antennas allowing transmission over a distance R. This implies that for any given node in the network, we can individuate its neighbors as those nodes within distance R from it, that is, within its transmission/interference range. In this model,

  ii) *a collision happens at a node x if two or more of its neighbors try to transmit at the same time.*

However, simultaneous transmissions among pair of nodes can successfully occur whenever conditions i) and ii) of the above interference model are respected. The time is slotted so that one–hop transmission of one data item consumes one time slot; the network is assumed to be synchronous. Moreover, following [12,15,26] and contrarily to [3,17], we assume that no buffering is done at intermediate nodes, that is each node forwards a message as soon as it receives it. Finally, it is assumed that the only traffic in the network is due to data to be collected, thus data transmissions can be completely scheduled.

Summarizing, the network can be represented by means of a direct graph $G = (V, A)$ where V represents the sensors (devices) nodes and A the set of possible calls; i.e. an arc $(u, v) \in A$ if v is in the transmission/interference range of u. Throughout this paper we assume that all nodes have the same transmission range, hence the graph G is a directed symmetric graph, e.g., $(u, v) \in A$ if and only if $(v, u) \in A$. The fact that there is no collision can be expressed by the fact that two calls (u, v) and (u', v') are compatible (can be done in the same time slot) iff $d(u, v') \geq 2$ and $d(u', v) \geq 2$ (i.e., both u and v', and u' and v have not to be neighbors, by ii)).

The collision–free data gathering problem can be then stated as follows [26].

Data Gathering. *Given a graph $G = (V, A)$ and a base station (BS) s, for each $v \in V - \{s\}$, schedule the multi-hop transmission of the data items sensed at v to s so that the whole process is collision–free, and the time when the last data is received by s is minimized.*

We will actually study the related one–to-all personalized broadcast problem in which the BS wants to communicate different data items to each other node in the network.

One–to-all personalized broadcast: *Given a graph G and a BS s, for each node $v \neq s$, schedule the multi-hop transmission from s to v of the data items destined to v so that the whole process is collision-free, and the time when the last data item is received at the corresponding destination node is minimized.*

Solving the above dissemination problem is equivalent to solve data gathering in sensor networks. Indeed, let T denote the delay, that is, the largest time–slot used by a dissemination algorithm; a gathering schedule with delay T consists in scheduling a transmission from node y to x during slot t iff the broadcasting algorithm schedules a transmission from node x to y during slot $T - t + 1$, for any t with $1 \leq t \leq T$.

It should be noticed that our algorithms are centralized requiring the BS perform a distinct topology learning phase and schedule broadcasting. When requirements are more stringent, these algorithms may no longer be practical. However, they still continue to provide a lower bound on the data collection time of any given collection schedule.

1.2   Related Work

Much effort has been devoted to the study of efficient data gathering algorithms taking into consideration various aspects of sensor networks [8]. The problem of minimizing the delay of the gathering process has been recently recognized and studied. The authors of [12] first afford such a problem; they use the same model for sensor networks adopted in this paper. The main difference with our work is that [12] mainly deals with the case when nodes are equipped with directional antennas, that is, only the designed neighbor of a transmitting node receives the signal while its other neighbors can safely receive from different nodes. Under this assumption, [12] gives optimal gathering schedules for trees. An optimal algorithm for general networks has been presented in [15] in the case that each node has one packet of sensed data to deliver.

The work in [26] also deals with the latency of data gathering under the assumption of unidirectional antennas; the difference with [12] is the assumption of the possibility to have multiple channels between adjacent nodes. By adopting this model an approximation algorithm with performance ratio 2 is obtained.

Fast gathering with omnidirectional antennas is considered in [1,3,4,5,7], under the assumption of possibly different transmission and interference ranges, that is, when a node transmits, all the nodes within a fixed distance d_T in the graph can receive while nodes within distance d_I $(d_I \geq d_T)$ cannot listen to other transmissions due to interference (in our paper $d_I = d_T$). Lower bounds on the time to gather and NP-hardness proofs are given in [3]; an approximation algorithm with approximation factor 4 is also presented. Paper [7] presents an on–line gathering algorithm under the described model.

The case where $d_T = 1$ and where each node has one packet to transmit is solved for the line in [1], for the uniform grids in [4] and for trees when furthermore $d_I = 1$ in [5]. All the above papers allow buffering at intermediate nodes.

Several papers deal with the problem of maximize the lifetime of the network through topology aware placement [10,13], data aggregation [16,19,20,21], or efficient data flow [11,18,23]. Papers [9,22,25] consider the minimization of the gathering delay in conjunction with the energy spent to complete the process.

1.3   Paper Overview

We consider the model introduced in Section 1.1 and give optimal gathering schedules in case the graph modeling the network is a line or a tree.

In Section 3 we shortly illustrate an optimal algorithm in case G is a line with the BS s as one of its endpoints. The result was first presented in [12]; we report it in our settings as a starting point for our result on trees.

In Section 4 we give an optimal algorithm in case the graph is a tree T with *one data item at each node*, apart the source s.

We look at T as rooted at s and denote by $T_1, T_2, \cdots, T_m$ the subtrees of T rooted at the sons of s.

Definition 1. *For each $i = 1, \ldots, m$, we denote by:*
- *s_i the son of s which is the root of T_i and is at level 1 in T_i,*
- *α_i the number of nodes at level 2 in T_i,*
- *β_i the number of nodes at level 3 or more in T_i.*

Moreover, we define the shade *of subtree T_i, for $1 \leq i \leq m$, as $\tau_i = 1 + 2\alpha_i + 3\beta_i$.*

Let $|T_i|$ represent the number of nodes in the subtree T_i, for $i = 1, \ldots, m$.

Definition 2. *Given $i, j = 1, \ldots, m$ with $i \neq j$, we say that*
- *$T_i \prec T_j$ if either $\tau_i > \tau_j$ or $\tau_i = \tau_j$ and $|T_i| > |T_j|$,*
- *$T_i = T_j$ if $\tau_i = \tau_j$ and $|T_i| = |T_j|$.(they are not necessarily isomorphic)*

Definition 3. *Assume that $T_1 \preceq T_2 \preceq \ldots \preceq T_m$. Define*

$$\epsilon_T = \begin{cases} 1 & \text{if } T_1 = T_2 \\ 0 & \text{otherwise} \end{cases} \quad \text{and} \quad \Delta_{i,j} = |T_i| + |T_j| + \beta_i - 1,$$

for each $i, j = 1, \ldots, m$, with $i \neq j$.

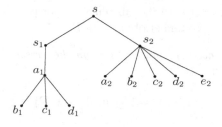

Fig. 1.

Theorem 1. *Suppose each node of a tree T has one data item and let n denote the number of vertices of T. Assuming that $T_1 \preceq T_2 \preceq \ldots \preceq T_m$, we have that the optimal gathering time is*

$$T^*(T) = \max\{n - 1, \ \tau_1 + \epsilon_T, \ \Delta_{1,2}, \ \Delta_{2,1}, \ \Delta_{1,3}\}.$$

Example 1. Let T be the tree in Fig.1 with BS s. We have: $m = 2$, $|T_1| = 5$, $|T_2| = 6$, $\alpha_1 = 1$, $\beta_1 = 3$, $\tau_1 = 12$, $\alpha_2 = 5$, $\beta_2 = 0$, $\tau_2 = 11$, $\epsilon_T = 0$, $\Delta_{1,2} = 13$, and $\Delta_{2,1} = 10$. Hence, $T_1 \prec T_2$ and, by Theorem 1, $T^*(T) = \Delta_{1,2} = 13$.

Due to space limits some proofs are omitted from this extended abstract; a full version is in [2].

2   Mathematical Formulation

We now formally formulate the one-to-all personalized broadcast problem. Let $G = (V, A)$ be the directed symmetric graph that is obtained by replacing each edge connecting two nodes u and v of the network with two directed arcs (u, v) and (v, u). Furthermore, let $s \in V$ be a special node that will be called the *source*.

Each node $v \in V - \{s\}$ is associated with an integer weight $w(v) \geq 0$ that represents the number of data items destined to node v. The set $\mathbf{w} = \{w(v) \mid v \in V - \{s\}\}$ represents the set of the weights of the nodes in V.

We need to schedule (time–label) the transmissions in order to create $w(v)$ collision–free routes from s to node v, for each $v \in V - \{s\}$.

Definition 4. *Let $\mathbf{p} = (u_0, \cdots, u_h)$ be a path in G. An increasing labeling L of $\mathbf{p}$ is an assignment of integers, $L_{\mathbf{p}}(u_0, u_1) \ldots, L_{\mathbf{p}}(u_{h-1}, u_h)$, to the arcs of $\mathbf{p}$ such that for $j = 1, \ldots, h - 1$.*

$$L_{\mathbf{p}}(u_j, u_{j+1}) = L_{\mathbf{p}}(u_{j-1}, u_j) + 1$$

The labeling is called t-increasing, for some integer $t \geq 1$, if it is increasing and $L_{\mathbf{p}}(u_0, u_1) = t$.

Consider any set $\mathcal{P}$ of paths in G from s to (not necessarily pairwise distinct) nodes in $V - \{s\}$ together with the labellings $L_{\mathbf{p}}$, for $\mathbf{p} \in \mathcal{P}$. Notice that any arc $a \in A$ can belong to any number of paths in $\mathcal{P}$.

Definition 5. *The labeling induced by $\mathcal{P}$ on the arcs of G consists, for each $(u, v) \in A$ of the multisets*

$$L(u, v) = \{L_{\mathbf{p}}(u, v) \mid \mathbf{p} \in \mathcal{P}\}.$$

Let $N(u)$ be the set of neighbors of u in G, that is, $N(u) = \{x \mid (u, x) \in A\} = \{x \mid (x, u) \in A\}$.

Definition 6. *The labeling L induced by $\mathcal{P}$ on the arcs of G is called strictly collision–free (SCF) if L is increasing and, for each $(u, v) \in A$ it holds:*
- *$L(u, v)$ is a set (e.g, any integer has at most one occurrence in $L(u, v)$),*
- *$L(u, v) \cap L(w, u) = \emptyset$, for each $w \in N(u)$,*
- *$L(u, v) \cap L(w, z) = \emptyset$, for each $w \in N(v) \cup \{v\}$, $z \in N(w)$.*

Definition 7. *An instance of SCF labeling is a triple $\langle G, \mathbf{w}, s \rangle$ where G is the graph, s is the source, and $\mathbf{w}$ is the set of weights of the nodes in G.*
A feasible solution for $\langle G, \mathbf{w}, s \rangle$ is a pair $(\mathcal{P}, L)$ where:

- *$\mathcal{P}$ is a set of $w(v)$ paths (not necessarily distinct) from s to v in G, for each $v \in V - \{s\}$;*
- *L is a SCF–labeling induced by $\mathcal{P}$.*

An optimal solution $(\mathcal{P}^*, L^*)$ is a feasible solution minimizing the largest label given to any arc of G.
The value attained by the optimal solution $(\mathcal{P}^*, L^*)$ for $\langle G, \mathbf{w}, s \rangle$ is denoted by $T^*(\langle G, \mathbf{w}, s \rangle)$ (or simply by $T^*(G)$ when $\mathbf{w}$ and s are clear from the context).

Example 2. In the tree T of Fig.1, let $w(u) = 1$ for each $u \neq s$. A feasible solution for $\langle T, \mathbf{w}, s \rangle$ is the pair $(\mathcal{P}, L)$ where $\mathcal{P} = \{\mathbf{p}_u \mid \mathbf{p}_u$ is the unique path from s to u in T, $u \neq s\}$ and the SCF labeling L is such that each path $\mathbf{p}_u$ is labeled with a t_u-increasing labeling as follows: $t_{b_1} = 1$, $t_{a_2} = 2$, $t_{b_2} = 4$, $t_{c_1} = 5$, $t_{c_2} = 6$, $t_{d_1} = 8$, $t_{s_2} = 9$, $t_{d_2} = 10$, $t_{a_1} = 11$, $t_{e_2} = 12$, $t_{s_1} = 13$.

As an example, we have
$\mathbf{p}_{b_1} = (s, s_1, a_1, b_1)$, with $L_{\mathbf{p}_{b_1}}(s, s_1) = t_{b_1} = 1$, $L_{\mathbf{p}_{b_1}}(s_1, a_1) = 2$, $L_{\mathbf{p}_{b_1}}(a_1, b_1) = 3$ $L(s, s_1) = \{L_{\mathbf{p}_{b_1}}(s, s_1), L_{\mathbf{p}_{c_1}}(s, s_1), L_{\mathbf{p}_{d_1}}(s, s_1), L_{\mathbf{p}_{a_1}}(s, s_1), L_{\mathbf{p}_{s_1}}(s, s_1)\} = \{1, 5, 8, 11, 13\}$, $L(s, s_2) = \{L_{\mathbf{p}_{a_2}}(s, s_2), L_{\mathbf{p}_{b_2}}(s, s_2), L_{\mathbf{p}_{c_2}}(s, s_2), L_{\mathbf{p}_{d_2}}(s, s_2), L_{\mathbf{p}_{e_2}}(s, s_2), L_{\mathbf{p}_{s_2}}(s, s_2)\} = \{2, 4, 6, 9, 10, 12\}$.

Notice that minimizing the largest label is equivalent to minimize the time needed by the algorithm. Indeed, one can just consider solutions where all labels in $\{1, \cdots, T\}$ are used: If some integer c is never used, we can decrease by 1 the value of each label $c' \geq c + 1$ in the considered feasible solution.

3   Lines

In this section we present an optimal algorithm to solve the SCF–labeling problem for an instance $\langle G, \mathbf{w}, s \rangle$, where G is a line, s is one of its end points, and node weights are arbitrary non negative integers, that is, $w(v) \geq 0$ for each $v \neq s$. The optimal value given in Theorem 2 was already given in [12] (theorem 4.1); however, we restate the algorithm in our notation since it is a starting point for the algorithm on trees given in the next section.

Let G be the line of length n with nodes $0, 1, \cdots, n$ and let $(i, i + 1)$, for $i = 0, \cdots, n - 1$, be the connection between subsequent nodes. Assume that the source node is $s = 0$ and $w(n) > 0$ (otherwise delete the end vertices of the line with weight 0).

Property 1. *A solution $(\mathcal{P}, L)$ of $\langle G, \mathbf{w}, s \rangle$ is feasible iff*
1) The labeling L induced by $\mathcal{P}$ is increasing,
2) for each $\mathbf{p}, \mathbf{q} \in \mathcal{P}$ with $L_{\mathbf{q}}(s, 1) \geq L_{\mathbf{p}}(s, 1)$: if $\mathbf{p}$ leads from s to node h, with $1 \leq h \leq n$, then

$$L_{\mathbf{q}}(s, 1) \geq L_{\mathbf{p}}(s, 1) + \min\{3, h\}. \tag{1}$$

The following notation will be used in the algorithm description.

- *Set a path (resp. a t–path) to node v*: establish a path from s to v together with its increasing labeling (resp. t–increasing labeling);
- A node $v \neq s$ is *completed*: if $w(v)$ paths from s to v have been set.

Theorem 2. [12] *For a line G with nodes $\{s = 0, 1, \ldots, n\}$ and and $w(i) \geq 0$ for $i = 1, \ldots, n$, it holds*

$$T^*(G) = \max_{1 \leq i \leq n} M_i,$$

where $M_1 = w(1) + 2w(2) + 3\sum_{j \geq 3} w(j)$, $M_2 = 2w(2) + 3\sum_{j \geq 3} w(j)$, and $M_i = i - 3 + 3\sum_{j \geq i} w(j)$ if $i \geq 3$.

When each sensor node of the line has at least one request to be completed, Theorem 2 provides a simpler form of the optimal label (i.e. minimum time) .

Corollary 1. *If $w(i) \geq 1$, for $i = 1, \ldots, n$, then $T^*(G) = M_1 = w(1) + 2w(2) + 3\sum_{j=3}^{n} w(j)$.*

Table 1. The SCF labeling algorithm on a line

LINE-labeling $(G, \mathbf{w}, s)$
- Set $\mathcal{P} = \emptyset$, $k = 1$.
- **while** there is a non completed node, **do**
  - Let i be the largest node which is not completed
    (e.g $i = \max\{j \mid 1 \leq j \leq n, w(j) > 0\}$.
  - Set a k–path to i in G, call it $\mathbf{p}_i$.
  - Let $\mathcal{P} = \mathcal{P} \cup \{\mathbf{p}_i\}$.
  - Let $w(i) = w(i) - 1$.
  - Set $k = k + \min\{3, i\}$.
- **return** $(\mathcal{P}, L)$, where L is the labeling induced by $\mathcal{P}$.
---

4   Trees

Let $T = (V, E)$ be any tree and s be a fixed node in T. We assume that each node has exactly one path to be set, i.e., $w(v) = 1$ for each $v \in V - \{s\}$ (recall that the source has weight $w(s) = 0$). We will show how to obtain an optimal labeling for $\langle T, \mathbf{w}, s \rangle$. The extention to the case $w(v) \geq 1$ can be easly obtained.

Definition 8. *Given a tree T. We shall denote by $|T|$ the size of T in terms of the weights of the nodes in T, that is*

$$|T| = \sum_{v \in V(T)} w(v).$$

Notice that $|T|$ represents the number of paths to be set in T. Since we assume that $w(v) = 1$ for each $v \in V - \{s\}$ then the algorithms starts with $|T| = |V| - 1$. Root T at s and let $T_1, T_2, \cdots, T_m$ be the subtrees of T rooted at the sons of s.

We also notice that in case $m = 1$, then T consists of a root of degree 1 and T_1 as the only subtree. A one-to-all personalized optimal broadcasting in T is obtained by applying the optimal algorithm LINE-labeling to the line L obtained from T by replacing the $w(j)$ vertices at distance j in T by a vertex j with weight $w(j)$ in L. Then by Corollary 1 the number of steps is $T^*(L) = 1 + 2\alpha_1 + 3\beta_1$.

The main idea of the algorithm consists in setting, whenever that is possible, a path to a node in the subtree T_i having the largest shade value $\tau_i = 1 + 2\alpha_i + 3\beta_i$ (Definition 1). However, we have to be careful and, even if the algorithm is relatively simple, the proof of the value of gathering time in Theorem 1 is involved.

4.1   The Algorithm

In order to describe the SCF labeling algorithm, we introduce the following terminology. Unless otherwise stated, in the following we assume that the subtrees are numbered according to the ranking given in Definition 2, that is $T_1, \cdots, T_m$ is a reordering of the subtrees of T such that $T_1 \preceq \cdots \preceq T_m$.

- *One step:* one time–slot.
- A node $v \neq s$ *is completed* if a path from s to v has been set.
- *Set a path (resp. a t–path) to T_i:* set a path (resp. a t-path) to a node v in T_i which is the furthest from s among all nodes in T_i which are not yet completed.
  When we set a path to some T_i the corresponding value $|T_i|$ of the remaining weights in T_i will be decreased by one and also α_i and β_i if they are non zero.
- T_i *is completed:* if a path has been set to each node in T_i, that is $|T_i| = 0$.
- Step t is called *idle* if no t-path is set.
- T_i is *available* at step t (e.g. a t-path to T_i can be set) only if no path was set to a node v in T_i at some step t' s.t. $t' < t < t' + \min\{3, \ell(v)\}$, where $\ell(v)$ is the level of v in T. Said otherwise, if at some step t' we set a path to a node v in T_i, then T_i is not available at step $t' + j$ where $1 \leq j < \min\{3, \ell(v)\}$. in particular if v is at a level at least 3, then T_i is not available at steps $t' + 1$ and $t' + 2$.

The SCF labeling algorithm is given in Tab.2. Following is an informal description of the behavior of the algorithm during a generic step $t \geq 1$: Let T_i be an available subtree that precedes all the other available subtrees of T according to the order relation $\preceq$; set a t-path to T_i; update the shade of T_i.

Table 2. The SCF labeling algorithm on trees

TREE-labeling $(T, \mathbf{w}, s)$ *[T has non empty subtrees $T_1, \ldots, T_m$ and root s]*
1. Set $\mathcal{P} = \emptyset$ and $t = 1$
   Let $t_i = 1$ for $i = 1, \ldots, m$ *[t_i is the minimum step to set a path to T_i]*
   Set $M = \{1, \ldots, m\}$ *[M represents the set of subtrees not yet completed]*
2. **while** $M \neq \emptyset$
   2.1   Rename the indices in M so that for the permuted subtrees $T_1 \preceq \ldots \preceq T_{|M|}$
   2.2   **if** there exists $i \leq |M|$ with $t_i \leq t$ **then**
               Let i be the smallest such index (e.g. $t_1, \ldots, t_{i-1} > t$, $T_i \preceq \ldots \preceq T_{|M|}$).
               if NOT $(|M| = 2, i = 1, \beta_1 = 1, \alpha_2 > \beta_2 = 0, t_2 \leq t + 1)$ **then**
                  [Execute the generic step of the algorithm]
                  - Set a t-path to T_i and call it $\mathbf{p}$
                  - $\mathcal{P} = \mathcal{P} \cup \{\mathbf{p}\}$.
                  - If T_i is completed then $M = M - \{i\}$.
                  - $t_i = t + \min\{3, \ell\}$, where ℓ is the length of $\mathbf{p}$,
                  - Update T_i, eg.: $\tau_i = \tau_i - \min\{3, \ell\}$, $w(s_i) = w(s_i) - \begin{cases} 1 & \text{if } \ell = 1 \\ 0 & \text{oth.} \end{cases}$
                  $\alpha_i = \alpha_i - \begin{cases} 1 & \text{if } \ell = 2 \\ 0 & \text{oth.} \end{cases}$, $\beta_i = \beta_i - \begin{cases} 1 & \text{if } \ell \geq 3 \\ 0 & \text{oth.} \end{cases}$.
   2.3   **else** *[The special case: $|M| = 2$, $i = 1$, $\beta_1 = 1$, $\alpha_2 > \beta_2 = 0, t_2 \leq t + 1$]*
                  - Set a t-path to T_1 and call it $\mathbf{p}$
                  - Set a $t + 1$-path to s_2 and call it $\mathbf{q}_1$
                  - Set a $t + 2$-path to T_2 and call it $\mathbf{q}_2$
                  - $\mathcal{P} = \mathcal{P} \cup \{\mathbf{p}, \mathbf{q}_1, \mathbf{q}_2\}$.
                  - $t_1 = t + 3$ and $t_2 = t + 4$.
                  - Update T_1 and T_2
                      (e.g. $\tau_1 = \tau_1 - 3$, $\beta_1 = 0$, $\tau_2 = \tau_2 - 3$, $w(s_2) = 0$, $\alpha_2 = \alpha_2 - 1$).
                  - If $\alpha_2 = 0$ then $M = \{1\}$.
                  - $t = t + 2$.
   2.4 $t = t + 1$.
3. **return** $(\mathcal{P}, L)$

Example 3. The solution $(\mathcal{P}, L)$ given in Example 2 is the same one gets by applying the TREE-labeling algorithm on the tree T of Fig.1. Notice that steps $t = 8, 9, 10$ correspond to the special case of point 2.3 of the algorithm.

The TREE-labeling algorithm sets, at step t, a t-path to T_i only if T_i is available. We can then conclude that

Lemma 1. *The solution $(\mathcal{P}, L)$ returned by algorithm TREE-labeling on $\langle T, \mathbf{w}, s \rangle$ is feasible.*

4.2   Preliminary Results

We establish now some facts that will be used to prove the optimality of the proposed algorithm.

Fact 1. *For any subtree T_i with $|T_i| > 1$ it holds that $2|T_i| - 1 \leq \tau_i \leq 3|T_i| - 3$.*

Fact 2. *Let $T_i \preceq T_j$.*

- *If $\tau_i = \tau_j$ and $T_i \prec T_j$ then $\alpha_i > \alpha_j$ and $\beta_i < \beta_j$.*
- *$T_i = T_j$ (e.g., $\tau_i = \tau_j$ and $|T_i| = |T_j|$) iff $\alpha_i = \alpha_j$ and $\beta_i = \beta_j$.*

Fact 3. *If $T_i \preceq T_j$ then $\beta_j \leq \begin{cases} |T_i| - 2 & \text{if } |T_i| \geq 2 \\ 0 & \text{otherwise} \end{cases}$.*

The quantities $\Delta_{i,j}$ introduced in Definition 3 satisfy the following properties.

Fact 4. *For any i, j it holds $\Delta_{i,j} - \tau_i = |T_j| - |T_i|$*

Fact 5. *$\Delta_{i,j} \geq \max\{|T|, \tau_1 + \epsilon_T\}$ only if either $i = 1$ and $j = 2, 3$ or $i = 2$ and $j = 1$.*

Proof. Assume first either $i \geq 3$ or $i = 2$ and $j \geq 3$. We have $|T| - |T_i| - |T_j| \geq |T_1|$ or $|T| - |T_i| - |T_j| \geq |T_2|$. By Fact 3 we know that $\beta_i < \min\{|T_1|, |T_2|\} - 1$. Hence, in any case we get

$$|T| - \Delta_{i,j} = |T| - |T_i| - |T_j| - \beta_i + 1 > 2,$$

which implies $\Delta_{i,j} < |T| \leq \max\{|T|, \tau_1 + \epsilon_T\}$.

Assume now $i = 1$ and $j \geq 4$; supposing, by contradiction, $\Delta_{1,j} \geq |T|$ and $\Delta_{1,j} \geq \tau_1 + \epsilon_T$, we have

$$|T_2| + |T_3| \leq |T| - |T_1| - |T_j| = |T| - \Delta_{1,j} + \beta_1 - 1 \leq \beta_1 - 1 \leq |T_1| - 3. \quad (2)$$

From the assumption that $\Delta_{1,j} \geq \tau_1 + \epsilon_T$ and by Fact 4 we get $|T_1| \leq |T_j|$. This, (2), and Fact 1 imply

$$\tau_j \geq 2|T_j| - 1 \geq 2|T_1| - 1 \geq 2(|T_2| + |T_3|) + 5 > \frac{2}{3}(\tau_2 + \tau_3) + 5 \geq \frac{4}{3}\tau_3 + 5 > \tau_3$$

thus contradicting the assumption $T_3 \preceq T_j$ for any $j \geq 4$. □

4.3 A Lower Bound

Let T be such that $T_1 \preceq T_2 \preceq \ldots \preceq T_m$. Define

$$Max(T) = \max\{|T| = |V| - 1, \ \tau_1 + \epsilon_T, \ \Delta_{1,2}, \ \Delta_{2,1}, \ \Delta_{1,3}\}$$

Lemma 2. *Assuming that $T_1 \preceq T_2 \preceq \ldots \preceq T_m$, we have $T^*(T) \geq Max(T)$.*

Proof. Any algorithm needs to set a path to each node, hence $T^*(T) \geq |T|$.

By Definition 1 and Corollary 1, the shade τ_i of T_i is the minimum label that can be assigned when only paths to the nodes in T_i are set. Since paths must be set to all nodes in each T_i, for $i = 1, \cdots, m$, and $\tau_1 \geq \tau_2 \geq \cdots \geq \tau_m$ we have that $T^*(T) \geq \tau_1$.

Furthermore, if $T_1 = T_2$ then at least $\tau_1 + 1$ labels are necessary.

Consider now $\Delta_{i,j}$. For each path to a node at level at least 3 in T_i no path to some other node in T_i can be set in the following 2 steps. Moreover, at most one of the following two steps can be used to set a path to T_j, except for the eventual step in which a path to the root of T_j is set and immediately after a path to some other node in T_j is set. The remaining step can be used to set a path to some T_ℓ with $\ell \neq i, j$. Hence, any algorithm has at least $\beta_i - 1 - \sum_{\ell \neq i,j} |T_\ell|$ idle steps, which implies $T^*(T) \geq |T| + \beta_i - 1 - \sum_{\ell \neq i,j} |T_\ell| = \Delta_{i,j}$. By Fact 5, we get that $Max(T)$ lower bounds $T^*(T)$. □

4.4 Optimality

We show now that the SFC–labeling algorithm for trees is optimal, that is, the maximum label assigned to any arc of T is $\mathcal{T}(T) \leq Max(T)$ thus matching the lower bound of Theorem 2.

We first recall that we are in the hypothesis that the weight of each node is 1. The order in which nodes are chosen as end–points of the paths set by the algorithm implies that the largest label assigned to an arc of T is always to be searched among those assigned to the arcs outgoing the root s of T. Therefore, it coincides with the largest t for which a t–path is set in T.

Lemma 3. *Let t denote the largest integer such that a t–path is set in T during the execution of the SFC–labeling algorithm. The largest label assigned by the algorithm to any arc of T is $\mathcal{T}(T) = t$.*

By the above Lemma, we need to show that the largest t such that a t–path is set in T is upper bounded by $Max(T)$. The proof will proceed by induction. We will consider the first steps of the algorithm mainly those which send to different subtrees (before the step where we send again to a subtree to which we already sent) and we will apply the induction on the tree T' obtained by deleting the vertices completed in these first steps. For that we give the following definition.

Definition 9. *We denote the fact that the algorithm on T starts by setting k paths to pairwise different subtrees of T (that is, it sets a t–path to some node v_i in T_i, for $i = 1, \ldots, k$) by*

$$\langle T_1 \ldots T_k \rangle$$

We denote by T' the updated tree, resulting from $\langle T_1 \ldots T_k \rangle$, that is, T' has subtrees $T'_1 \ldots, T'_k, T'_{k+1} \ldots, T'_m$, where

- *T'_i denotes the updated subtree T_i after the i–path to v_i has been set (that is, $w'(v_i) = 0$ and $|T'_i| = |T_i| - 1$, for $i = 1, \ldots, k$*
- *$T'_{k+1} = T_{k+1}, \ldots, T'_m = T_m$.*

Notice that the subtrees $T'_1 \ldots, T'_m$, are not necessarily ordered according to the relation $\preceq$. Let $i_1, i_2, \cdots, i_m$ be a permutation of $1, \ldots, m$ such that $T'_{i_1} \preceq \ldots \preceq T'_{i_m}$; we will always consider permutations that maintain the original order on equal subtrees, that is

$$\text{if } T'_{i_j} = T'_{i_\ell} \text{ then } i_j < i_\ell. \tag{3}$$

We denote by $\alpha'_i, \beta'_i, \tau'_i$ the parameters of T'_i. In particular (unless the special case $k = 2$, $\beta_1 = 1, \alpha_2 > \beta_2 = 0$, $T_3 = \emptyset$, and T_2 is available) we have for $i = 1, \ldots, k$:

$$\alpha'_i = \alpha_i - \begin{cases} 1 & \text{if } \beta_i = 0,\ \alpha_i \geq 1 \\ 0 & \text{otherwise} \end{cases}, \quad \beta'_i = \beta_i - \begin{cases} 1 & \text{if } \beta_i \geq 1 \\ 0 & \text{otherwise} \end{cases},$$

$$\tau'_i = \tau_i - \begin{cases} 3 & \text{if } \beta_i \geq 1 \\ 2 & \text{if } \beta_i = 0,\ \alpha_i \geq 1 \\ 1 & \text{if } |T_i| = 1 \\ 0 & \text{if } T_i = \emptyset \end{cases}.$$

Fact 6. *Assume* $\langle T_1 \ldots T_k \rangle$ *and that NOT (k = 2, $\beta_1 = 1, \alpha_2 > \beta_2 = 0$, $T_3 = \emptyset$, and T_2 is available). For any $1 \le i < j \le k$.*

1) *If $\tau_i > \tau_j$ then $\tau_i' \ge \tau_j'$;*
2) *if $T_i \prec T_j$ and $T_j' \prec T_i'$ then $\beta_j = 0$, $\beta_i \ge 2$, $|T_i| < |T_j|$, and $\tau_i = \tau_j + 1$.*

Fact 7. *Assume* $\langle T_1 \ldots T_k \rangle$ *with either $k \ge 4$ or $k = 3$ and $T_3 \preceq T_1', T_2'$:*

i) $|T| \ge \tau_1 + k - 2$;
ii) $|T_i| \ge \beta_1 + 1$, *for each $i = 2, \ldots, k$;*
iii) $|T_i| \ge \beta_2 + 1$, *for each $i = 3, \ldots, k$.*

The following lemma together with lower bound of Lemma 2 prove Theorem 1.

Lemma 4. *Assume $T_1 \preceq \ldots \preceq T_m$. The solution returned by algorithm TREE-labeling satisfies*

$$\mathcal{T}(T) \le Max(T) \tag{4}$$

Proof. At any step of the algorithm the tree can have any number $m \ge 1$ of subtrees of positive weight. When we say that the algorithm sets a t–path to a subtree T_i and $|T_i| = 0$ at step t, this means that no t–path is actually set (e.g. t is an idle step).

We first analyze the special case of the algorithm in which $m = 2$, $\beta_1 = 1$, $\beta_2 = 0$ and T_2 is available. So $\tau_1 > \tau_2$ and $\alpha_1 \ge \alpha_2 - 1$. The first two steps of the algorithm are $\langle T_1 T_2 \rangle$, where the path set to T_2 is a path to s_2 (the root of T_2). Let T' be the tree resulting after $\langle T_1 T_2 \rangle$, at the third step a path to T_2' is set. Hence, the first three steps of the algorithm are: $\langle T_1 T_2 \rangle \langle T_2' \rangle$

Let T^2 be the tree resulting after $\langle T_1 T_2 \rangle \langle T_2' \rangle$. Next the algorithm on T proceeds as follows

$$\langle T_1^2 T_2^2 \rangle \langle T_1^3 T_2^3 \rangle \ldots \langle T_1^\ell T_2^\ell \rangle \ldots \langle T_1^{\alpha_1+1} T_2^{\alpha_1+1} \rangle \langle T_1^{\alpha_1+2} \rangle.$$

where T^ℓ is the tree resulting from $T^{\ell-1}$ after the 2 steps $\langle T_1^{\ell-1} T_2^{\ell-1} \rangle$. To see this, we notice that in each T^ℓ it holds $T_1^\ell \prec T_2^\ell$, since $\tau_1^2 = \tau_1 - 3 > \tau_2 - 3 = \tau_2^2$ and $\tau_1^\ell = \tau_1^{\ell-1} - 2 > \tau_2^\ell = \max\{\tau_2^{\ell-1} - 2, 0\}$, for $\ell > 2$. Moreover, in the hypothesis of this case $\alpha_1 \ge \alpha_2 - 1$, which implies that $T_2^\ell = \emptyset$ for $\ell > \alpha_2$. Finally, by the hypothesis we have

$$\epsilon_T = 0, \quad |T| = 3 + \alpha_1 + \alpha_2 \le 3 + 2\alpha_1 + 1 = \tau_1, \quad \text{and } \Delta_{1,2}, \Delta_{2,1} \le |T|.$$

Hence, $Max(T) = \tau_1$; but $\mathcal{T}(T) = 3 + 2\alpha_1 + 1 = \tau_1 = Max(T)$.

The rest of the proof is devoted to show that $\mathcal{T}(T) \le Max(T)$ for each tree. The proof is by induction on the shade of T_1, (recall that $T_1 \preceq T_2 \preceq \ldots \preceq T_m$). As a base consider the trees of the special case above and trees T such that $\tau_1 = 1$; in the latter case, we have $|T_i| = 1$ for each $i = 1, \ldots, m$ and $\mathcal{T}(T) = |T| = Max(T)$.

Suppose now that (4) holds for any tree in which the shade of the first subtree (according to the relation $\preceq$) is at most $\tau_1 - 1$; we prove that (4) holds for T.

Notice that we are assuming that T does not belong to the special case (e.g., $m = 2$, $\beta_1 = 1$, $\beta_2 = 0$, and T_2 is available) and that $|T_1| \ge 2$.

We separate four cases according to the value attaining $Max(T)$.

Case 1: $Max(T) = \Delta_{1,2} > \max\{\tau_1 + \epsilon_T, |T|\}$.

In such a case we know that $\beta_1 > 1$, otherwise $\Delta_{1,2} = |T_1| + |T_2| + \beta - 1 \leq |T|$; hence, the first tree steps of the algorithm are (including the case $|T_3| = 0$) $\langle T_1 T_2 T_3 \rangle$.

Let T' be the tree resulting after $\langle T_1 T_2 T_3 \rangle$. We will show that after the first 3 steps $\langle T_1 T_2 T_3 \rangle$, the algorithm on T proceeds as on input T' and

$$Max(T') \leq Max(T) - 3. \qquad (5)$$

This implies the desired inequality $T(T) = 3 + T(T') \leq 3 + Max(T') = Max(T)$.

By definition of $\Delta_{1,2}$ and using $\Delta_{1,2} > |T|$, we get

$$|T| - |T_1| - |T_2| < \beta_1 - 1. \qquad (6)$$

By Fact 4 and using $\Delta_{1,2} > \tau_1$, we get

$$|T_1| < |T_2|. \qquad (7)$$

By (6) and Fact 1, we get

$$\tau_3 < 3|T_3| \leq 3(|T| - |T_1| - |T_2|) < 3(\beta_1 - 1) = (3\beta_1 + 2\alpha_1 + 1) - (2\alpha_1 + 4),$$

from which, since $\alpha_1 \geq 1$, it follows

$$\tau_3 < \tau_1 - 6 = \tau_1' - 3. \qquad (8)$$

Moreover, by (6) and (7) we have

$$|T_2| \geq |T_1| + 1 \geq \beta_1 + \alpha_1 + 2 \geq \beta_1 + 3 > (|T| - |T_1| - |T_2|) + 4 \geq |T_3| + |T_4| + 4; \qquad (9)$$

which, by Fact 1, implies $\tau_2 \geq 2|T_2| - 1 > 2(|T_3| + |T_4|) + 7 \geq 4 \min\{|T_3|, |T_4|\} + 7$.

Noticing that Fact 1 implies $4|T_4| > \tau_4$ and $4|T_3| > \tau_3 \geq \tau_4$, we get

$$\tau_2 \geq \tau_4 + 8. \qquad (10)$$

From (8) and (10) and recalling that $\tau_1 \geq \tau_2$, we obtain that in the tree T', resulting after $\langle T_1 T_2 T_3 \rangle$:

$$T_1' \prec T_3', \quad T_1' \prec T_4' = T_4, \quad T_2' \prec T_4' = T_4.$$

Moreover, we have

$$T_2' \prec T_3';$$

indeed, if we assume $T_2' \succeq T_3'$ we have either $|T_2| = |T_3|$ or, by Fact 6, $|T_3| > |T_2|$ contradicting (9).

We notice that $T_1' \neq T_2'$, since by (7) they have different weights. Hence, by the definition of $\prec$ (cfr. Definition 2), we get that the only possible orderings on the the subtrees of T' are:

$$T_1' \prec T_2' \prec T_3', \quad T_1' \prec T_2' \prec T_4', \quad T_2' \prec T_1' \prec T_3', \quad T_2' \prec T_1' \prec T_4'.$$

Moreover, both sequences of steps $\langle T_1 T_2 T_3 \rangle \langle T_1' T_2' \rangle$ and $\langle T_1 T_2 T_3 \rangle \langle T_2' T_1' \rangle$ are possible during the execution of the algorithm on T; in particular if $T_2' \prec T_1'$ we know by Fact 6 that $\beta_2 = 0$.

Hence, after the first 3 steps, the algorithm on T proceeds as on input T'. For T' we have: $\tau_1' = \tau_1 - 3$ (since $\beta_1 > 1$),

$$|T'| = |T| - \begin{cases} 3 & \text{if } |T_3| > 0 \\ 2 & \text{otherwise} \end{cases}, \quad \epsilon_{T'} = \epsilon_T = 0 \text{ (since } |T_1| < |T_2|).$$

In case $T_1' \prec T_2'$, it holds $\Delta_{1,2}' = \Delta_{1,2} - 3$,

$$\Delta_{2,1}' = \begin{cases} \Delta_{2,1} - 3 & \text{if } \beta_2 > 0 \\ |T_2'| + |T_1'| - 1 < |T'| & \text{if } \beta_2 = 0 \end{cases}, \quad \Delta_{1,3}', \Delta_{1,4}' < \Delta_{1,2} - 3,$$

where the last inequality follows from (9).

In case $T_2' \prec T_1'$, by Fact 6 we have $\beta_2 = 0$, $\beta_1 \geq 1$ and $\tau_1 > \tau_2$; hence $\tau_2' = \tau_2 - 2 = \tau_1 - 3$ and

$$\Delta_{1,2}' = \Delta_{1,2} - 3, \qquad \Delta_{2,i}' = |T_2'| + |T_i'| + \beta_2' - 1 = |T_2'| + |T_i'| - 1 < |T'| \ (i = 1, 3, 4).$$

Summarizing, in both cases $T_1' \prec T_2'$ and $T_2' \prec T_1'$, inequality (5) holds.

Due to space limits, the proofs of the other cases:
$Max(T) = \Delta_{2,1} > \max\{\tau_1 + \epsilon_T, |T|\}$, $Max(T) = \Delta_{1,3} > \max\{\tau_1 + \epsilon_T, |T|\}$, and $Max(T) = \max\{\tau_1 + \epsilon_T, |T|\}$ are omitted from this extended abstract.   □

5   Conclusion

In this paper we give a relatively simple protocol for trees with $w(u) = 1$ packet to transmit. The results can be easily extended to the case where all the $w(u)$ are positive (or at least there is no more than two consecutive nodes with weights 0). It might be that the algorithm is optimal for any weight function by replacing τ_i with M_i (see Theorem 2); but the proof seems more complicated. It will be also interesting to find the complexity of the gathering problem for general graphs without buffering (with buffering it is known to be NP-hard).

References

1. Bermond, J.-C., Corrêa, R., Yu, M.: Gathering algorithms on paths under interference constraints. In: Calamoneri, T., Finocchi, I., Italiano, G.F. (eds.) CIAC 2006. LNCS, vol. 3998, pp. 115–126. Springer, Heidelberg (2006)
2. Bermond, J.-C., Gargano, L., Rescigno, A.: Gathering with Minimum Delay in Sensor Networks, INRIA TR (2008), http://hal.inria.fr/inria-00256896/fr/
3. Bermond, J.-C., Galtier, J., Klasing, R., Morales, N., Pérennes, S.: Hardness and approximation of gathering in static radio networks. Parallel Processing Letters 16(2), 165–183 (2006)
4. Bermond, J.-C., Peters, J.: Efficient gathering in radio grids with interference. In: AlgoTel 2005, pp. 103–106, Presqu'île de Giens (2005)
5. Bermond, J.-C., Yu, M.: Optimal gathering algorithms in multi-hop radio tree-networks with interferences (manuscript, 2008)

6. Bertin, P., Bresse, J.-F., Le Sage, B.: Accès haut débit en zone rurale: une solution "ad hoc". France Telecom R&D 22, 16–18 (2005)
7. Bonifaci, V., Korteweg, P., Marchetti-Spaccamela, A., Stougie, L.: An Approximation Algorithm for the Wireless Gathering Problem. In: Arge, L., Freivalds, R. (eds.) SWAT 2006. LNCS, vol. 4059. Springer, Heidelberg (2006)
8. Chong, C.-Y., Kumar, S.P.: Sensor networks: Evolution, opportunities, and challenges. Proc. of the IEEE 91(8), 1247–1256 (2003)
9. Coleri, S., Varaiya, P.: Energy Efficient Routing with Delay Guarantee for Sensor Networks. Wireless Networks (to appear)
10. Dasgupta, K., Kukreja, M., Kalpakis, K.: Topology-aware placement and role assignment for energy-efficient information gathering in sensor networks. In: Proc. IEEE ISCC 2003, pp. 341–348 (2003)
11. Falck, E., Floreen, P., Kaski, P., Kohonen, J., Orponen, J.P.: Balanced data gathering in Energy-constrained sensor networks. In: Nikoletseas, S.E., Rolim, J.D.P. (eds.) ALGOSENSORS 2004. LNCS, vol. 3121, pp. 59–70. Springer, Heidelberg (2004)
12. Florens, C., Franceschetti, M., McEliece, R.J.: Lower Bounds on Data Collection Time in Sensory Networks. IEEE J. on Sel. Ar. in Com. 22(6), 1110–1120 (2004)
13. Ganesan, D., Cristescu, R., Beferull-Lozano, B.: Power-efficient sensor placement and transmission structure for data gathering under distortion constraints. In: IPSN 2004, pp. 142–150 (2004)
14. Gargano, L.: Time Optimal Gathering in Sensor Networks. In: Prencipe, G., Zaks, S. (eds.) SIROCCO 2007. LNCS, vol. 4474, pp. 7–10. Springer, Heidelberg (2007)
15. Gargano, L., Rescigno, A.A.: Optimally Fast Data Gathering in Sensor Networks. In: Královič, R., Urzyczyn, P. (eds.) MFCS 2006. LNCS, vol. 4162, pp. 399–411. Springer, Heidelberg (2006)
16. Gupta, H., Navda, V., Das, S.R., Chowdhary, V.: Efficient gathering of correlated data in sensor networks. In: Proc. of ACM MobiHoc 2005, pp. 402–413 (2005)
17. Gasieniec, L., Potapov, I.: Gossiping with Unit Messages in Known Radio Networks. In: IFIP TCS 2002, pp. 193–205 (2002)
18. Ho, B., Prasanna, V.K.: Constrained flow optimization with application to data gathering in sensor networks. In: Nikoletseas, S.E., Rolim, J.D.P. (eds.) ALGOSENSORS 2004. LNCS, vol. 3121, pp. 187–200. Springer, Heidelberg (2004)
19. Intanagonwiwat, C., Govindan, R., Estrin, D., Heidemann, J., Silva, F.: Directed diffusion for wireless sensor networking. IEEE/ACM Trans. Netw. 11(1), 2–16 (2003)
20. Krishnamachari, B., Estrin, D., Wicker, S.: Modeling data-centric routing in wireless sensor networks. In: Proc. of IEEE INFOCOM (2002)
21. Lindsey, S., Raghavendra, C.: Pegasis: Power-efficient gathering in sensor wireless networks. In: Proc. of IEEE Aerospace Conference (2002)
22. Lindsey, S., Raghavendra, C., Sivalingam, K.M.: Data gathering algorithms in sensor networks using energy metrics. IEEE Trans. on Par. and Distr. Sys. 13(9), 924–935 (2002)
23. Padmanabh, K., Roy, R.: Multicommodoty flow fased maximum lifetime routing in wireless sensor network. In: Proc. of IEEE ICPADS 2006, pp. 187–194 (2006)
24. Shen, C., Srisathapornphat, C., Jaikaeo, C.: Sensor information networking architecture and applications. IEEE Personal Communications, 52–59 (2001)
25. Yu, Y., Krishnamachari, B., Prasanna, V.: Energy-latency tradeoffs for data gathering in wireless sensor networks. In: Proc. of IEEE INFOCOM 2004 (2004)
26. Zhu, X., Tang, B., Gupta, H.: Delay efficient data gathering in sensor networks. In: Jia, X., Wu, J., He, Y. (eds.) MSN 2005. LNCS, vol. 3794, pp. 380–389. Springer, Heidelberg (2005)

Centralized Communication in Radio Networks with Strong Interference

František Galčík*

Institute of Computer Science,
P.J. Šafárik University, Faculty of Science,
Jesenná 5, 041 54 Košice, Slovak Republic
frantisek.galcik@upjs.sk

Abstract. We study communication in known topology radio networks with the presence of interference constraints. We consider a real-world situation, when a transmission of a node produces an interference in the area that is larger than the area, where the transmitted message can be received. For each node, there is an area, where a signal of its transmission is too low to be decoded by a receiver, but is strong enough to interfere with other incoming simultaneous transmissions. Such a setting is modelled by a newly proposed interference reachability graph that extends the standard graph model based on reachability graphs. Further, focusing on the information dissemination problem in bipartite interference reachability graphs, we introduce interference ad-hoc selective families as an useful combinatorial tool. They are a natural generalization of ad-hoc selective families. Adopting known algorithms and techniques, we show how to construct small interference ad-hoc selective families in the case when, for each node, the ratio of the only-interfering neighbors to the other neighbors is bounded. Finally, taking into account the maximum degree in an underlying interference reachability graph, we study the broadcasting problem in general radio networks.

1 Introduction

A *radio network* is a collection of autonomous stations that are referred to as *nodes*. The nodes communicate via sending messages. Each node is able to receive and transmit messages. However, a node can transmit messages only to the nodes, which are located within its *transmission range*. We say that a node w belongs to the transmission range of a node v ($w \in T(v)$) if and only if a message transmitted by v can reach the node w. Hence, the transmission range of v is a set of the network nodes that are located at positions, where the signal transmitted by v has enough intensity and quality to be successfully decoded. All nodes of the radio network operate at the same frequency. Owing to properties of the radio communication medium, simultaneous transmissions of two or more nodes cause interference in the area that is in the range of those transmitted

---

* Research of the author is supported in part by Slovak VEGA grant number 1/3129/06.

A. Shvartsman and P. Felber (Eds.): SIROCCO 2008, LNCS 5058, pp. 277–290, 2008.
© Springer-Verlag Berlin Heidelberg 2008

signals. However in some practical applications, a transmitted signal can reach an area, where decoding of the signal is not possible due to its low intensity, but the signal is intensive enough to interfere with other simultaneous transmissions. We define an *interference range* $I(v)$ of a node v as follows. A node w belongs to the interference range of a node v ($w \in I(v)$) if and only if a transmission of v can interfere with other transmissions reaching the node w. It is natural to assume that $T(v) \subseteq I(v)$. Indeed, if a signal is intensive enough to be decoded, it is intensive enough to cause interference with other transmissions.

Most of the literature concerning communication in radio networks (e.g. [10], [5], [9]) assume that the transmission range $T(v)$ and the interference range $I(v)$ of a node v are the same, i.e. $T(v) = I(v)$ for each network node v. Such a communication network can be modelled by a *reachability graph*. A reachability graph is a directed graph $G = (V, E)$. The vertex set V corresponds to the network nodes and two vertices $u, v \in V$ are connected by an edge $e = (u, v)$ if and only if the node v is in the transmission range of a node u, i.e. $v \in T(u)$. This model is also referred to as the *graph model* of radio networks. If the transmission power of all nodes is the same, then a network can be modelled by an undirected graph.

A communication model, in which the interference range of a node is larger than its transmission range, was considered by Bermond et al. in [2]. The authors studied time complexity of the gathering task in known topology radio networks. They defined transmission and interference range of a node with respect to distances in an underlying communication graph. Particularly, denote by $dist_G(u, v)$ the length (the number of edges) of a shortest path between nodes u and v in the graph G. Fix the numbers d_T and d_I. The number d_T, $d_T \geq 1$, is called a *transmission distance* and the number d_I, $d_I \geq d_T$, is called an *interference distance*. The transmission range $T(v)$ of a node v is defined as $T(v) = \{w | dist(v, w) \leq d_T\}$ and the interference range $I(v)$ of a node v as $I(v) = \{w | dist(v, w) \leq d_I\}$. Note that the standard graph model corresponds to the case when $d_T = d_I = 1$.

It is easy to see that there are such settings, where the model introduced by Bermond et al. is not appropriate, e.g. due to large obstacles or signal reflexes. That means, that there are settings, for which it is difficult or even impossible to express the transmission and interference ranges of the nodes with respect to distances in an underlying communication graph. In this paper we focus on a new model of the communication environment. Particularly, we shall assume that $I(v)$ is an arbitrary set of network nodes satisfying $T(v) \subseteq I(v)$. In such a setting, the communication network can be modelled by a directed graph $G = (V, E_T \cup E_I)$, called an *interference reachability graph* (IRG), such that $E_T \cap E_I = \emptyset$. Two vertices $u, v \in V$ are connected by an edge $e = (u, v) \in E_T$ (a *transmission edge*) if and only if the node v is in the transmission range of the node u, i.e. $v \in T(u)$. The node u is referred to as a *transmission neighbor* of the node v. Similarly, two vertices $u, v \in V$ are connected by an edge $e = (u, v) \in E_I$ (an *interference edge*) if and only if the node v is in the interference range of the node u but not in its transmission range, i.e. $v \in I(u) \setminus T(u)$. The node u is referred to as an *interference neighbor* of the node v. Note that no message can be brought forward by an interference edge in compare to a transmission edge. We shall denote the spanning subgraph $G(E_T)$ as a *transmission subgraph*. For practical reasons, we assume that the transmission subgraph $G(E_T)$ is strongly connected. Hence there is an oriented path, using only transmission edges, from each network node to any other network node. Observe that each radio network modelled by the model introduced by Bermond et al. can be described by an IRG. It implies that the proposed model is more general.

Communication in radio networks is considered to be synchronous. In particular, the network nodes work in synchronized steps (time slots) called *rounds*. In every round, a node can act either as a *receiver* or as a *transmitter*. A node u acting as transmitter sends a message, which can be potentially received by every node in its transmission range. In a given round, a node, acting as a receiver, receives a message if and only if it is located in the transmission range of exactly one transmitting node and in the interference range of none transmitting node. Otherwise, no message is received by the receiving node. The received message is the same as the message transmitted by the transmitting neighbor.

2 Centralized Broadcasting

We focus on the *broadcasting* - one of the most studied and important communication primitives. The goal of broadcasting is to distribute a message, called a *source message*, from one distinguished node, called a *source*, to all other nodes. Remote nodes of the network are informed via intermediate nodes. The time (number of rounds), that is required to complete an operation, is an important efficiency measure and is a widely studied parameter of mostly every communication task.

It is known that assumptions about initial knowledge of nodes significantly influence the time required to complete a communication task. In this paper we shall assume that each node possesses full knowledge of the network topology, i.e. every node possesses a labelled copy of an underlying IRG. Communication in radio networks with full knowledge of nodes (known topology radio networks) is referred to as *centralized communication*. In this setting, an execution of a broadcasting algorithm can be seen as a process controlled by a central monitor. Thus the goal is to design a polynomial time (tractable) algorithm, that for a given (interference) reachability graph G and a source node s produces a schedule of transmissions, referred to as a *radio broadcast schedule*, that is as short as possible and disseminates the source message to all network nodes.

Centralized broadcasting in radio networks modelled by a reachability graph, i.e. in the standard graph model, has been intensively studied. In [9], Kowalski and Pelc presented an algorithm that produces a radio broadcast schedule of the length $O(D + \log^2 n)$, where D is the diameter of a given underlying reachability graph and n denotes the number of nodes. In the view of the lower bound $\Omega(\log^2 n)$ for graphs with diameter 2 given by Alon et al. [1] and, a trivial lower bound D, this algorithm is asymptotically optimal.

The rest of the paper is devoted to the information dissemination problem in known topology radio networks modelled by IRGs.

2.1   Difficulty of Fast Broadcasting in IRG

It is easy to see that the broadcasting task can be always completed in $O(n)$ rounds, where n is the number of network nodes. Hence there is a radio broadcast schedule such that each node is informed at most $O(n)$ rounds after first transmission of the source node. Indeed, in each round, we select one informed node that is a transmission neighbor of at least one uninformed node. Only the selected node is allowed to transmit in this round. Thus all uninformed nodes in its transmission range receive the source message and become informed. On the other hand, there is an IRG with diameter 2 such that the broadcasting time is bounded by $\Omega(n)$ rounds. The $(2 \cdot m + 1)$-node graph $G_m = (V_m, E_T \cup E_I)$ is defined as follows:

- $V(G_m) = \{s, a_1, a_2, \ldots, a_m, b_1, b_2, \ldots, b_m\}$
- $E_T(G_m) = \{(s, a_i), (a_i, b_i) | 1 \leq i \leq m\}$
- $E_I(G_m) = \{(a_i, b_j) | 1 \leq i \neq j \leq m\}$

Let s to be a source of the broadcasting. Each node a_i becomes informed after first transmission of s. The node b_i can be informed only by a transmission of a_i. However, if a node a_i transmits, no other node b_j, $j \neq i$, can receive the message due to the presence of interference edges. It follows, that at least $m + 1 = \Omega(n)$ rounds are necessary to complete the broadcasting.

Therefore, in order to study the time complexity of the broadcasting task in the proposed interference model, we should consider other parameters of IRG, or introduce new appropriate parameters expressing the presence of interference edges in the IRG.

3   Interference Ad-Hoc Selective Families

Following the work of Clementi et al. [3], that is devoted to selective structures related to the standard model of radio networks, we define the notion of *interference ad-hoc selective family* and show some useful properties of it. As we will discuss later, this notion is closely related to the considered interference model of radio networks. In the case, when a considered collection of set-pairs satisfies a specific property (defined later), we show the existence of small interference ad-hoc selective families by a probabilistic argument. Finally, we design a deterministic polynomial-time algorithm that computes small interference ad-hoc selective family for a given input collection of set-pairs. Algorithms presented in this section extends the work [3] of Clementi et al.

The (interference) ad-hoc selective families are related to intensively studied combinatorial structures called *selectors* (see e.g. [6], [4], or [8]). One of their applications is in communication algorithms under the standard graph model of radio networks [7] in the case when the nodes are not aware of the network topology. The *k-selectors*, defined and investigated by Chrobak et al. in [7], can be seen as a weaker variant of interference ad-hoc selective families introduced in this section.

Definition 1. *Let* $\mathcal{F} = \{(T_1, I_1), (T_2, I_2), \ldots, (T_m, I_m)\}$ *to be a collection of set-pairs such that* $T_i \cap I_i = \emptyset$ *and* $T_i \neq \emptyset$, *for all* $i = 1, \ldots, m$. *Denote* $U(\mathcal{F}) = \bigcup_{i=1}^{m} T_i \cup I_i$. *A family* $\mathcal{S} = \{S_1, S_2, \ldots, S_k\}$ *of subsets of* $U(\mathcal{F})$ *is said to be selective for* $\mathcal{F}$ *if and only if for any* (T_i, I_i) *there is a set* S_j *such that* $|T_i \cap S_j| = 1$ *and* $I_i \cap S_i = \emptyset$. *We say that the set* S_j *is selective for* (T_i, I_i).

There is a relationship between interference ad-hoc selective families and the broadcasting task. To see it, suppose that a proper subset of network nodes is already informed. Initially, only the source is informed. Let V_S to be a set of informed network nodes that have an uninformed node within transmission range. Let V_R to be a set of uninformed nodes that are located within transmission ranges of informed nodes. Interference ad-hoc selective families can be utilized to construct a schedule of transmissions such that all nodes in V_R become informed by transmissions of nodes in the set V_S. Indeed, consider a collection $\mathcal{F} = \{(T_v, I_v) | v \in V_R\}$ such that $T_v = \{u \in V_S | v \in T(u)\}$ and $I_v = \{u \in V_S | v \in I(u)\}$. Note that $U(\mathcal{F}) \subseteq V_S$. Let $\mathcal{S} = \{S_1, S_2, \ldots, S_k\}$ to be an interference ad-hoc selective family for $\mathcal{F}$. Observe, that if exactly the nodes of S_i transmit in the i-th round of a schedule, all nodes in V_R become informed in at most $k = |\mathcal{S}|$ rounds. Hence it seems useful to search for small selective families. Obviously, we can always construct a selective family of the size $min\{|\mathcal{F}|, |U(\mathcal{F})|\}$ by a trivial construction. On the other hand, for any n, there is an instance $\mathcal{F}$, $|\mathcal{F}| = |U(\mathcal{F})| = n$, such that it is not possible to construct interference ad-hoc selective family of the size smaller than n. These instances correspond to the example of "slow" IRG in the section 2.1. It follows, as for the broadcasting in IRG, that a new parameter characterizing collection $\mathcal{F}$ should be introduced and considered.

Definition 2. *Let* $\mathcal{F} = \{(T_1, I_1), (T_2, I_2), \ldots, (T_m, I_m)\}$ *to be a collection of set-pairs such that* $T_i \cap I_i = \emptyset$ *and* $T_i \neq \emptyset$, *for all* $i = 1, \ldots, m$. *We say that* r *is an interference ratio of the pair* (T_i, I_i) *if and only if* $|I_i| \leq r \cdot |T_i|$. *Analogously, we say that* $r(\mathcal{F})$ *is an interference ratio of the collection* $\mathcal{F}$, *if and only if* $|I_i| \leq r(\mathcal{F}) \cdot |T_i|$, *for all* $i = 1, \ldots, m$.

Intuitively, the notion of the interference ratio is introduced in order to express a ratio of the interference edges to the transmission edges of a node. Now, using a probabilistic argument, we show that there are small interference ad-hoc selective families.

Theorem 1. *Let* $\mathcal{F} = \{(T_1, I_1), (T_2, I_2), \ldots, (T_m, I_m)\}$ *to be a collection of set-pairs such that* $T_i \cap I_i = \emptyset$, $T_i \neq \emptyset$, *and* $\Delta_{min} \leq |T_i| + |I_i| \leq \Delta_{max}$, *for all* $i = 1, \ldots, m$. *There is a family* $\mathcal{S}$ *of the size* $O((1 + r(\mathcal{F})) \cdot ((1 + \log(\Delta_{max}/\Delta_{min}))) \cdot \log |\mathcal{F}|)$ *that is selective for* $\mathcal{F}$.

Proof. Let us define $\mathcal{F}' = \{(T_i, I_i) \in \mathcal{F}, |T_i| + |I_i| = 1\}$. Since $T_i \neq \emptyset$, we have that $|T_i| = 1$ and $|I_i| = 0$, for all members of $\mathcal{F}'$. It is easy to see that the set $S_0 = \bigcup_{(T_i, I_i) \in \mathcal{F}'} T_i$ is selective for $\mathcal{F}'$. Therefore, in what follows we can assume that $\Delta_{min} \geq 2$.

For each $j \in \{\lceil \log \Delta_{min} \rceil, \ldots, \lceil \log \Delta_{max} \rceil\}$, consider a family $\mathcal{S}_j$ of l sets, where an unknown parameter l will be determined at the end of the proof. Each set $S \in \mathcal{S}_j$ is constructed by picking each element of $U(\mathcal{F})$ independently with the probability $1/2^j$.

Fix a pair $(T_i, I_i) \in \mathcal{F}$. Let j to be an integer such that $2^{j-1} \leq |T_i| + |I_i| < 2^j$. Consider a set $S \in \mathcal{S}_j$. Let us estimate the probability that the set S is selective for (T_i, I_i):

$$Pr[|T_i \cap S| = 1 \wedge I_i \cap S = \emptyset] = |T_i| \cdot \frac{1}{2^j} \cdot \left(1 - \frac{1}{2^j}\right)^{|T_i| + |I_i| - 1}$$

$$> |T_i| \cdot \frac{1}{2^j} \cdot \left(1 - \frac{1}{2^j}\right)^{2^j} \overset{(1)}{\geq} \frac{1}{2 \cdot (r(\mathcal{F}) + 1)} \cdot \left(1 - \frac{1}{2^j}\right)^{2^j} \overset{(2)}{\geq} \frac{1}{8 \cdot (r(\mathcal{F}) + 1)}$$

The inequality (1) holds because $2^{j-1} \leq |T_i| + |I_i| \leq (r(\mathcal{F}) + 1) \cdot |T_i|$. The inequality (2) follows from the fact that $\left(1 - \frac{1}{t}\right)^t \geq \frac{1}{4}$, for $t \geq 2$.

The sets in $\mathcal{S}_j$ are constructed independently. Thus the probability that none of l sets of the family $\mathcal{S}_j$ is selective for (T_i, I_i) is upper-bounded by the expression

$$\left(1 - \frac{1}{8 \cdot (r(\mathcal{F}) + 1)}\right)^l \leq e^{-\frac{l}{8 \cdot (r(\mathcal{F}) + 1)}}$$

due to the inequality $(1 - x)^y \leq e^{-x \cdot y}$, for $0 < x < 1$ and $y > 1$.

Finally, let us define a family $\mathcal{S}$ as the union of the families $\mathcal{S}_j$, for $j \in \{\lceil \log \Delta_{min} \rceil, \ldots, \lceil \log \Delta_{max} \rceil\}$. Now we estimate the probability that $\mathcal{S}$ is not selective for $\mathcal{F}$:

$$Pr[\mathcal{S} \text{ is not selective for } \mathcal{F}] \leq \sum_{(T_i, I_i) \in \mathcal{F}} Pr[\mathcal{S} \text{ is not selective for } ((T_i, I_i))]$$

$$\leq \sum_{(T_i, I_i) \in \mathcal{F}} e^{-\frac{l}{8 \cdot (r(\mathcal{F}) + 1)}} = |\mathcal{F}| \cdot e^{-\frac{l}{8 \cdot (r(\mathcal{F}) + 1)}}$$

It follows, that the probability of $\mathcal{S}$ not being selective for $\mathcal{F}$, is less than 1 for $l > 8 \cdot (r(\mathcal{F}) + 1) \cdot \ln |F|$. It implies the existence of an interference ad-hoc selective family $\mathcal{S}$ of the size $O((1 + r(\mathcal{F})) \cdot ((1 + \log(\Delta_{max}/\Delta_{min}))) \cdot \log |\mathcal{F}|)$. $\square$

Using de-randomization method of conditional probabilities, we show that a selective family of the size $O((1 + r(\mathcal{F})) \cdot ((1 + \log(\Delta_{max}/\Delta_{min}))) \cdot \log |\mathcal{F}|)$ can be constructed deterministically in the polynomial time.

At first, we fix an ordering of elements of $U(\mathcal{F})$, i.e. $U(\mathcal{F}) = \{u_1, u_2, \ldots, u_n\}$. For $S \subseteq U(\mathcal{F})$, let us denote $\delta_i(S) = S \cap \{u_i, u_{i+1}, \ldots, u_n\}$. Finally, let us fix a pair $(T, I) \in \mathcal{F}$ and let Δ to be a power of 2 (i.e. $\Delta = 2^j$, for some j) such that $\Delta/2 \leq |T| + |I| < \Delta$. For a fixed set $S \subseteq \{u_1, u_2, \ldots, u_{j-1}\}$, we define the conditional probabilities

$$Y_j(S, (T, I)) = Pr[S \cup X \cup \{u_j\} \text{ is selective for } (T, I)]$$

$$N_j(S, (T, I)) = Pr[S \cup X \text{ is selective for } (T, I)]$$

where X is a subset of $\{u_{j+1}, \ldots, u_n\}$ constructed by picking each element of $\{u_{j+1}, \ldots, u_n\}$ independently at random with the probability $1/\Delta$, that is, $Pr[u_k \in X] = 1/\Delta$.

Lemma 1. *The conditional probabilities $Y_j(S, (T, I))$ and $N_j(S, (T, I))$ can be computed in $O(n)$ time.*

Utilizing those conditional probabilities, we design an algorithm that computes an interference ad-hoc selective family for a given input collection of set-pairs $\mathcal{F}$. Algorithm is based on the procedure *IASF*. It produces an interference ad-hoc selective family for a given collection $\mathcal{F} = \{(T_1, I_1), (T_2, I_2), \ldots, (T_m, I_m)\}$ satisfying the property that there is a power of 2 denoted as Δ (i.e. $\Delta = 2^j$, for some $j \geq 2$) such that the condition $\Delta/2 \leq |T_i| + |I_i| < \Delta$ is valid for all members of $\mathcal{F}$.

Input : $\Delta = 2^j$, $\mathcal{F} = \{(T_1, I_1), (T_2, I_2), \ldots, (T_m, I_m)\}$
Output: $\mathcal{S} = \{S_1, S_2, \ldots, S_k\}$
let n to be the number of elements of $U(\mathcal{F}) = \{u_1, u_2, \ldots, u_n\}$;
while $\mathcal{F} \neq \emptyset$ do
  $S \leftarrow \emptyset$;
  for $i \leftarrow 1$ to n do
    $Y_i \leftarrow \sum_{(T,I) \in \mathcal{F}} Y_i(S, (T, I))$;
    $N_i \leftarrow \sum_{(T,I) \in \mathcal{F}} N_i(S, (T, I))$;
    if $N_i < Y_i$ then $S \leftarrow S \cup \{u_i\}$;
  end
  $\mathcal{F} \leftarrow \mathcal{F} \setminus \{(T, I) \in \mathcal{F} \mid S \text{ is selective for } (T, I)\}$;
  $\mathcal{S} \leftarrow \mathcal{S} \cup S$;
end
return $\mathcal{S}$

Algorithm 1. Procedure *IASF*

Theorem 2. *Let $\mathcal{F} = \{(T_1, I_1), (T_2, I_2), \ldots, (T_m, I_m)\}$ to be a collection of set-pairs such that $T_i \cap I_i = \emptyset$, $T_i \neq \emptyset$, and $\Delta_{min} \leq |T_i| + |I_i| \leq \Delta_{max}$, for all $i = 1, \ldots, m$. There is a deterministic algorithm that produces an interference ad-hoc selective family $\mathcal{S}$ of the size $O((1 + r(\mathcal{F})) \cdot ((1 + \log(\Delta_{max}/\Delta_{min}))) \cdot \log |\mathcal{F}|)$ for the given collection $\mathcal{F}$. Computation takes polynomial time, more precisely $O((1 + \log(\Delta_{max}/\Delta_{min})) \cdot r(\mathcal{F}) \cdot (\log |\mathcal{F}|) \cdot |\mathcal{F}| \cdot |U(\mathcal{F})|^2)$.*

Proof. The goal of the procedure *IASF* is to compute an interference ad-hoc selective family for a specific subset of the input collection $\mathcal{F}$. For each $j \in \{\lceil \log \Delta_{min} \rceil, \ldots, \lceil \log \Delta_{max} \rceil\}$, the procedure *IASF* is executed with the two input parameters: $\Delta = 2^j$ and a subset of the collection $\mathcal{F}$ (denoted as $\mathcal{F}_j$), that is restricted to those set-pair (T, I) satisfying $\Delta/2 \le |T| + |I| < \Delta$. The resulting selective collection is the union of all selective families returned by executions of *IASF*. As in the proof of the theorem 1, we focus only on j such that $j \ge 2$. Indeed, for $j = 1$ the construction of a selective set is trivial.

At first, we show that each execution of *IASF* produces a selective family of the size at most $O((1 + r(\mathcal{F}_j)) \cdot \log |\mathcal{F}_j|)$. Let us fix considered input parameters of *IASF*: $\Delta = 2^j$ and the collection $\mathcal{F}_j$. In the following, we shall analyze a single execution of the *while* loop in the procedure *IASF*. Hence, the symbols $\mathcal{F}$ and S will correspond to the variables of the algorithm. Note that the variable (collection) $\mathcal{F}$ remains unchanged during the analyzed part of the execution. It is modified only at the end of each iteration of the *while* loop.

Let W to be a set constructed by picking each element of $U(\mathcal{F})$ independently with the probability $1/\Delta$. Denote as $E(X)$ the expected number of set-pairs $(T, I) \in \mathcal{F}$ that are selected by W. Analogously, for a set Y, $Y \subseteq U(\mathcal{F})$, and an integer i, $i \ge 1$, satisfying $Y \cap \delta_i(U(\mathcal{F})) = \emptyset$, we denote as $E(X|(Y, i))$ the expected number of set-pairs $(T, I) \in \mathcal{F}$ that are selected by a random set $W_{Y,i}$. The set $W_{Y,i}$ is the union of the set Y and a set of independently (with probability $1/\Delta$) picked elements of the set $\delta_i(U(\mathcal{F}))$. Clearly, it follows from the proof of the theorem 1 that

$$E(X|(\emptyset, 1)) = E(X) \ge \frac{|\mathcal{F}|}{8 \cdot (r(\mathcal{F}) + 1)}.$$

Now we prove by induction on i that the inequality $E(X|(S, i + 1)) \ge E(X)$ is valid after i ($i \in \{0, \ldots, |U(\mathcal{F})|\}$) iterations of the *for* loop in *IASF*:

- For $i = 0$, it holds $S = \emptyset$. Since $E(X|(\emptyset, 1)) = E(X)$, the claim is true.
- Suppose that the claim is true for all j, $j < i$. Recall that the symbol S corresponds to the set variable S (containing a subset of $U(\mathcal{F})$) in the procedure *IASF* after i iterations of the *for* loop. Denote $S' = S \setminus \{u_i\}$. Due to the definition of the expected value, it holds for $i > 0$ that

$$E(X|(S', i)) = \frac{1}{\Delta}E(X|(S' \cup \{u_i\}, i + 1)) + \left(1 - \frac{1}{\Delta}\right)E(X|(S', i + 1)).$$

Obviously, $A = qB + (1 - q)C \Rightarrow A \le max\{B, C\}$, for $A, B, C \ge 0$ and $0 \le q \le 1$. Thus it follows

$$E(X|(S', i)) \le max\{E(X|(S' \cup \{u_i\}, i + 1)), E(X|(S', i + 1))\}.$$

Moreover, the definition of the expected value implies $Y_i = E(X|(S' \cup \{u_i\}, i + 1))$ and $N_i = E(X|(S', i + 1))$. The choice, between adding the element u_i to S or not, depends on the values Y_i and N_i. Since the larger value is chosen, it follows that $E(X|(S, i + 1)) = max\{Y_i, N_i\} = max\{E(X|(S' \cup \{u_i\}, i + 1)), E(X|(S', i + 1))\} \ge E(X|(S', i))$. Finally, the inductive hypothesis implies

$$E(X|(S, i + 1)) \ge E(X|(S', i)) \ge E(X).$$

It follows from the previous claim for $i = |U(\mathcal{F})|$, that in each iteration of the *while* loop such a set S is constructed that at least $\lceil \frac{|\mathcal{F}|}{8 \cdot (r(\mathcal{F})+1)} \rceil$ set-pairs of the collection variable $\mathcal{F}$ (in execution of *IASF*) are selected by S. Thus after k iterations of the *while* loop, the number of unselected set-pairs in the collection variable $\mathcal{F}$ can be upper-bounded by the expression $\left(1 - \frac{1}{8 \cdot (r(\mathcal{F}_j)+1)}\right)^k \cdot |\mathcal{F}_j|$, where $\mathcal{F}_j$ is input of *IASF*. Since, for $z \geq 1$, it holds $\ln\left(1 - \frac{1}{z}\right) \leq -\frac{1}{z}$, this expression is lower than 1 for k at least $(8 \cdot r(\mathcal{F}_j) + 1) \cdot |\mathcal{F}_j|$. Finally, we get that at most $O((r(\mathcal{F}_j)+1) \cdot |\mathcal{F}_j|)$ iterations of the *while* loop are sufficient to select all set-pairs of the collection $\mathcal{F}_j$. Hence the interference ad-hoc selective family constructed by one execution of *IASF* has the size $O((1 + r(\mathcal{F}_j)) \cdot \log |\mathcal{F}_j|)$.

There are $\lceil 1 + \log(\Delta_{max}/\Delta_{min}) \rceil$ executions of *IASF*. Considering the definition of the interference ratio $r(\mathcal{F})$, it is easy to see, that if $\mathcal{F}_j \subseteq \mathcal{F}$ then $r(\mathcal{F}_j) \leq r(\mathcal{F})$. This concludes the proof that the constructed interference ad-hoc selective family has the size $O((1 + r(\mathcal{F})) \cdot ((1 + \log(\Delta_{max}/\Delta_{min}))) \cdot \log |\mathcal{F}|)$. □

4 Centralized Broadcasting and Interference Ad-Hoc Selective Families

Now, we sketch a simple algorithm producing a radio broadcast schedule for radio networks modelled by arbitrary IRGs.

The algorithm works as follows. At first, we split the network nodes into layers $L_0, L_1, \ldots, L_{ecc}$ with respect to their distances to a fixed source s, where $L_i = \{v \in V | dist(s, v) = i\}$ and ecc is the eccentricity of the source s in the transmission subgraph. The source message is disseminated in phases layer by layer. During the i-th phase, the source message is received by the nodes of the layer L_i owing to transmissions of the nodes in the layer L_{i-1}. Particularly, for each node v of L_i, we construct a set-pair (T_v, I_v) such that $T_v = \{w \in L_{i-1} | (w, v) \in E_T\}$ and $I_v = \{w \in L_{i-1} | (w, v) \in E_I\}$. Furthermore, for a collection $\mathcal{F}^i = \{(T_v, I_v) | v \in L_i\}$, an interference ad-hoc selective family $\mathcal{S}^i = \{S_1^i, \ldots, S_m^i\}$ is obtained as an output of the algorithm described in the proof of the theorem 2. Finally, transmissions of the phase i are scheduled in such a way, that in the j-th round of the phase i exactly the nodes $S_j^i \subseteq L_{i-1}$ transmit the source message. Thus the i-th phase takes totally $|\mathcal{S}^i|$ rounds.

Theorem 3. *Let $G = (V, E_T \cup E_I)$ to be an IRG. There is a deterministic polynomial time algorithm that for a given source node s produces a schedule of the length $O((1 + \log(\Delta_{max}/\Delta_{min})) \cdot R(s))$, where $R(s) = \sum_{i=0}^{ecc-1}((1 + r(\mathcal{F}^i)) \cdot \log |\mathcal{F}^i|)$ with ecc standing for the eccentricity of the node s in the transmission subgraph of G.*

Note, that there are IRGs for which the utilized layer by layer information dissemination approach is not suitable. For instance, consider the following undirected IRG $G_m = (V, E_T \cup E_I)$, where

- $V = \{s, v_1, \ldots, v_m, w_1, \ldots, w_m\}$
- $E_T = \{(s, v_i), (v_i, w_i)|i = 1, \ldots, m\} \cup \{(v_i, v_j), (w_i, w_j)|1 \leq i \neq j \leq m\}$
- $E_I = \{(v_i, w_j)|1 \leq i \neq j \leq m\}$

Observe, that the ratio of the incident interference edges to the incident transmission edges is at most 1, for each node of the constructed graph G_m. It is easy to see, that it is not possible to complete the broadcasting task from the node s in less then $m + 1$ rounds utilizing the layer-by-layer approach. Indeed, all nodes of the layer $L_1 = \{v_1, \ldots, v_m\}$ have to transmit in separate rounds. On the other hand, broadcasting with the source s can be completed in 3 rounds:

1. the source s transmits and informs all nodes of the layer L_1
2. the node v_1 transmits and informs the node w_1
3. the node w_1 transmits and informs the remaining nodes.

Although this simple algorithm is not suitable for all IRGs, its combination with some graph analysis or heuristics can lead to algorithms that produce radio broadcast schedules of sufficient length (for practical applications), at least for a large subclass of radio networks.

5   Time Complexity of the Centralized Broadcasting in IRG with Respect to the Maximum Degree

Let Δ to be the maximum degree of IRG, i.e. the largest degree (the total number of incident, transmission and interference, edges of a node) over all networks nodes. In this section, we shall investigate the impact of the parameter Δ to the time complexity of the centralized broadcasting in IRGs.

Theorem 4. *Let $G = (V_S \cup V_R, E_T \cup E_I)$ to be a directed bipartite IRG, where E_T are the transmission edges and E_I are the interference edges. Suppose that all nodes in V_S are informed, i.e. they possess the source message, and the nodes in V_R are uninformed. Let Δ to be the maximum degree in the IRG G, i.e. $\Delta = max\{deg_T(v)+deg_I(v)|v \in V_S \cup V_R\}$, where $deg_T(v) = |\{(u,v)|(u,v) \in E_T \vee (v,u) \in E_T\}|$ and $deg_I(v) = |\{(u,v)|(u,v) \in E_I \vee (v,u) \in E_I\}|$. If $deg_T(v) \geq 1$ for all $v \in V_R$, then the following holds:*

- *all nodes in V_R can be informed in at most Δ^2 rounds,*
- *if $deg(v) = 1$ for all $v \in V_S$, then all nodes in V_R can be informed in at most $2 \cdot \Delta$ rounds.*

Proof. At first, we prove that the first part of the claim holds. Assignment of transmission rounds for the nodes of the set V_S can be obtained by a simple greedy algorithm *Greedy-rounds-assignment* (Algorithm 2). The goal of the algorithm is to assign a non-colliding round number $r(v) \in \{1, \ldots, \Delta^2\}$ to each node v. Two invariants are valid during the computation of assignments:

- For the nodes of the set V_S, the number $r(v)$ denotes a round, in which the node v transmits the source message.

Input   : $G = (V_S \cup V_R, E_T \cup E_I)$ - bipartite IRG, nodes in V_S are informed,
            nodes in V_R are uninformed
Output: round assignment $r : (V_S \cup V_R) \longrightarrow \{1, \ldots, \Delta^2\}$
initially, $r(v)$ is unassigned for each $v \in V_S \cup V_R$;
$A_S \leftarrow \emptyset$;
$A_R \leftarrow \emptyset$;
while $A_S \neq V_S$ **do**
    pick a random node v from $V_S \setminus A_S$;
    $B_I \leftarrow \{r(u)|u \in A_R \wedge (v, u) \in E_T \cup E_I\}$;
    $N \leftarrow \{u|u \in V_R \setminus A_R \wedge (v, u) \in E_T\}$;
    $B_N \leftarrow \{r(u)|u \in A_S \wedge \exists w \in N, (u, w) \in E_I \cup E_T)\}$;
    $B \leftarrow B_I \cup B_N$;
    $r(v) \leftarrow$ any element of the set $\{1, \ldots, \Delta^2\} \setminus B$;
    $A_S \leftarrow A_S \cup \{v\}$;
    foreach $u \in V_R \setminus A_R$ such that $(v, u) \in E_T$ **do**
        $r(u) \leftarrow r(v)$;
        $A_R \leftarrow A_R \cup \{u\}$;
    end
end
return *assignment r*

Algorithm 2. Algorithm *Greedy-rounds-assignment*

– For the nodes of the set V_R, the value $r(v)$ denotes a round, in which the node v receives the source message.

Note that a node $v \in V_R$ can receive the source message also in other rounds than $r(v)$. The algorithm GRA produces an assignment r as a result of the following computation. In each iteration, we pick a node $v \in V_S$ such that $r(v)$ is unassigned. The set variables A_S and A_R contain only the nodes of the sets V_S and V_R, respectively. Each node w, that is a member of V_S or V_R, has already defined the value $r(w)$. For the picked node v, a set of colliding transmission rounds B is computed. The set B contains the round numbers of all nodes with assigned round number that are in the transmission or interference range of the node v. These round numbers are colliding due to the invariant property defined for the nodes in the set V_R. Particularly, the round number means a round when the source message is received for sure. Moreover, in order to inform all uninformed nodes in the transmission range of the picked node v, i.e. the nodes of set variable N with unassigned round numbers, we have to guaranty that none of their (transmission or interference) neighbors transmits in the round $r(v)$. This achieved by adding their round numbers to the set of colliding rounds B. The picked node v is a neighbor of at most Δ other nodes. Each of them adds to the set B at most $\Delta - 1$ colliding round numbers. Hence, it holds that $|B| \leq \Delta(\Delta - 1)$, and we can pick a non-colliding round number from the set $\{1, \ldots, \Delta^2\} \setminus B$. It is easy to see, that if each node $v \in V_S$ transmits the source message in the round $r(v)$ then all nodes in V_R become informed in at most Δ^2 rounds. It concludes the proof of the first part of the claim.

Now we prove the second part of the claim. Observe, that if $deg_T(v) = 1$ for all $v \in V_S$, then in the algorithm GRA the set B of colliding round number contains at most $2 \cdot \Delta - 1$ rounds. Thus, we can modify the algorithm GRA in such a way that assigned round number is picked from the set $\{1, \ldots, 2 \cdot \Delta\} \setminus B$. Finally, transmissions according to the computed assignment ensure that all nodes in V_R become informed in at most $2 \cdot \Delta$ rounds.   □

In [5], Gąsieniec et al. discussed centralized communication in radio networks (assuming the standard model without extended interference). They presented algorithms that produce a radio broadcast schedule of the length $O(D + \Delta \cdot \log n)$ and $D + O(\log^3 n)$, for a reachability graph G. Algorithms are based on the gathering spanning tree and can be reformulated as follows:

Theorem 5. *Let $G' = (V_S \cup V_R, E)$ to be an undirected bipartite graph. Suppose that all nodes in V_S are informed (possess the source message) and the nodes in V_R are uninformed. Let $n = |V_S \cup V_R|$ and denote as*

- *$A_S(n)$ the maximal length of a schedule produced by an algorithm A_S ensuring that all nodes in V_R become informed due to transmissions of the nodes in V_S,*
- *$A_F(n)$ the maximal length of a schedule produced by an algorithm A_F ensuring that all nodes in the set V_R become informed due to transmissions of the nodes in V_S under the following assumption: $deg(v) = 1$, for all $v \in V_S \cup V_R$.*

Let $G = (V, E)$ to be a reachability graph (no extended interference). There is an algorithm (schema) that produces a radio broadcast schedule of the length $O(A_F(n) \cdot D + A_S(n) \cdot \log n)$, where $n = |V|$.

Now, we show how to realize centralized broadcasting in radio networks (with extended interference) modelled by an IRG.

Theorem 6. *Let $G = (V, E_T \cup E_I)$ to be an IRG with the maximum degree Δ. There is a deterministic polynomial time algorithm that for a given source node s produces a radio broadcast schedule of the length $O(\Delta D + min\{\Delta, \log \Delta \cdot \log n\} \cdot \Delta \log n)$.*

Proof. We utilize the schema of the theorem 5 for the transmission subgraph of a given IRG G. Due to the presence of interference edges, we cannot apply algorithms for the standard "non-interference" model. However observe, that the algorithm presented in the second part of proof of the theorem 4 produces a schedule such that $A_F(n) = O(\Delta)$. Moreover, the algorithm presented in the first part of the proof can be used as the algorithm A_S, in order to produce schedules such that $A_S(n) = O(\Delta^2)$. Another choice for the algorithm A_S is to apply the algorithm presented in the theorem 3 that leads to a schedule such that $A_S(n) = O(\Delta \cdot \log \Delta \cdot \log n)$. Note that this algorithm provides shorter schedules than the former algorithm in the case when $\Delta = \Omega(\log^2 n)$.   □

One can easily show that it is possible to construct an IRG with the maximum degree Δ such that the broadcasting time is lower-bounded by $\Omega(\Delta \cdot D)$ rounds.

6   Conclusion

In this paper, we introduced a new model (an extension of the standard graph model) of radio networks that reflects a situation when a transmission of a node causes interference in an area where the decoding of this transmission is impossible. We focused on the broadcasting problem in the newly proposed model. Designed algorithms, one of them based on the introduced notion of interference ad-hoc selective families, can be seen as a first step to study the efficiency of communication in this model.

The evident open problem is design of optimal communication (broadcasting, gossiping, etc.) algorithms with respect to parameters of an underlying IRG that express the presence of interference edges in an appropriate way. This could answer the question how the presence of interference edges makes the communication process more difficult (e.g. slower) in compare to the communication under the standard graph model.

References

1. Alon, N., Bar-Noy, A., Linial, N., Peleg, D.: A lower bound for radio broadcasting. Journal of Computer and System Sciences 43, 290–298 (1991)
2. Bermond, J.-C., Galtier, J., Klasing, R., Morales, N., Perennes, S.: Hardness and approximation of gathering in static radio networks. Parallel Processing Letters 16, 165–184 (2006)
3. Clementi, A.E.F., Crescenzi, P., Monti, A., Penna, P., Silvestri, R.: On computing ad-hoc selective families. In: Goemans, M.X., Jansen, K., Rolim, J.D.P., Trevisan, L. (eds.) RANDOM 2001 and APPROX 2001. LNCS, vol. 2129, pp. 211–222. Springer, Heidelberg (2001)
4. De Bonis, A., Gąsieniec, L., Vaccaro, U.: Generalized framework for selectors with applications in optimal group testing. In: Baeten, J.C.M., Lenstra, J.K., Parrow, J., Woeginger, G.J. (eds.) ICALP 2003. LNCS, vol. 2719, pp. 81–96. Springer, Heidelberg (2003)
5. Gąsieniec, L., Peleg, D., Xin, Q.: Faster communication in known topology radio networks. In: Proc. 24th Annual ACM Symposium on Principles of Distributed Computing (PODC 2005), pp. 129–137 (2005)
6. Chlebus, B., Kowalski, D.: Almost optimal explicit selectors. In: Liśkiewicz, M., Reischuk, R. (eds.) FCT 2005. LNCS, vol. 3623, pp. 270–280. Springer, Heidelberg (2005)
7. Chrobak, M., Gąsieniec, L., Rytter, W.: Fast broadcasting and gossiping in radio networks. Journal of Algorithms 43(2), 177–189 (2002)
8. Indyk, P.: Explicit constructions of selectors and related combinatorial structures, with applications. In: Proc. 13th ACM-SIAM Symposium on Discrete Algorithms (SODA), pp. 697–704 (2002)
9. Kowalski, D., Pelc, A.: Optimal deterministic broadcasting in known topology radio networks. Distributed Computing 19, 185–195 (2007)
10. Pelc, A.: Broadcasting in radio networks. In: Stojmenovic, I. (ed.) Handbook of Wireless Networks and Mobile Computing, pp. 509–528. John Wiley and Sons, Inc., New York (2002)

A   Appendix

Lemma 1. *The conditional probabilities $Y_j(S,(T,I))$ and $N_j(S,(T,I))$ can be computed in $O(n)$ time.*

Proof. Evaluation of the conditional probabilities $Y_j(S,(T,I))$ and $N_j(S,(T,I))$ is based on the following equalities. In these equalities, we use α to denote $|\delta_j(T)| + |\delta_j(I)|$.

- $\delta_j(T) = 0$

$$Y_j(S,(T,I)) = N_j(S,(T,I)) = \begin{cases} \left(1 - \frac{1}{\Delta}\right)^{|\delta_j(I)|} & |T \cap S| = 1 \wedge I \cap S = \emptyset \\ 0 & \text{otherwise} \end{cases}$$

- $\delta_j(T) \geq 1$
  - $u_j \in T$

$$Y_j(S,(T,I)) = \begin{cases} \left(1 - \frac{1}{\Delta}\right)^{\alpha-1} & (T \cup I) \cap S = \emptyset \\ 0 & \text{otherwise} \end{cases}$$

$$N_j(S,(T,I)) = \begin{cases} 0 & |(T \cup I) \cap S| \geq 2 \\ 0 & |T \cap S| = 0 \wedge |I \cap S| = 1 \\ \left(1 - \frac{1}{\Delta}\right)^{\alpha-1} & |T \cap S| = 1 \wedge |I \cap S| = 0 \\ 0 & |(T \cup I) \cap S| = 0 \wedge \\ & \quad |\delta_j(T)| = 1 \\ (|\delta_j(T)| - 1) \cdot \frac{1}{\Delta} \cdot \left(1 - \frac{1}{\Delta}\right)^{\alpha-2} & |(T \cup I) \cap S| = 0 \wedge \\ & \quad |\delta_j(T)| \geq 2 \end{cases}$$

  - $u_j \in I$

$$Y_j(S,(T,I)) = 0$$

$$N_j(S,(T,I)) = \begin{cases} 0 & |(T \cup I) \cap S| \geq 2 \\ 0 & |T \cap S| = 0 \wedge |I \cap S| = 1 \\ \left(1 - \frac{1}{\Delta}\right)^{\alpha-1} & |T \cap S| = 1 \wedge |I \cap S| = 0 \\ |\delta_j(T)| \cdot \frac{1}{\Delta} \cdot \left(1 - \frac{1}{\Delta}\right)^{\alpha-2} & |(T \cup I) \cap S| = 0 \end{cases}$$

  - $u_j \notin T \cup I$

$$\begin{aligned} Y_j(S,(T,I)) \\ N_j(S,(T,I)) \end{aligned} = \begin{cases} 0 & |(T \cup I) \cap S| \geq 2 \\ 0 & |T \cap S| = 0 \wedge |I \cap S| = 1 \\ \left(1 - \frac{1}{\Delta}\right)^{\alpha} & |T \cap S| = 1 \wedge |I \cap S| = 0 \\ |\delta_j(T)| \cdot \frac{1}{\Delta} \cdot \left(1 - \frac{1}{\Delta}\right)^{\alpha-1} & |(T \cup I) \cap S| = 0 \end{cases}$$

$\square$

Fast Radio Broadcasting with Advice

David Ilcinkas[1,*], Dariusz R. Kowalski[2], and Andrzej Pelc[3,**]

[1] CNRS, LaBRI, Université Bordeaux I, France
david.ilcinkas@labri.fr
[2] Department of Computer Science, The University of Liverpool, United Kingdom
darek@csc.liv.ac.uk
[3] Département d'informatique, Université du Québec en Outaouais, Canada
pelc@uqo.ca

Abstract. We study deterministic broadcasting in radio networks in the recently introduced framework of network algorithms with *advice*. We concentrate on the problem of trade-offs between the number of bits of information (size of advice) available to nodes and the time in which broadcasting can be accomplished. In particular, we ask what is the minimum number of bits of information that must be available to nodes of the network, in order to broadcast very fast. For networks in which constant time broadcast is possible under complete knowledge of the network we give a tight answer to the above question: $O(n)$ bits of advice are sufficient but $o(n)$ bits are not, in order to achieve constant broadcasting time in all these networks. This is in sharp contrast with geometric radio networks of constant broadcasting time: we show that in these networks a constant number of bits suffices to broadcast in constant time. For arbitrary radio networks we present a broadcasting algorithm whose time is inverse-proportional to the size of advice.

Keywords: radio network, distributed algorithm, deterministic broadcasting, advice.

1 Introduction

The Framework and the Problem

We study deterministic broadcasting in radio networks in the recently introduced [17] framework of network algorithms with *advice*. This paradigm permits to investigate the minimum amount of information (size of advice) that nodes of the network have to be given in order to accomplish some distributed task with a given efficiency. In our present context the task is broadcasting in radio networks and the measure of efficiency is time.

A *radio network* is a collection of sites (stations) equipped with wireless transmission and receiving capabilities, with a distinguished node s called the source.

---

* This work was done during the stay of David Ilcinkas at the Research Chair in Distributed Computing of the Université du Québec en Outaouais and at the University of Ottawa, as a postdoctoral fellow.
** Research partially supported by NSERC discovery grant and by the Research Chair in Distributed Computing at the Université du Québec en Outaouais.

A. Shvartsman and P. Felber (Eds.): SIROCCO 2008, LNCS 5058, pp. 291–305, 2008.

The topology of a radio network is modeled as a directed graph $G = (V, E)$, where nodes in V represent sites of the network and oriented edges in E correspond to wireless connections. It is assumed that there is a directed path from the source to every other node. The existence of an edge (u, v) means that v is within the reach of u. We say that u is an *in-neighbor* of v and v is an *out-neighbor* of u. Nodes that are not neighbors must communicate via intermediate (relaying) nodes. Similarly as in most papers in the literature on radio networks, we assume that communication is synchronous, i.e., all nodes have internal clocks that tick at the same rate, measuring consecutive time steps, referred to as *rounds*. All clocks show the same round number at any given time.

At any round every node can be either in the *transmitting* or in the *receiving* mode, i.e., a node cannot transmit and receive messages during the same round. When a node v transmits in round i, its message is delivered during this round to all out-neighbors of v. However, if w is an out-neighbor of v, this message is *heard* by w, i.e., w receives the message correctly, if and only if the node v is the only in-neighbor of w that transmits during the round i. Otherwise a *collision* occurs at w and the message is not heard. An important property of radio networks is the *collision detection* capability, i.e., the ability of a node to differentiate collision from silence in a given round. All our results hold both with this assumption and without it. Indeed, our positive results (algorithms) are valid even without collision detection, and our impossibility results are valid even assuming this capability.

Among the large class of (arbitrary) radio networks, an important subclass consists of *geometric radio networks* (GRN). In the case of an approximately flat region without large obstacles, nodes that can be reached from u are those within a circle of radius r centered at u, and the positive real r, called the *range* of u, depends on the power of the transmitter located at u. Reachability graphs corresponding to such radio networks are called geometric radio networks. More precisely, they are defined as follows. We assume that there is a constant number ρ of possible powers of transmitters, thus we fix a set $R = \{r_1, ..., r_\rho\}$ of positive reals, $r_1 < ... < r_\rho$, called *ranges*. Let C be a set of points in the plane with a distinguished source. Points of C are nodes of the graph (representing radio stations). Each point $u \in C$ is assigned a range $r(u) \in R$ and a directed edge (u, v) exists in the graph, if and only if the Euclidean distance between u and v does not exceed $r(u)$.

The number of nodes of a radio network is denoted by n, and the eccentricity of the source (the maximum length of all shortest paths in the graph from the source to all other nodes) is denoted by D. Throughout the paper, log denotes the logarithm with base 2 and ln denotes the natural logarithm. Nodes of a radio network have distinct labels from the set $\{1, ..., N\}$, where $N \in O(n)$. Moreover, nodes of a geometric radio network have also their (x, y) coordinates. A priori, each node of a (general) radio network knows only its own label, and each node of a GRN knows only its own label and its (x, y) coordinates, as well as the set R of available ranges (which has constant size). All other information about the network must be given to nodes as advice, to be defined below.

One of the most studied communication primitives in networks is *broadcasting*, also known as one-to-all communication. The source has a message that should be distributed to all other nodes in the network. The *time* of a deterministic broadcasting algorithm is the number of rounds in which all the nodes get the source message. With every radio network G we associate its optimal broadcasting time $Opt(G)$. This is the minimum time in which broadcasting in this network can be accomplished, if nodes have full information about the network. Establishing optimal broadcasting time for a given radio network is an NP-hard problem [5].

It remains to formalize the framework of advice (cf. [17]) in our present context. All additional knowledge available to the nodes of the network (in particular knowledge concerning the rest of the network), is modeled by an oracle providing advice. An *oracle* is a function $\mathcal{O}$ whose arguments are labeled networks (in the case of geometric radio networks these arguments are actual sets of points in the plane, together with the assigned ranges and labels), and the value $\mathcal{O}(G)$, for a network $G = (V, E)$, called the *advice* provided by the oracle to this network, is in turn a function $f : V \rightarrow \{0, 1\}^*$ assigning a binary string to every node v of the network. Intuitively, the oracle looks at the entire labeled network and assigns to every node some information, encoded as a string of bits. The *size* of the advice given by the oracle to a given network G is the sum of the lengths of all the strings it assigns to nodes. Hence this size is a measure of the amount of information about the network, available to its nodes. Solving the broadcasting problem in radio networks using advice provided by oracle $\mathcal{O}$ consists in designing an algorithm that is *unaware* of the network G at hand but accomplishes broadcasting in it, as long as every node v of the network G is provided with the string of bits (advice) $f(v)$, where $f = \mathcal{O}(G)$.

The main interest of this framework is the significance of lower bounds on the size of advice. If we have a broadcasting algorithm using some advice of size $O(g(n))$ and achieving time $O(T(n))$, in n-node networks, and at the same time we prove that $\Omega(g(n))$ is the lower bound on the size of advice needed to achieve time $O(T(n))$, this implies optimality in a very strong sense: smaller amount of information of *any* type cannot help to achieve broadcasting time $O(T(n))$ using *any* algorithm. In other words, changing the type of information provided to nodes cannot help to achieve the same efficiency of broadcasting at lower information cost.

This paper is the first to consider communication in radio networks in the framework of algorithms with advice. Our research is motivated by the following problems:

- What is the minimum size of advice permitting to achieve broadcasting time $O(Opt(G))$ for a radio network G?
- What are the trade-offs between the size of advice and the time of broadcasting in radio networks?

Our Results
Our main focus is on radio networks with constant optimal broadcasting time, i.e., on networks in which deterministic broadcast in constant time is possible

under complete knowledge of the network. For this class of networks we establish the minimum size of advice sufficient to achieve constant broadcasting time. We show that $O(n)$ bits of advice are sufficient and $o(n)$ bits are not sufficient, in order to achieve constant broadcasting time in all these networks. The main contribution of this part of the paper is the above tight lower bound on the size of advice. This is in sharp contrast with geometric radio networks of constant broadcasting time: we show that in these networks a *constant* number of bits of advice suffices to broadcast in constant time.

For arbitrary radio networks we show a trade-off between the size of advice and the time of deterministic broadcasting, by presenting a broadcasting algorithm whose time is inverse-proportional to the size of advice. More precisely, for any $q \in O(n)$ we show an oracle which gives advice of size q to the nodes of a network, and an algorithm using this advice, which performs broadcasting in time $O(\frac{nD}{q} \log^3 n)$ in any n-node network with source eccentricity D. As a corollary we get that for "short" networks, i.e., with D polylogarithmic in n, an advice of sublinear size suffices to achieve polylogarithmic broadcasting time.

Related Work

The paradigm of distributed computing with advice has been recently introduced in [17] and used there to study the task of broadcasting with a linear number of messages, in the message passing model. Subsequently, this approach has been used in [18] to study efficient exploration of networks by mobile agents, in [19] to study distributed graph coloring, in [20] to study the distributed minimum spanning tree construction, and in [30] to study graph searching.

Broadcasting in radio networks is a topic extensively studied in the last twenty years. Most of the papers represented radio networks as arbitrary (undirected or directed) graphs. Models used in the literature about algorithmic aspects of radio communication, starting from the paper [5], differ mostly in the amount of information about the network that is assumed available to nodes. However, assumptions about this knowledge concern particular items of information, such as the knowledge of the size of the network, its diameter, maximum degree, or some neighborhood around the nodes, rather than limiting the total number of bits available to nodes, regardless of their meaning, as is the case with the advice approach.

Deterministic centralized broadcasting assuming complete knowledge of the network was considered, e.g., in [6], where a polynomial-time algorithm constructing a $O(D \log^2 n)$-time broadcasting scheme was given for all n-node networks of radius D. Subsequent improvements by many authors [15,21,22] were followed by the polynomial-time algorithm from [27] constructing a $O(D + \log^2 n)$-time broadcasting scheme, which is optimal. On the other hand, in [1] the authors proved the existence of a family of n-node networks of radius 2, for which any broadcast requires time $\Omega(\log^2 n)$.

One of the first papers to study deterministic distributed broadcasting in radio networks whose nodes have only limited knowledge of the topology, was [2]. The authors assumed that nodes know only their own label and labels of their neighbors. Many authors [4,7,8,10] studied deterministic distributed broadcasting in radio networks under the assumption that nodes know only their own label (but not labels of their neighbors). Increasingly faster broadcasting algorithms working on arbitrary radio networks were constructed, the currently fastest being the $O(n \log^2 D)$-time algorithm from [11] and the $O(n \log n \log \log n)$ algorithm from [12]. On the other hand, in [10] a lower bound $\Omega(n \log D)$ on broadcasting time was proved for n-node networks of radius D.

Randomized broadcasting algorithms in radio networks were studied, e.g., in [2,28,26]. For these algorithms, no topological knowledge of the network and no distinct identities of nodes were supposed.

Broadcasting in geometric radio networks and some of their variations was considered, e.g., in [13,14,29]. In [29] the authors proved that scheduling optimal broadcasting is NP-hard even when restricted to such graphs, and gave an $O(n \log n)$ algorithm to schedule an optimal broadcast when nodes are situated on a line. In [14] broadcasting with restricted knowledge was considered but the authors studied only the special case of nodes situated on the line. In [13], the authors investigated the impact of the size of the part of the geometric radio network known to nodes, on the efficiency of broadcasting. In particular they showed that with the full knowledge of the network broadcasting can be accomplished in (optimal) time $O(D)$, and if all nodes know only their own label, range and coordinates, broadcasting in time $O(n)$ is possible. For symmetric geometric radio networks, time $O(D + \log n)$ was proved optimal under this restricted knowledge, if collision detection is available. If it is not, the same broadcasting time was achieved if nodes know positions, labels and ranges of all nodes within a constant (arbitrarily small) positive radius. In a recent paper [16] the authors considered broadcasting in radio networks represented by unit disk graphs. They compared broadcasting time in two models: the model allowing spontaneous transmissions of nodes that have not yet gotten the source message, and the model in which only nodes that already obtained the source message can transmit.

2   Broadcasting in Constant Time

In this section we focus on radio networks with constant optimal broadcasting time, i.e., on the class of networks in which broadcasting in constant time is possible if nodes have complete knowledge of the network. Such networks must of course have bounded source eccentricity D. However, this is not a sufficient condition. Indeed, there are n-node networks with $D = 2$, whose minimum broadcasting time is $\Omega(\log^2 n)$, even if the network is completely known to all nodes (cf. [1]).

Networks with constant optimal broadcasting time may require a very long broadcasting time if their topology is unknown and in the absence of any advice. In [25] a family of such n-node networks was proved to require time $\Omega(n)$. In fact, even for the more restricted class of geometric radio networks, strong lower bounds of this type can be proven. Using techniques from [16] a class of geometric radio networks with constant optimal broadcasting time can be shown to require time $\Omega(\sqrt{n})$, if nodes know only their own label and coordinates. Therefore it is natural to ask how sensitive to advice is broadcasting time in networks (geometric or not) with constant optimal broadcasting time. More precisely, how much advice is needed to achieve constant broadcasting time in such networks.

First observe that for networks of the considered class, $O(n)$ bits of advice are sufficient in order to achieve constant broadcasting time.

Proposition 1. *For any positive constant c let $\mathcal{C}$ be the class of n-node radio networks whose optimal broadcasting time is at most c. There exists an oracle which gives advice of size $O(n)$ to the nodes of networks of class $\mathcal{C}$ and an algorithm using this advice, which performs broadcast in time at most c, for any network in class $\mathcal{C}$.*

Proof. Fix a network $C \in \mathcal{C}$ and consider an algorithm having complete knowledge of the network and broadcasting in time at most c. For any fixed node v of C, let $t_1, \ldots, t_k$ be numbers of rounds in which v has to transmit, according to this algorithm. The oracle gives this information, encoded as a string of bits of bounded length, to node v. Hence the total size of advice is $O(n)$. Now the broadcasting algorithm simply makes node v transmit in rounds $t_1, \ldots, t_k$. □

2.1 Lower Bounds

The main result of this section shows that the above upper bound on the size of advice needed to achieve constant broadcasting time is tight, i.e., that $o(n)$ bits of advice are not sufficient to broadcast in constant time.

Theorem 1. *For every integer function $k^* \in o(n)$ there exist an integer function c^* such that $c^*(n) \to \infty$ and a family of n-node networks with constant optimal broadcasting time, such that every algorithm using at most $k^*(n)$ bits of advice requires time $c^*(n)$ on some of them, for sufficiently large n.*

We will use the following lemmas whose proofs are omitted.

Lemma 1. *If $k^* \in o(k)$ then for any integer $0 \le \ell \le k^*$ and for sufficiently large k*

$$\binom{k}{k^* - \ell} \le e^{-2\ell} \cdot \left(\frac{ke}{k^*}\right)^{k^*} .$$

Lemma 2. *Let $x, x_1, \ldots, x_a$ be non-negative integers satisfying $x \ge x_1 \ge \ldots \ge x_a \ge 0$ and $x_1 + \ldots + x_a = 2x$, where $2 \le a \le x$. The number of permutations of the set $X = \{1, \ldots, 2x\}$ satisfying the following condition:*

for any $1 \le i \le a$ and $1 \le j \le x$, no two elements from the interval $X_i = [1 + \sum_{i'=1}^{i-1} x_{i'}, \sum_{i'=1}^{i} x_{i'}]$ are placed in positions $2j-1, 2j$ (called group j)

is at most

$$\beta(2x, a) = \sqrt{2} \cdot (2x)! \cdot e^{2a^2 \ln(2ex/a) - x/a} .$$

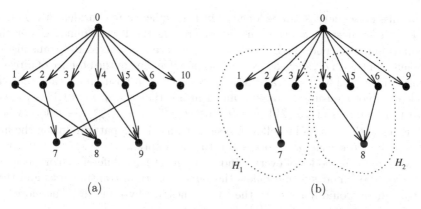

Fig. 1. (a) Example of the network in the class used in the proof of Theorem 1, for $n = 11$ and $k = 3$; (b) Example of the network in the class used in the proof of Theorem 2, for $n = 10$, $n' = 3$, $k = 2$, $S_1 = \{2\} \subseteq \{1, 2, 3\}$ and $S_2 = \{4, 6\} \subseteq \{4, 5, 6\}$

Proof of Theorem 1. Fix n. Consider the following family $\mathcal{C}$ of n-node directed networks, see Fig. 1(a). Let $k = \lfloor (n-1)/3 \rfloor$. (We omit parameter n from the arguments of functions k, k^*, c^* since it is known from the context.) Each network in $\mathcal{C}$ is composed of three layers. Layer L_0 consists of the source with label 0. Layer L_1 consists of $2k$ nodes with labels from $\{1, \ldots, 2k\}$, partitioned into k disjoint *groups* of size 2, and of $n - (3k+1)$ nodes with labels from $\{3k+1, \ldots, n\}$. Layer L_2 consists of k nodes with labels from $\{2k+1, \ldots, 3k\}$. All nodes of layer L_1 are out-neighbors of the source, and each node $2k+i$ from layer L_2 is the out-neighbor of both nodes from the ith group. There are no other edges in networks from $\mathcal{C}$. There are $\alpha = (2k)!/2^k$ different networks in this family. The optimal broadcasting time of any network from $\mathcal{C}$ is clearly 2. Let $c^* = \log \frac{k - k^*}{2k^* \log(ke/k^*)}$. Clearly, $c^* \in \omega(1)$ for $k^* \in o(k)$. In view of $k = \lfloor (n-1)/3 \rfloor$, it is enough to prove that every algorithm using at most $k^* \in o(k)$ bits of advice requires time larger than c^* on some network in $\mathcal{C}$. We fix n such that k is sufficiently large for the purpose of Lemma 1 and Fact 1, and assume that $k^* \in o(k)$ and $k^* \in \omega(k^{4/5})$ (if we show that the time is $\omega(1)$ for functions $k^* \in \omega(k^{4/5})$, the same remains true also for all smaller functions k^*).

The proof is by contradiction. Fix an oracle giving advice of size at most k^* to networks from the family $\mathcal{C}$, and an algorithm using this oracle and completing the broadcast in all these networks in time at most c^*. Let $\mathcal{C}(\ell)$ be the subfamily of $\mathcal{C}$ for which the oracle gives ℓ bits of advice to the source, for $0 \leq \ell \leq k^*$, and gives the remaining bits to some other nodes of the network. For a sequence $\hat{y}$ of ℓ bits, let $\mathcal{C}(\ell)_{\hat{y}}$ be the subfamily of $\mathcal{C}(\ell)$ containing those networks for which the oracle gives the advice $\hat{y}$ to the source.

Fix $0 \leq \ell \leq k^*$. There are 2^ℓ possible advice sequences in the source, and at least $k - k^* + \ell$ groups none of whose nodes has any advice. We call these groups *blind groups*. Fix a sequence $\hat{y}$ of ℓ bits of advice in the source and consider the transmission sequence of length at most c^* for each node in $\{1, \ldots, 2k\}$ assuming that it has no bit of advice and that the source has the advice $\hat{y}$. Formally, the transmission sequences can be longer, but it is sufficient to consider only prefixes of length at most c^* for the purpose of proving the lower bound. Under this assumption, each node in L_1 without any advice has a fixed 0-1 transmission sequence of length c^*, since no feedback is possible, due to the absence of directed cycles in the graph. (In a transmission sequence, 0 in position i means that the node does not transmit in round i, and 1 means that it transmits.) This yields a partition of nodes $\{1, \ldots, 2k\}$ into at most $a = 2^{c^*} = \frac{k-k^*}{2k^* \log(ke/k^*)}$ colors, where all nodes of the same color follow the same transmission pattern during the first c^* rounds. Note that $a^4 \in o(k^*)$, by the assumption $k^* \in \omega(k^{4/5})$. In every network in $\mathcal{C}(\ell)_{\hat{y}}$, nodes of every blind group must have different colors, because otherwise both nodes would follow the same pattern of transmissions and their out-neighbor would not receive the source message by round c^*. Therefore, the number of networks in class $\mathcal{C}(\ell)_{\hat{y}}$ is at most

$$\binom{k}{k^* - \ell}\binom{2k}{2(k^* - \ell)}\frac{(2k^* - 2\ell)!}{2^{k^* - \ell}} \cdot 2^{-(k-k^*+\ell)} \cdot \beta(2k - 2k^* + 2\ell, a) \ ,$$

where $\beta(2x, a)$ is the upper bound from Lemma 2. In the above formula the first factor corresponds to the number of choices of non-blind groups (and possibly some blind ones, since the number of non-blind groups is *at most* $k^* - \ell$) among all groups. The second factor corresponds to the number of choices of the $2(k^* - \ell)$ elements to be allocated to the above groups. The third factor corresponds to the number of ways of allocating these elements to these groups. The last two factors form an upper bound on the number of different configurations of the remaining nodes such that the remaining (blind) groups are not monochromatic (i.e., the number of permutations without remaining monochromatic blind groups, divided by the number of possible flips of elements inside those groups—there are $2^{k-k^*+\ell}$ such flips).

Using Lemma 2 and the properties $k^* \in \omega(k^{4/5})$ and $a^4 \in o(k^*)$, we get the following fact whose proof is omitted.

Fact 1. For sufficiently large k,

$$|C(\ell)_{\hat{y}}| \leq \binom{k}{k^* - \ell} \cdot \alpha \cdot e^{-k^* \log(ke/k^*) - 1} \ .$$

Finally, using Fact 1 and Lemma 1, we can bound

$$|C| = \sum_{\ell=0}^{k^*} \sum_{\hat{y}} |C(\ell)_{\hat{y}}| \leq \sum_{\ell=0}^{k^*} \sum_{\hat{y}} \left(\alpha \cdot \binom{k}{k^* - \ell} \cdot e^{-k^* \log(ke/k^*) - 1} \right)$$

$$\leq \sum_{\ell=0}^{k^*} \left(2^\ell \cdot \alpha \cdot e^{-2\ell} \cdot e^{k^* \ln(ke/k^*)} \cdot e^{-k^* \log(ke/k^*) - 1} \right)$$

$$\leq (\alpha/e) \cdot \sum_{\ell=0}^{k^*} e^{-\ell} < \alpha \ ,$$

for sufficiently large k. This is a contradiction which completes the proof of Theorem 1. $\square$

Our next result shows that if the advice is of sublogarithmic size then the time required for broadcasting is not only unbounded but sometimes quite large.

Theorem 2. *Fix any constant $\delta < 1$. There exists a constant $c > 0$ such that, for sufficiently large n, there exists a family of n-node networks with constant optimal broadcasting time, for which every algorithm using at most $c \log n$ bits of advice requires time at least n^δ on some of them.*

Proof. Fix any $0 < \delta < 1$. For a positive integer n, we set $n' = \lceil n^\delta \rceil$ and $k = \lfloor \frac{n-1}{n'+1} \rfloor$. For n large enough, there exists $0 < \epsilon < 1$ such that $k \geq n^\epsilon$. For any k-tuple $S = (S_1, S_2, \ldots, S_k)$, where each S_i, $1 \leq i \leq k$, is an arbitrary non-empty subset of $\{1, \ldots, n'\}$, we define the directed graph G_S as follows. The source is node 0. It has directed edges to $k \cdot n'$ nodes labelled from 1 to $k \cdot n'$. For any $1 \leq i \leq k$, if $j \in S_i$ then node $(i-1)n' + j$ has a directed edge to node $k \cdot n' + i$. Finally, in order to have exactly n nodes, the source has directed edges to the nodes from $k(n'+1) + 1$ to $n - 1$, if any. Hence the graph has k disjoint $(n'+1)$-node subgraphs $H_1, \ldots, H_k$, attached to the source. More precisely, the subgraph H_i is induced by the nodes $(i-1)n' + 1, \ldots, i \cdot n', k \cdot n' + i$. The directed edges inside a subgraph H_i are determined by the set S_i. The set of graphs G_S, for all possible S, is denoted $\mathcal{G}$. See Fig. 1(b).

We prove that there is no algorithm using advice of size $q \leq \frac{1}{2} \log k$ that achieves broadcast in the family $\mathcal{G}$ in time smaller than n'. Fix an algorithm using advice of size $q \leq \frac{1}{2} \log k$. Let $s_1, \ldots, s_Q$, for $Q = 2^{q+1} - 1$ be an enumeration of all binary sequences of length at most q (including the empty sequence). First note that $Q \cdot (q+1) \leq k$, for sufficiently large n. Consider the following property:

For any $1 \leq i \leq Q \cdot (q + 1)$, there exists a non-empty subset S_i of $\{1, \ldots, n'\}$ such that for any k-tuple S containing S_i as the i-th element we have that, in the graph G_S, either
(1) the source has advice different from s_j, where $j = \lfloor \frac{i-1}{q+1} \rfloor$, or
(2) at least one node of the subgraph H_i receives at least one bit of advice.

This implies that for a k-tuple S such that the $Q \cdot (q + 1)$ first elements are the above mentioned sets S_i, there exist at least $q + 1$ different subgraphs H_i receiving at least one bit. Indeed, if the advice given to the source is s_j, each of the graphs H_i, for $i = (j - 1)(q + 1) + 1, \ldots, j(q + 1)$, gets at least one bit. This contradicts the fact that the total size of advice is at most q.

Therefore, the property does not hold. This means that there exists an integer $i \leq k$ such that for any non-empty subset S_i of $\{1, \ldots, n'\}$, there exists a k-tuple S containing S_i as the i-th element such that, in the graph G_S, the source has advice s_j, where $j = \lfloor \frac{i-1}{q+1} \rfloor$, and the subgraph H_i receives no bit of advice. In other words, there exists an index i and a subfamily $\mathcal{G}'$ of $\mathcal{G}$ such that for each graph in $\mathcal{G}'$ the source always receives the same string while the subgraph H_i never receives any advice from the oracle; moreover, for any non-empty subset S_i of $\{1, \ldots, n'\}$, there exists a graph in $\mathcal{G}'$ where the graph H_i is constructed from S_i. Therefore, for this subgraph H_i, the situation is identical as if it were alone (the graph is directed) and as if there were no oracle. Since there are no directed cycles in the graph, no node can receive any feedback, and hence any broadcasting algorithm in such a graph is oblivious. Therefore, using the argument from the proof of Theorem 2.2. in [23], for some graph H_i the time of informing node $k \cdot n' + i$ is at least n'. This implies that there exists a graph in $\mathcal{G}'$ in which the algorithm does not achieve broadcast in time less than n'. Since $n' \geq n^\delta$ and $\frac{1}{2} \log k \geq c \log n$, for $c = \epsilon/2$, this proves the theorem. □

2.2   Geometric Radio Networks

We finally show that the large advice requirements established in the previous section do not hold in the more restricted class of geometric radio networks. Indeed, for these networks we have the following result which should be contrasted with Theorems 1 and 2.

Theorem 3. *For any positive constant c let $\mathcal{G}$ be the class of geometric radio networks whose optimal broadcasting time is at most c. There exists an oracle which gives advice of constant size to the nodes of networks of class $\mathcal{G}$ and an algorithm using this advice, which performs broadcast in constant time c', for any network in class $\mathcal{G}$.*

To prove Theorem 3 we will use the following construction. Fix the ranges $r_1 < \ldots < r_\rho$. (Recall that both the number ρ of ranges and the ranges themselves are constants.) Partition the plane into a mesh of squares of side $z = r_1/\sqrt{2}$, called *tiles*, with the bottom-left corner of one of them in $(0,0)$. Include the left and bottom sides and exclude the top and right sides from every square. Knowing its position, every node knows to which tile it belongs. The tile to which the source belongs is called *central*. Observe that any two nodes belonging to the same tile are within each other's range. For any positive integer x, the *x-block* is a square consisting of $B(x) = (2x + 1)^2$ tiles with the central tile in the center of this square.

A configuration of points in the plane yielding a geometric radio network with optimal broadcasting time at most c must have the property that the most distant points are at distance at most $2cr_\rho$ and hence all points are contained in a d-block, for some positive constant d. Take the smallest such integer d. Order all the $B(d)$ tiles of the d-block in a fixed way, giving them indices $1, \ldots, B(d)$ and then order the $p(d) = B(d)(B(d) - 1)$ ordered pairs of these indices in a fixed way, giving them indices $1, \ldots, p(d)$. Let $\lambda(a, b)$ denote the index of the pair (a, b), where a, b are (indices of) distinct tiles.

Advice. We now describe the oracle, called Geometric Oracle in the sequel. Consider an ordered pair (a, b) of distinct tiles of the d-block. If there is a pair (u, v) of nodes in tiles a and b, respectively, such that v is in the range of u, choose one such a pair. The oracle gives advice $(\lambda(a, b), out)$ to u and advice

$(\lambda(a,b), in)$ to v. Clearly, the same node can get many pieces of advice, however, for constant d, the total number of bits of advice is constant. Moreover, any node that received the above advice, gets additionally the integer d.

We now describe the algorithm using the advice obtained from Geometric Oracle. It uses global round numbers which are transmitted from node to node appended to the source message.

Algorithm GRN-Broadcasting-with-Advice. The algorithm lasts $1 + 2p(d)B(d)$ rounds. After round 1 it is divided into $B(d)$ identical stages, each lasting $p(d)$ 2-round periods. The pseudo-code follows:

> in round 1 the source transmits;
> starting in round 2, **repeat** $B(d)$ times procedure **Stage**

where **Stage** is the following subroutine:

> **if** u has advice (i, out), for $1 \leq i \leq p(d)$, and got the source message
>   **then** it transmits in the first round of period i of this stage
> **if** v has advice (i, in), for $1 \leq i \leq p(d)$, and got the source message
>   **then** it transmits in the second round of period i of this stage

Theorem 3 follows from the following lemma whose proof is omitted.

Lemma 3. *Algorithm GRN-Broadcasting-with-Advice, using the Geometric Oracle, is correct and has constant running time.*

3   The General Algorithm

In this section we design and analyze a broadcasting algorithm working for arbitrary radio networks, whose running time is inverse-proportional to the size of advice given to nodes. We prove the following theorem.

Theorem 4. *For any $q \in O(n)$ there exists an oracle which gives advice of size q to the nodes of a network and an algorithm using this advice, which performs broadcast in time $O(\frac{nD}{q} \log^3 n)$ in any n-node network with source eccentricity D.*

We prove Theorem 4 by constructing an appropriate oracle and algorithm. First assume that $q \in O(D \log n + \log^2 n)$. In this case we can use the broadcasting algorithm from [11] running in time $O(n \log^2 D)$ without using any advice, since $O(n \log^2 D) \subseteq O(\frac{nD}{q} \log^3 n)$, for this range of q. Therefore, in the sequel, we can assume $q \geq 6(D \log n + \log^2 n)$.

Given the directed graph $G = (V, E)$ with source s, let $L_1, \ldots, L_D$ be BFS layers in G, i.e., sets of nodes at distance exactly i from the source, for $1 \leq i \leq D$. Let T be the smallest power of 2 greater or equal to $1152\frac{n}{q} \log^2 n$. For each $1 \leq i \leq D-1$ we will need sets $L_i(j) \subseteq L_i$, for $j = \log T, \log T+1, \ldots, \lfloor \log |L_i| \rfloor$, such that for every such j the following properties hold:

(i) every node in L_{i+1} having at least 2^j and less than 2^{j+1} neighbors in L_i, has at least 1 and at most $144 \log n$ neighbors in $L_i(j)$;
(ii) $|L_i(j)| < 144|L_i| \log n / 2^j$.

The following lemma justifies the existence of such sets (the proof is omitted).

Lemma 4. *There exist sets $L_i(j) \subseteq L_i$, for $j = \log T, \log T + 1, \ldots, \lfloor \log |L_i| \rfloor$ with the above properties.*

Advice. We now describe the advice given by the oracle. The advice given to the source consists of integers N, n, q and of the sizes of layers $L_1, \ldots, L_D$. This can be encoded using $3D \log n \leq q/2$ bits of advice. Moreover, to every node in set $L_i(j)$, for $1 \leq i \leq D - 1$ and $\log T \leq j \leq \log |L_i|$, the oracle gives the integer j. (Note that, since sets $L_i(j)$ are not necessarily disjoint, a node may get several integers as advice.) This costs a total of at most

$$2 \cdot \sum_{i=1}^{D-1} \sum_{j=\log T}^{\lfloor \log |L_i| \rfloor} (144|L_i| \log n / 2^j \cdot \log j) \leq 4 \cdot 144 \cdot (n/T) \log^2 n \leq q/2$$

bits, by property (ii) of sets $L_i(j)$. Hence the total size of advice is at most q.

Algorithm Radio-Broadcasting-with-Advice. We now describe the algorithm using the above advice. It uses global round numbers which are transmitted from node to node appended to the source message. First we define the additional information attached to the source message. We will use the notion of a (N, x)-selective family. This is a family $\mathcal{F}$ of subsets of $\{1, \ldots, N\}$, such that, for any set $X \subseteq \{1, \ldots, N\}$ of size at most x, there exists a set $F \in \mathcal{F}$, for which $|F \cap X| = 1$. For any x, fix a (N, x)-selective family $S(N, x)$ of size $s(N, x)$. By [10] there exist (N, x)-selective families of size $O(x \log(N/x)) \subseteq O(x \log n)$, thus we can assume that $s(N, x) \leq b \cdot x \log n$ for some constant $b > 0$. Fix an order $(F_1, \ldots, F_{s(N,x)})$ of the family $S(N, x)$. Knowing T, sizes $|L_i|$ of layers and the constant b, the source computes the sequence of rounds $t_1 < \ldots < t_{D-1}$ recursively as follows:
$$t_0 = 0, \quad t_{i+1} = t_i + s(N, T) + (\log |L_i| - \log T + 1) \cdot s(N, \lceil 144 \log n \rceil), \text{ for}$$
$1 \leq i \leq D - 1$.

Then the source broadcasts the source message together with the sequence $t_1, \ldots, t_{D-1}$ and $|L_1|, \ldots, |L_{D-1}|$ in round 0. A node that receives this message for the first time in round t, where $t_{i-1} < t \leq t_i$ for some $1 \leq i \leq D - 1$, waits till round t_i and starts transmitting according to the (N, T)-selective family $S(N, T)$, starting in round $t_i + 1$ until round $t_i + s(N, T)$. More precisely, a node with label u transmits in round $t_i + y$, if u is in F_y, where F_y is the y-th set of the family $S(N, T)$. Additionally, if a node has the integer j in its advice string then it transmits according to the family $S(N, \lceil 144 \log n \rceil)$ in the time interval from $t_i + s(N, T) + (j - \log T) \cdot s(N, \lceil 144 \log n \rceil) + 1$ to $t_i + s(N, T) + (j + 1 - \log T) \cdot s(N, \lceil 144 \log n \rceil)$, for any $\log T \leq j \leq \log |L_{i+1}|$. A node without the integer j in its advice string waits during this period. A node that receives the source message for the first time in round at most t_i does not transmit in rounds beyond t_{i+1}. We omit the proof of the following lemma.

Lemma 5. *Assume $q \in O(n)$ and $q \geq 6(D \log n + \log^2 n)$. Our algorithm Radio-Broadcasting-with-Advice performs broadcasting in any n-node network with source eccentricity D in time $O(\frac{nD}{q} \log^3 n)$ using at most q bits of advice.*

Since, as we noticed before, for $q \in O(D \log n + \log^2 n)$, the time $O(\frac{nD}{q} \log^3 n)$ of broadcasting can be achieved even without advice, Lemma 5 concludes the proof of Theorem 4.

Corollary 1. *For n-node networks with source eccentricity D polylogarithmic in n, there exists advice of size $o(n)$ sufficient to achieve polylog(n) broadcasting time.*

The above corollary should be contrasted with the lower bound from [10], were it is shown that (without advice) some n-node networks with source eccentricity D require time $\Omega(n \log D)$.

4   Conclusion

We studied the impact of the size of information (advice) given to nodes of a radio network on the time of broadcasting. Our approach was quantitative, i.e., we were concerned with the total number of bits, as opposed to particular items of information, such as the knowledge of neighborhood, or of the size of the network, whose impact on broadcasting time was previously studied in the literature. While our algorithm is a first step towards grasping the trade-off between the size of advice and the time of broadcasting, establishing the exact trade-offs, for any number of bits of advice, remains an open problem. Its general formulation is: What is the minimum time to broadcast in radio networks, with advice of size q? A more specific question is: What is the minimum size of advice permitting to achieve broadcasting time $O(Opt(G))$ for any radio network G. We answered this question for networks with constant optimal time.

Establishing trade-offs between the size of advice and broadcasting time is also open for geometric radio networks. For these networks time $O(D)$, where D is the eccentricity of the source, is optimal under full knowledge of the network. It is easy to show that $O(\min(n, D^2))$ bits of advice are sufficient to achieve this time. Is this size of advice also necessary?

Another interesting problem is to compare the size of arbitrary advice permitting given broadcasting time with the size of advice of given type, e.g., concerning the immediate neighborhood. It was proved in [24] that giving to all nodes information about their immediate neighborhood (a total of $\Theta(|E| \log n)$ bits) permits broadcasting in time $O(n^{2/3} \log n)$ in networks with source eccentricity 2. In [3] it was proved that time $\Omega(\sqrt{n})$ is necessary for these networks with this information. This should be contrasted with the algorithm from the present paper which, e.g., permits broadcasting in these networks in the same time $O(n^{2/3} \log n)$ using only $O(n^{1/3} \text{polylog}(n))$ bits of advice, provided that the advice is of non-restricted type. On the other hand, $O(\sqrt{n} \text{polylog}(n))$ bits of advice suffice to beat time $\Theta(\sqrt{n})$ for these networks. These examples suggest that using advice of non-restricted type may be much more efficient than that of a particular type.

The paradigm of radio broadcasting with advice also suggests related problems for randomized algorithms: What is the minimum number of random bits provided to the nodes of a radio network of unknown topology that is sufficient to achieve randomized broadcasting in optimal expected time? The lower bound on the expected broadcasting time obtained by Kushilevitz and Mansour [28] can be directly applied to the class of graphs G defined as follows: G consists of three layers, the only directed connections are from a layer to the subsequent layer, the first layer consists of the source, and each node in the middle layer has at most one out-neighbor in the last layer. In view of the result from [28], the number of random bits provided to the system must be $\Omega(n \log n)$ in order to guarantee $O(\log n)$ expected time. By contrast, $O(n)$ bits of advice suffice to achieve constant deterministic broadcast time for these networks. This means that randomization is sometimes more costly than advice by a logarithmic factor, in terms of the number of bits. The precise trade-off between randomized broadcasting time and the number of random bits used by a distributed randomized broadcasting algorithm remains open.

References

1. Alon, N., Bar-Noy, A., Linial, N., Peleg, D.: A lower bound for radio broadcast. Journal of Computer and System Sciences 43, 290–298 (1991)
2. Bar-Yehuda, R., Goldreich, O., Itai, A.: On the time complexity of broadcast in radio networks: an exponential gap between determinism and randomization. Journal of Computer and System Sciences 45, 104–126 (1992)
3. Brito, C., Gafni, E., Vaya, S.: An information theoretic lower bound for broadcasting in radio networks. In: Diekert, V., Habib, M. (eds.) STACS 2004. LNCS, vol. 2996, pp. 534–546. Springer, Heidelberg (2004)
4. Bruschi, D., Del Pinto, M.: Lower bounds for the broadcast problem in mobile radio networks. Distributed Computing 10, 129–135 (1997)
5. Chlamtac, I., Kutten, S.: On broadcasting in radio networks - problem analysis and protocol design. IEEE Transactions on Communications 33, 1240–1246 (1985)
6. Chlamtac, I., Weinstein, O.: The wave expansion approach to broadcasting in multihop radio networks. IEEE Transactions on Communications 39, 426–433 (1991)
7. Chlebus, B., Gąsieniec, L., Gibbons, A., Pelc, A., Rytter, W.: Deterministic broadcasting in unknown radio networks. Distributed Computing 15, 27–38 (2002)
8. Chlebus, B., Gąsieniec, L., Östlin, A., Robson, J.M.: Deterministic radio broadcasting. In: Welzl, E., Montanari, U., Rolim, J.D.P. (eds.) ICALP 2000. LNCS, vol. 1853, pp. 717–728. Springer, Heidelberg (2000)
9. Chrobak, M., Gąsieniec, L., Kowalski, D.: The wake-up problem in multi-hop radio networks. In: Proc. 15th ACM-SIAM Symposium on Discrete Algorithms (SODA 2004), pp. 985–993 (2004)
10. Clementi, A.E.F., Monti, A., Silvestri, R.: Selective families, superimposed codes, and broadcasting on unknown radio networks. In: Proc. 12th Ann. ACM-SIAM Symposium on Discrete Algorithms (SODA 2001), pp. 709–718 (2001)
11. Czumaj, A., Rytter, W.: Broadcasting algorithms in radio networks with unknown topology. In: Proc. 44th Symposium on Foundations of Computer Science (FOCS 2003), pp. 492–501 (2003)

12. De Marco, G.: Distributed broadcast in unknown radio networks. In: Proc. 19th ACM-SIAM Symp. on Discrete Algorithms (SODA 2008) (2008)
13. Dessmark, A., Pelc, A.: Broadcasting in geometric radio networks. Journal of Discrete Algorithms 5, 187–201 (2007)
14. Diks, K., Kranakis, E., Krizanc, D., Pelc, A.: The impact of knowledge on broadcasting time in linear radio networks. Theoretical Computer Science 287, 449–471 (2002)
15. Elkin, M., Kortsarz, G.: Improved broadcast schedule for radio networks. In: Proc. 16th ACM-SIAM Symposium on Discrete Algorithms(SODA 2005) (2005)
16. Emek, Y., Gasieniec, L., Kantor, E., Pelc, A., Peleg, D., Su, C.: Broadcasting time in UDG radio networks with unknown topology. In: Proc. 26th Ann. ACM Symposium on Principles of Distributed Computing (PODC 2007), pp. 195–204 (2007)
17. Fraigniaud, P., Ilcinkas, D., Pelc, A.: Oracle size: a new measure of difficulty for communication problems. In: Proc. 25th Ann. ACM Symposium on Principles of Distributed Computing (PODC 2006), pp. 179–187 (2006)
18. Fraigniaud, P., Ilcinkas, D., Pelc, A.: Tree exploration with an oracle. In: Královič, R., Urzyczyn, P. (eds.) MFCS 2006. LNCS, vol. 4162, pp. 24–37. Springer, Heidelberg (2006)
19. Fraigniaud, P., Ilcinkas, D., Gavoille, C., Pelc, A.: Distributed computing with advice: Information sensitivity of graph coloring. In: Arge, L., Cachin, C., Jurdziński, T., Tarlecki, A. (eds.) ICALP 2007. LNCS, vol. 4596, pp. 231–242. Springer, Heidelberg (2007)
20. Fraigniaud, P., Korman, A., Lebhar, E.: Local MST computation with short advice. In: Proc. 19th Annual ACM Symposium on Parallelism in Algorithms and Architectures (SPAA 2007), pp. 154–160 (2007)
21. Gaber, I., Mansour, Y.: Centralized broadcast in multihop radio networks. Journal of Algorithms 46, 1–20 (2003)
22. Gąsieniec, L., Peleg, D., Xin, Q.: Faster communication in known topology radio networks. In: Proc. 24th Annual ACM Symposium on Principles Of Distributed Computing (PODC 2005), pp. 129–137 (2005)
23. Gąsieniec, L., Pelc, A., Peleg, D.: The wakeup problem in synchronous broadcast systems. SIAM Journal on Discrete Mathematics 14, 207–222 (2001)
24. Kowalski, D., Pelc, A.: Time of deterministic broadcasting in radio networks with local knowledge. SIAM Journal on Computing 33, 870–891 (2004)
25. Kowalski, D., Pelc, A.: Time complexity of radio broadcasting: adaptiveness vs. obliviousness and randomization vs. determinism. Theoretical Computer Science 333, 355–371 (2005)
26. Kowalski, D., Pelc, A.: Broadcasting in undirected ad hoc radio networks. Distributed Computing 18, 43–57 (2005)
27. Kowalski, D., Pelc, A.: Optimal deterministic broadcasting in known topology radio networks. Distributed Computing 19, 185–195 (2007)
28. Kushilevitz, E., Mansour, Y.: An $\Omega(D \log(N/D))$ lower bound for broadcast in radio networks. SIAM Journal on Computing 27, 702–712 (1998)
29. Sen, A., Huson, M.L.: A new model for scheduling packet radio networks. In: Proc. 15th Annual Joint Conference of the IEEE Computer and Communication Societies (IEEE INFOCOM 1996), pp. 1116–1124 (1996)
30. Soguet, D., Nisse, N.: Graph searching with advice. In: Prencipe, G., Zaks, S. (eds.) SIROCCO 2007. LNCS, vol. 4474, pp. 51–65. Springer, Heidelberg (2007)

Author Index

Lecture Notes in Computer Science

Sublibrary 1: Theoretical Computer Science and General Issues

For information about Vols. 1–4711
please contact your bookseller or Springer